LEARNING TO
TEACH MATHEMATICS

LEARNING TO TEACH MATHEMATICS

Randall J. Souviney
University of California, San Diego

Merrill Publishing Company
A Bell & Howell Information Company
Columbus Toronto London Melbourne

Published by Merrill Publishing Company
A Bell & Howell Information Company
Columbus, Ohio 43216

This book was set in Garamond.

Administrative Editor: Jeff Johnson
Developmental Editor: Amy Macionis
Production Coordinator: Carol Sykes
Art Coordinator: Lorraine Woost
Cover Designer: Cathy Watterson

All photos by Walter Lilly.

Library of Congress Catalog Card Number: 88-60987
International Standard Book Number: 0-675-20687-1
Printed in the United States of America
1 2 3 4 5 6 7 8 9—93 92 91 90 89

For my wife,
Barbara Laura Miller-Souviney
Together, the world is a joy. . .

Preface

When asked by a tourist if he was making any money selling his lobster traps, an old Maine fisherman thought a moment and said, "Well, on Monday I didn't sell any. On Tuesday a feller came in and took ten, but on Wednesday he brought 'em back. I didn't sell any on Thursday, and on Friday I went fishin' . . . guess you could say that Tuesday was my best day."

Teaching often has its ups and downs as well. To be effective, teachers must be highly motivated individuals with outstanding communication skills; they must have a solid grasp of the subject matter, a practical understanding of how children learn, and, like the Down East fisherman, an ability to see something positive in virtually any situation. As an elementary teacher, you will be responsible for introducing young minds to the field of mathematics. By helping children explore mathematics and its many useful applications, you can play an important role in their development of the problem-solving skills they need to make informed decisions in life.

The purpose of this book is to help beginning teachers with one aspect of this complex enterprise—the design of effective mathematics instruction for elementary school children. Regardless of your mathematics background, the activities described in this book can heighten your level of curiosity about mathematics and broaden your understanding of elementary mathematics content and teaching practices. Because I believe that high-quality mathematics instruction requires the thoughtful integration of *content* and *instructional methods*, emphasis has been placed on the introduction of mathematics concepts together with the presentation of appropriate pedagogy. The instructional approach is developmental in nature, drawing on research in cognitive development and classroom practices. A range of instructional techniques is presented and applications for each are demonstrated using step-by-step, graphic examples. Suggested course activities are included to encourage the reader to practice the procedures with peers and, whenever possible, in actual classroom settings.

The review of intellectual development and learning presented in this book is strongly influenced by the developmental view of cognitive growth. The developmental view concludes that learning takes place as the result of complex interactions between an individual and the environment—our surroundings affect the ways in which our minds are organized, and this organization, in turn, influences our understanding of how the environment is structured. It is the interactive nature of the learning enterprise that enables individuals to construct order out of chaos, organize noise into music, and rationalize patterns to generate meaning.

Chapter 1 provides an overview of the goals of mathematics education into the next century. Chapters 2 through 4 address mathematics concept development, skill attainment, and problem solving in mathematics education. Affective issues related to learning, such as confidence, motivation, and student and teacher attitudes, are also explored. Lesson planning, evaluation procedures, and techniques for accommodating the special-needs learner are also highlighted. Chapter 5 introduces techniques for using calculators and computers in mathematics education. Chapters 6 through 14 each address a specific strand of the elementary mathematics curriculum—geometry, measurement, number, whole and rational number operations, patterns and functions, graphing, statistics, and probability. Each chapter provides an overview of significant cognitive development issues, a description of teaching practices appropriate for each mathematics concept and skill, and a set of problem-solving activities. Calculator and computer applications and examples of ways to adapt instruction for special-needs children are also included. Practice exercises are introduced throughout each chapter, as well as course activities, lists of relevant microcomputer software, references, and additional readings.

Many people have made significant contributions to the development of this book. I wish to thank Jeff Johnston, Amy Macionis, and the other creative people at Merrill Publishing Company for their good advice, patience, and support as these pages were taking shape. To Norma Allison and Alma Salcido I extend my deepest appreciation for their efforts in deciphering the original coffee-stained manuscript and emerging victorious over an occasionally reluctant word processor. I also owe a debt of gratitude to Michelle Souviney for her diligence in producing the hundreds of computer-designed graphics and Walter Lilly for sharing his skills with a camera. I thank the faculty and students at Laurel and Garrison Elementary Schools in Oceanside, California, for allowing us to photograph their classroom activities to include as chapter openings. Also, apologies to Jim Brunelle for my adaptation of some of his Down East stories that appear from time to time throughout the book. I extend my deepest appreciation to Barbara Miller-Souviney, James Levin, Doris Randall Souviney, Hugh Mehan, Miriam Gamoran, and, in particular, to the following reviewers for their perceptive comments and helpful suggestions, which contributed significantly to the final manuscript: Otto Bassler, Vanderbilt University; Jack Beal, University of Washington; William Bush, University of Kentucky; Roy Callahan, State University of New York at Buffalo; Warren Crown, Rutgers University; Jon Engelhardt, Arizona State University; Thomas Gibney, The University of Toledo; Alan Hoffer, Boston University; Michael Hynes, University of Central Florida; Hiram Johnston, Georgia State University; Betty Myers,

Shepherd College; Alan Osborne, Ohio State University; Jack Ott, University of South Carolina; Margaret Riel, University of California, San Diego; and C. Winston Smith, Jr., Ohio University. Finally, I wish to acknowledge the contribution of all the elementary, secondary, and university students I have known over the past 22 years, without whom this book would not have been possible.

Contents

Sample Process Problems and Solutions; Problem-Solving Lesson Plan; Adapting Instruction for Children with Special Needs; Summary; Course Activities; References and Readings

SECTION ONE

MATHEMATICS TEACHING AND LEARNING

The five chapters in this section address general issues associated with teaching and learning mathematics in grades K–6. The chapters contain

1 An overview of the psychological factors involved in learning mathematics.

2 An introduction to appropriate methods for teaching concepts, skills, and problem solving.

3 A review of relevant research.

4 An overview of how special-needs children learn and specialized instructional techniques.

5 A discussion of the special role calculators and computers play.

6 Course activities for chapter review.

7 Lists of related resource books, software, and references.

1

Mathematics Education into the Twenty-First Century

I don't know what's the matter with people: they don't learn by un-derstanding; they learn by some other way—by rote, or something. Their knowledge is so fragile!

—Richard P. Feynman

Upon completing Chapter 1, the reader will be able to

1. Identify eight goals for mathematics education into the twenty-first century.
2. Identify nine steps a teacher can take to minimize mathematics anxiety in the classroom.
3. Identify personal teaching goals.

I remember how I felt during math lessons in elementary school. I enjoyed the order of the ideas and the predictability of the results, but I didn't like the attention paid to accuracy and neatness. I felt powerful when I was able to see relationships and patterns in the mathematics ideas we were learning. My earliest memory involved discovering the pattern of sums equal to 9. We were filling in a worksheet, and the two sums **5 + 4** and **4 + 5** were written next to each other:

$$\begin{array}{cc} 5 & 4 \\ +\,4 & +\,5 \end{array}$$

I always had trouble remembering facts and in self-defense, I suppose, constantly looked for patterns to use as memory cues. I noticed that when I continued the pattern to the right, the sum of each pair of values equaled 9:

$$\begin{array}{cccccc} \mathbf{5} & \mathbf{4} & 3 & 2 & 1 & 0 \\ \mathbf{4} & \mathbf{5} & 6 & 7 & 8 & 9 \end{array}$$
$$\longrightarrow$$

Similarly, continuing the pattern to the left created additional pairs of values whose sum was 9:

$$\begin{array}{cccccccccc} 9 & 8 & 7 & 6 & \mathbf{5} & \mathbf{4} & 3 & 2 & 1 & 0 \\ 0 & 1 & 2 & 3 & \mathbf{4} & \mathbf{5} & 6 & 7 & 8 & 9 \end{array}$$
$$\longleftarrow$$

I remember thinking that it was probably important that there were exactly ten ways to make 9 (I later discovered more when I learned about the integers). Few of my classmates thought that this was a particularly important discovery. I now know that the rules and shortcuts we learned during our math lessons can be explained using predictable patterns and relationships, for example: regrouping when adding and subtracting; indenting a column when multiplying multi-digit numbers; dividing into the first digit or two when the dividend is large; moving the decimal point the same number of places in the divisor and the dividend before dividing; dividing fractions by inverting the divisor and multiplying; solving proportions by cross-multiplying and solving for the unknown.

After many years learning mathematics (and even more teaching it), I now realize that many people do not take a problem-solving approach to math. I remember my elementary students saying, after what I thought to be an absolutely crystal clear explanation, *Please, just tell us what to do!*

Looking back, I realize how much I learned during those first few years in the classroom. To teach mathematics meaningfully requires more than a good knowledge of the subject—it requires a careful self-analysis of how *you* came to know the mathematics you know and an understanding of why so many people find it difficult to learn the subject.

This book demonstrates how mathematics can be *meaningfully* taught and learned. Each sample activity allows children (and the reader) opportunities to think about, discuss, and write about the mathematics they are learning. Manipulative materials are introduced to encourage the direct observation of mathematical patterns and relationships. Children are also encouraged to work together on problems before they are required to complete work on their own.

One of the most important goals of public education is to help *all* children become independent actors in the world, fully capable of making informed decisions about matters important to themselves and society. In order to accomplish this goal, schools must foster the development of appropriate intellectual tools for analyzing information to solve problems. As an elementary teacher, you can contribute significantly to this effort by helping children see the beauty in the structure of mathematics and learn its many useful applications.

ATTITUDES TOWARD MATHEMATICS LEARNING

Children seem to readily learn the attitudes, prejudices, and values of their parents and teachers. Though we do not know for sure if adult recalcitrance towards mathematics infects children's attitudes about the subject, many students do come to view mathematics as a difficult subject to learn. For example, which of the following statements do you consider to be true?

1. Mathematicians are especially good at mental computation.
2. When doing mathematics, the most important goal is to find correct answers.
3. Learning mathematics depends more on logic than intuition.
4. To be good at mathematics, you need a good memory.
5. In general, males are better than females at mathematics.
6. The best mathematics students solve problems quickly and accurately.

7. The advent of cheap calculators and computers makes learning mathematics less important today than it once was.

Most math educators regard these statements as myths about mathematics and learning (Kogelman & Warren, 1978). Why, then, do such ideas about the nature of mathematics persist? John Ernest (1976) reported that 40 percent of the teacher candidates at the University of California, Santa Barbara, did not express positive attitudes toward mathematics and mathematics teaching. Based on similar results from other studies (Becker, 1986), it is not unreasonable to suspect that uninformed attitudes like these are being transmitted to children at home and in the classroom. If prospective teachers recognize such prejudices in themselves and strive actively to overcome them, they will become better mathematics instructors.

The widely held perspective that learning mathematics consists of memorizing long sequences of isolated rules creates an intimidating environment in the mathematics classroom, an experience so frustrating for some children that they become victims of **mathematics anxiety**. Some studies have reported that more than half the adult population suffers to some degree from the effects of such anxiety. For example, in severe cases, people have reported that they carry only large bills so they never have to count out change or that they stay at home rather than being put in the position of estimating tips or sales tax.

Teachers who subscribe to such myths about mathematics are more likely to organize instruction that overemphasizes memorization, speed, and independent work. Being required to memorize sequences of poorly understood mathematical rules increases the potential for unnecessarily anxious behavior. Though mathematics anxiety generally manifests itself in adolescence, its roots can often be traced to such elementary school experiences.

Equity in Mathematics Education

While there is no definitive evidence that mathematics anxiety disproportionately affects one segment of the population over another, fewer female and minority students do choose to take advanced mathematics classes in secondary school (Fennema, 1979). Other explanations for the lack of female and minority participation in advanced mathematics courses include biased academic counseling, negative peer pressure, low parental (and teacher) expectations, and sociocultural stereotypic shaping of independent, risk-taking males and dependent, security-oriented females (Fennema, 1982). Regardless of what causes the relatively small number of female and minority students to take advanced mathematics classes, many otherwise qualified students are filtered from access to interesting jobs that depend on mathematical competence (Sells, 1980).

Mathematics Equity in the Classroom

Elementary teachers can take several steps to minimize practices that are known to lead to mathematics anxious behavior in students (Brush, 1980; Kaseberg, Kreinberg,

& Downie, 1980; Tobias, 1981). Teachers should consider each of the following recommendations when planning mathematics instruction.

1. Provide an **intellectually safe environment** for learning. When children feel confident that intuitive responses and incorrect answers may lead to important findings, they are more likely to enter into discussions of concepts and take intellectual risks when solving problems.

2. Emphasize all **strands**, the major classifications of mathematics content. Often, children equate mathematics with arithmetic. Other important strands such as geometry, measurement, statistics, and probability offer alternative entry points for children to excel. Avoid undue emphasis on number and computation.

3. Involve children in daily **problem-solving experiences**. Avoid extensive use of isolated drills by embedding independent practice within problem-solving activities whenever possible.

4. Employ **manipulative models** when appropriate to motivate key mathematics concepts. Allow ample time for students of different abilities to develop conceptual understanding prior to memorizing facts and practicing skills.

5. Arrange for **cooperative** small group work when developing concepts and solving problems. Peer interaction encourages verbalization and justification of one's ideas. Thinking about one's thinking has been shown to contribute to problem-solving ability.

6. Remember that the **process** involved in getting an answer may be at least as important as the answer itself. Children often think there is one correct way to find an answer. This belief is rarely true and, in fact, may lead to the mistaken impression that mathematics is merely a collection of step-by-step rules. When considered carefully, the process of arriving at an incorrect answer can lead to greater understanding of underlying concepts and perhaps suggest a new solution strategy.

7. Maintain **high expectations** for every child regardless of gender or cultural background. Distribute response opportunities among the entire class as evenly as possible. Count to five after asking a question to allow time for a wider range of considered responses. Reorganize working groups from time to time to allow leadership roles to vary.

8. Make sure children are aware of the growing **importance of mathematics competence** in our society. Integrate mathematics concepts and skills into other curriculum areas such as science, social studies, art, and music. Display materials showing the mathematics required for various career options. Make available biographical sketches of successful women and men from various cultural backgrounds who use mathematics in their work.

9. Before you begin teaching, take stock of your own **attitudes** toward mathematics and teaching. In elementary school, were you assigned extra math exercises as punishment? Do you think a woman who takes a calculus course must be a bit odd? Are you anxious when confronted with a mathematics problem? Discuss your concerns and fears openly with your peers. You are probably not alone. As an elementary teacher, you have a responsibility to demonstrate a positive attitude toward mathematics in your classroom. The readings and activities in this book are designed to help you attain this vital goal.

MATHEMATICS EDUCATION FOR A NEW CENTURY

In the United States, approximately 1.5 million teachers and more than 30 million children go to school each day. Each state establishes its own criteria for teacher certification and the public school curriculum. This diversity within the U.S. public school system is both its strength and its weakness. Diversity in curriculum and certification enables schools to address the unique needs of each local community more effectively. On the other hand, history has demonstrated that locally established policy can be designed to discriminate systematically against various segments of the population. Although the effectiveness of the past two decades of desegregation and compensatory education legislation continues to be debated, direct federal involvement in local school policy seems to be declining.

Innovation in Mathematics Education

In 1975, the National Advisory Committee on Mathematical Education (NACOME, 1975) reported on the effects of federally supported mathematics curriculum reform between 1955 and 1975. This period saw the development of a dizzying array of instructional programs at the elementary and secondary levels, collectively referred to as the **new math**. Though the content and pedagogy varied widely, most programs emphasized set concepts, number properties, multibase arithmetic, intuitive geometry, elementary statistics, and probability. Each of these strands was included in the curriculum in an attempt to enhance student understanding of the underlying principles of elementary mathematics topics and to improve computation performance and problem-solving ability.

The NACOME report observed that, due to inadequate teacher inservice training and poor distribution of teaching materials, these unifying concepts were often incorrectly taught as ends in themselves. For example, before children were allowed to move on to base-10 subtraction, many were required to master written exercises in base-8 arithmetic; or students were drilled on the correct spelling of *commutative* rather than focusing on applications of this important number property, for example, as an aid in learning the multiplication facts.

Reviewing research on classroom practice throughout this period, the NACOME authors concluded that the median elementary classroom in 1975 exhibited little effect from two decades of the massive, federally supported curriculum innovation. Teachers appeared to teach much as they had been taught a generation before. In general, students worked from a single textbook, read less than half of the pages of explanations, used the books primarily as a source for worked examples and exercises, and concentrated the majority of their efforts on learning arithmetic. There is little evidence to indicate that much has changed since then.

Of course, many teachers were successful in implementing innovative mathematics programs, particularly with college bound students. The NACOME conclusion was less an evaluation of the twenty-year effort to improve curriculum as it was an indication of the importance of the teacher in establishing what actually goes on inside classrooms.

Mathematics Education for the Eighties and Beyond

In the past decade, dramatic changes have been proposed for the content of the elementary mathematics curriculum. All evidence indicates that the demands placed on the mathematics curriculum will continue to increase through the turn of the century and beyond. The National Council of Supervisors of Mathematics (NCSM, 1977) published a list of basic skills recommended for mathematics education during the 1980s. The position statement was developed partly as a positive response to education critics who demanded a retreat to teaching only basic number facts and computation skills in elementary classrooms, but also as a means to reach some teachers who had retreated to this basic skills position.

The primary focus of the NCSM recommendations involved increasing the emphasis on **mathematical problem solving** at all grade levels. This recommendation suggests that problem-solving experiences should include not only story-type problems but also **nonroutine** problems requiring the organization of data, illustration and elaboration of problem components, and the use of problem-solving strategies. Such nonroutine problems are thought to be more relevant to everyday problem solving because they can often be solved in more than one way and may have more than one correct answer.

Other NCSM recommendations included increased emphasis on estimation, statistics, probability, measurement, and computation skills necessary to support effective problem solving. Their final recommendation stressed the need for every student to understand what computers can and cannot do and to engage in firsthand computing experiences.

The need to focus on problem solving also receives support from the results of several large-scale achievement studies. The report of the National Assessment of Educational Progress (Lindquist, Carpenter, Silver, & Mathews, 1983) indicated that while elementary students exhibited reasonable achievement on whole number operations, they had difficulty applying these computational skills in problem-solving situations.

In the late 1970s, the National Council of Teachers of Mathematics undertook a national survey of the educators and the general public, called *Priorities in School Mathematics* (1981), to establish priorities for mathematics education. The results of this survey were reported in *An Agenda for Action*, which offered eight recommendations for improving mathematics education (NCTM, 1980).

1. Problem solving must be the focus of school mathematics in the 1980s [and beyond].

2. The concept of basic skills in mathematics must encompass more than computational facility.

3. Mathematics programs must take full advantage of the power of calculators and computers at all grade levels.

4. Stringent standards of both effectiveness and efficiency must be applied to the teaching of mathematics.

5. The success of mathematics programs and student learning must be evaluated by a wider range of measures than conventional testing.

6. More mathematics must be required for all students and a flexible curriculum with a greater range of options should be designed to accommodate the diverse needs of the student population.

7. Mathematics teachers must demand of themselves and their colleagues a high level of professionalism.

8. Public support for mathematics instruction must be raised to a level commensurate with the importance of mathematical understanding to individuals and society.

An Agenda for Action also specified curriculum innovations designed to meet the needs of students to the year 2000. The key curriculum changes include an increased emphasis on teaching strategies for problem solving, estimation skills, and computing literacy; and decreased emphasis on isolated arithmetic drills and paper-and-pencil computations involving numbers larger than two digits.

An exciting challenge awaits the teacher of elementary mathematics throughout the remainder of this century. An important goal of mathematics instruction is to help students become expert solvers of nonroutine problems.

In order to apply memorized number facts and algorithms to nonroutine problem situations, students need to engage in skill practice within the context of real-world problems. For example, research indicates that intensive drills are useful immediately following the introduction of a new skill. However, once an average accuracy of about 90 percent is attained, additional drills offer diminishing returns (Jacobsen, 1975; Suppes & Morningstar, 1972).

Calculators and quality computer software help students with a wide range of abilities solve problems by providing support for tedious, or as yet unmastered, task components. A calculator or computer can be used to carry out complicated computations, freeing the student to focus on the underlying logical structure of the problem. Computing tools will play an increasing role in problem-solving instruction as we near the twenty-first century.

USING THIS BOOK

The purpose of this book is to help beginning teachers design effective mathematics lessons for elementary students. The organization is based on the assumption that mathematics teaching requires the thoughtful integration of mathematics **content** and **instructional methods**. Emphasis has been placed on presentation of teaching techniques that introduce the mathematics concepts and skills using **manipulative materials**. This instructional approach is developmental in nature, drawing on cognitive development research and proven classroom practices. Lesson design and teaching techniques are demonstrated throughout the book, using step-by-step examples. Course activities are included that encourage the reader to practice the recommended instructional procedures and the use of manipulatives with peers and, when possible, in an actual classroom setting.

Chapter 2 presents an overview of mathematics concept development and learning. The approach taken is strongly influenced by the developmental view, which argues that learning occurs as the result of complex interactions among an individual, other people, and the environment. In other words, our surroundings affect the way our minds are organized, and in turn, this organization influences our perception of how the world works. As the result of this interaction, individuals are able to construct order out of chaos, organize noise into music, and recognize patterns to generate meaning.

Chapter 3 introduces techniques for teaching mathematics facts and skills and highlights discussion of the structure of lessons, cognitive and affective evaluation procedures, and techniques for accommodating the special-needs learner.

An introduction to the special role of problem solving in mathematics instruction is presented in Chapter 4. Types of appropriate elementary mathematics problems are defined, and examples of specific problem-solving strategies are included.

Chapter 5 gives an overview of how computers and calculators are affecting elementary mathematics education today. Specific examples of instructional software and computing activities are included.

Chapters 6 through 14 each address a specific strand, or major content area, of the elementary mathematics curriculum—geometry, measurement, numbers, whole number operations, patterns and functions, fractions, decimals, statistics, and probability. Each chapter includes a discussion of related cognitive development issues, a description of mathematics concepts and associated teaching practices, and a selection of activity-oriented classroom exercises. Calculator and computer applications and suggestions for adapting instruction for the special-needs learner are also included.

TOWARD A PHILOSOPHY OF TEACHING AND LEARNING

Success in teaching stems from the effective integration of mathematics content, knowledge of the child, parent involvement, careful planning, and expert management skills. One develops expertise in teaching only if each year brings new experiences that challenge existing conceptions of the teaching/learning process. Remaining open to change encourages greater understanding of effective classroom practices and awakens a celebration of the complexity of the teaching act.

The discussion of teaching and learning explored in this book will help you define a personal philosophy of teaching. Such a philosophy will help guide your study of mathematics and mathematics teaching. To gain full benefit from this book, it is important to engage actively in working through the recommended mathematics experiences. Tap your ability to be flexible and creative as you attempt to elaborate your own understanding of elementary mathematics.

SUMMARY

Many students and adults exhibit unnecessarily anxious behavior when learning mathematics. In addition to inappropriate academic counseling, such experiences may discourage a disproportionate number of female and minority students from taking advanced secondary school mathematics courses. Many career options are closed to students who do not complete at least three years of high school mathematics. Elementary teachers can take steps to minimize mathematics anxiety by providing a relaxed mathematics learning environment, increasing emphasis on concept development, and encouraging small group work. Teachers must also honestly address their own anxiety toward mathematics and take steps to minimize its potential effects on students. The goals of mathematics education into the next century include an increased focus on problem solving, the appropriate use of technology and mental arithmetic, and a decreased emphasis on paper-and-pencil computations involving complex values. Manipulatives and cooperative learning groups are cited as particularly appropriate teaching techniques for elementary mathematics instruction.

COURSE ACTIVITIES

1. List classroom practices that lead to mathematics anxious behavior. Observe a mathematics class and make note of the occurrence of these practices. Discuss the results of your observation with your peers. List ways the lessons observed could be reorganized to minimize the potential for anxiety.

2. Review the NCSM *Position Paper on Basic Skills* and the NCTM *An Agenda for Action*. Develop a list of basic mathematical skills you consider necessary for effective participation in our society. Show your list to a parent, administrator, teacher, and student and ask them to rank the items in order of importance. Compare the results of your survey and discuss with your peers.

3. Review a K–6 Scope and Sequence Chart supplied with a textbook series. Compare the content described for each grade level with the NCSM and NCTM recommendations for mathematics education in the 1980s. Write a brief report describing where these documents agree and where they seem to conflict.

4. Read one of the *Arithmetic Teacher* articles listed in the reference section. Write a brief report summarizing the main ideas of the article and describe how the recommendations for instruction might apply to your own mathematics teaching.

REFERENCES AND READINGS

Becker, J. (1986). Mathematics attitudes of elementary education majors. *Arithmetic Teacher, 33*(5), 50–51.

Brush, L. (1980). *Encouraging girls in mathematics*. Cambridge, MA: Abt Books.

Crosswhite, J., & Reys, R. (Eds.). (1977). *Organizing for mathematics instruction: 1977 yearbook*. Reston, VA: National Council of Teachers of Mathematics.

Ernest, J. (1976). Mathematics and sex. *American Mathematical Monthly, 93*(8), 595–614.

Fennell, F. (1984). Mainstreaming in the mathematics classroom. *Arithmetic Teacher, 32*(3), 22–27.

Fennema, E. (1979). Women and girls in mathematics—Equity in mathematics education. *Educational Studies in Mathematics, 10*(4), 389–401.

Fennema, E. (1982). *Women and mathematics: State of the art review*. Paper presented at the Equity in Mathematics Core Conference, National Council of Teachers of Mathematics, Reston, VA.

Jacobsen, E. (1975). *The effect of different modes of practice on number facts and computational abilities*. Unpublished manuscript, University of Pittsburgh, Learning Research and Development Center.

Kaseberg, A., Kreinberg, N., & Downie, D. (1980). *Use equals to promote the participation of women in mathematics*. Berkeley, CA: Lawrence Hall of Science.

Kogelman, S., & Warren, J. (1978). *Mind over math*. New York: The Dial Press.

Lindquist, M., Carpenter, T., Silver, E., & Mathews, W. (1983). The Third National Mathematics Assessment: Results and implications for elementary and middle schools. *Arithmetic Teacher, 31*(4), 14–19.

National Advisory Committee on Mathematical Education. (1975). *Overview and analysis of school mathematics, grades K–12*. Washington, DC: Conference Board of the Mathematical Sciences, NACOME.

National Council of Supervisors of Mathematics. (1977). Position paper on basic skills. *Arithmetic Teacher, 25*(1), 19–22.

National Council of Teachers of Mathematics. (1980). *An agenda for action: Recommendations for school mathematics in the 1980s*. Reston, VA: National Council of Teachers of Mathematics.

National Council of Teachers of Mathematics. (1981). *Priorities in school mathematics: An executive summary of the PRISM project*. Reston, VA: National Council of Teachers of Mathematics.

Sells, L. (1978). Mathematics—A critical filter. *The Science Teacher, 45*(2), 28–29.

Sells, L. (1980). The mathematics filter and the education of women and minorities. In L. Fox, L. Brody, & D. Tobin (Eds.). *Women and the mathematical mystique*. Baltimore: The Johns Hopkins University Press.

Suppes, R., & Morningstar, M. (1972). *Computer-assisted instruction at Stanford, 1966–1968: Data, models, and evaluation of arithmetic programs*. New York: Academic Press.

Tobias, S. (1981). Stress in the math classroom. *Learning, 9*(6), 34, 37–38.

2

Teaching Mathematics Concepts

I am not a short adult!

—Marilyn Burns

Upon completing Chapter 2, the reader will be able to

1. Define a mathematical concept and give several examples.
2. Describe and contrast the developmental learning theories of Piaget and Vygotsky.
3. Describe and contrast the concept instruction theories of Bruner and Dienes.
4. Diagram a model sequence of instructional activities and give examples of each type of activity.
5. Write a sample lesson plan to teach a mathematics concept.

A Maine farmer who lived his entire life on the southern border of the state was surprised to discover that a new tax survey indicated that his farm was in New Hampshire instead of Maine. When asked how he felt about his new citizenship he responded, "Thank God, I don't think I could'a stood another one of them Maine winters."

Teaching requires a similar ability to shift one's point of view because each child will express a unique mixture of talents and attitudes. Coupled with the idea that mathematical ideas also come in different shapes and sizes, it is obvious that no single instructional principle will satisfy all children. Thus for the same instructional task, more than one approach may be required in order to give all children the opportunity to learn the mathematical concept.

Chapters 2, 3, and 4 explore several views about teaching and learning. The teacher must decide which of these perspectives on learning is appropriate for each student and instructional task. Chapter 2 explains how *concept learning* takes place. Chapter 3 introduces ways to foster the learning of mathematical *skills* and *basic facts*. Chapter 4 presents techniques for teaching *mathematical problem solving*. The general themes associated with teaching and learning developed in these first three chapters will recur throughout the text as specific teaching practices are introduced.

CONCEPT DEVELOPMENT

A mathematics **concept** can be defined as the underlying pattern that relates sets of objects or actions to one another. For example, an underlying pattern that defines the geometric concept of *triangle* is all closed figures that have exactly three straight sides. Teaching mathematics concepts can be a complex enterprise. Because each student brings a unique set of experiences and abilities to bear on each learning task, the teacher plays a critical role in planning effective instruction.

To make a thoughtful selection among the wide range of instructional methods available, you must first adopt a theory of learning to guide your planning. Such theory must serve the needs of a wide range of student abilities, take into account the characteristics of different types of content objectives, and place manageable organizational demands on your time.

The following sections examine the learning theories proposed by Piaget, Vygotsky, Bruner, Dienes, van Hiele, Ausubel, and Gagne. When reviewing each perspective on the learning process, it will be useful to consider the following points.

1. How do the theories complement one another, and in what ways do they seem contradictory?
2. What are the instructional implications associated with each theory of learning?
3. How successfully does each theory explain the development of mathematics concepts, skills, and problem solving?

JEAN PIAGET

The theory of intellectual development described by Jean Piaget (1896–1980) grew out of his early work on IQ testing with Binet. Although his work did not directly address issues related to learning specific concepts, it does characterize components of knowledge that could be considered universal among humans.

Through the use of careful observations and interviews with his own and others' children, Piaget concluded that children pass sequentially through four **stages** of development, each stage characterized by a qualitatively different ability to imagine and organize objects, symbols representing objects, and events in the world (Piaget, 1960). Piaget thought that successive stages are attained through changes in one's ability to internalize, or mentally organize, activities in the environment. He argued that intelligence is the ability to reorganize existing knowledge mentally in order to integrate new experiences. His position contrasted with the prevailing view of IQ as a static measure of mental capacity (Flavell, 1963).

The age at which individual children move from one developmental stage to another varies according to innate ability, personal history, and the richness of the learning environment (Flavell, 1963). According to Piaget, however, every child passes through the four stages *in the same order*. The four major stages described by Piaget are

1. *Sensorimotor* (birth to approximately one and one-half years old)
2. *Preoperational* (one and one-half to approximately seven years old)
3. *Concrete operational* (seven years old to adolescence)
4. *Formal operational* (adolescence through adult)

The ages listed indicate when the majority of children are likely to attain each stage of development. Recent studies show that these ages may vary widely depending on the assessment procedures used. Since reliance on a single measure of development may give inaccurate results, performance on the range of skills characteristic of each stage offers a more reliable indicator of intellectual development (Berry & Dasen, 1973; Cole & Scribner, 1974; Lancy, Souviney, & Kada, 1981; Serpell, 1976).

Sensorimotor Stage

Sensorimotor refers to the stage that begins with the reflex actions of infants and proceeds through the development of basic concepts of time, space, and causality. The end of the sensorimotor stage is characterized by the development of eye-hand coordination and spatial relationships, an understanding that objects still exist even when out of sight (object permanence), and the beginning of symbolic thought. At this point of development, children see themselves at the center of all actions in the world (egocentric). The effective coordination of these activity-dependent skills culminates in a concrete understanding of cause and effect and the internalization of sequences of actions that stand for, or symbolize, objects (e.g., seeing the table being set means dinner is on the way). This behavior marks the transition to the next stage of development.

Preoperational Stage

The **preoperational** stage is characterized by the development of symbolic thinking. Children start to view specific objects and events as represented by actions, symbols (including language), and other forms of language (facial expressions or hand signals). In Piaget's view, language is the systematic use of symbols. For example, a child might first pretend to be an animal, then use a wooden block to stand for a favorite pet, and finally use a name or other word to express symbolic understanding. As children gain experience, terms gradually take on more generalized meanings. For example, while the word *toy* might initially refer only to a frequently used truck, the set of objects represented by the term will later be expanded to include additional play objects. The child must also differentiate between types, or classes, of objects to determine membership. A picture book, for example, originally included in the set of *toys*, might later belong to another category called *books*.

While children are learning to deal with symbols (and during this stage they do learn the names of numbers in order), they have difficulty maintaining a consistent one-to-one matching between equivalent sets. The initial learning of symbols is still closely tied to perceived reality. This means the child is not able to manipulate symbols mentally. The following dialogue with Karla (five years old) was recorded during a study of cognitive development and arithmetic achievement (Souviney, 1980):

INTERVIEWER: Here are two rows of peanuts. Suppose I get to eat this row (pointing to row 2) and you have these (pointing to row 1).

Would we each have the same amount or would one of us have more?

KARLA: (After pointing at each one-to-one pairing of peanuts as shown in Figure 2.1a) They are the same.

INTERVIEWER: (The interviewer adjusts row 2 in full view of the child as shown in Figure 2.1b.) Now do we both have the same amount or does one of us have more peanuts?

KARLA: You have more.

INTERVIEWER: Why do you think I have more?

KARLA: Because yours goes from here to here (pointing to the ends of row 2) and mine only goes this far.

INTERVIEWER: (Interviewer adjusts row 2 to appear as in Figure 2.1a) Now do we both have the same amount or does one of us have more peanuts?

KARLA: Both the same.

INTERVIEWER: Why do you think we have the same amount now?

KARLA: They all line up now like before.

According to Piaget, Karla was unable to visualize both displays in order to compare them after a transformation. Piaget would claim that this preoperational child was unable to recognize a contradiction between her assertions that the rows contained the same amount in one configuration and different amounts in another configuration.

To understand that the two rows in Figure 2.1b continue to have the same number of counters requires the mental activity of *doing and undoing* physical actions. Piaget referred to the ability to maintain number invariance after moving a set of counters as **conservation**. The preoperational child is unable to conserve number, length, mass, or capacity. The ability to overcome contradictions caused by the reliance on perceptual cues marks the transition to Piaget's third stage of development (Labinowicz, 1980).

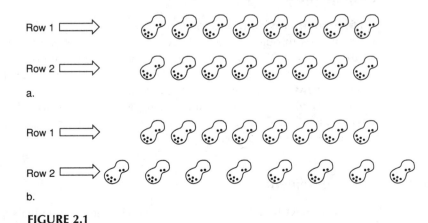

FIGURE 2.1

Concrete Operational Stage

Concrete operational reasoning is characterized by a significant increase in a child's ability to carry out *mental actions* related to physical objects and actual events. Children at this stage of development can accurately analyze the results of physical experiments by mentally reversing procedures, taking alternate points of view, and generalizing outcomes.

The ability to overcome perceptually guided conclusions enables concrete operational children to display conservation of number, length, mass, and capacity. The ability to group objects according to attributes, or **classification**, and order objects according to size, or **seriation**, improves dramatically during this period.

The ability to reverse actions mentally and hold attributes invariant after some transformation allows the development of new ideas. Much of elementary mathematics depends on the mental operations that develop during the concrete operational stage. Geometric and measurement topics are introduced, as are the concepts of number, place value, regrouping, and operations.

Piaget's theory suggests that the presence of perceptually salient objects (e.g., proportional base-10 blocks) is an important component in the development of mathematical concepts. He described the learning process at this stage as the visualization of the actual events encountered by students. While symbolic representations of such real situations are possible, propositions requiring the suspension of reality are difficult for concrete operational children. In geometry, for example, it is possible to show that opposite sides of a rectangle are parallel by employing a *proof by contradiction*. This method requires that one first assume the opposite sides are not parallel. Based on this assumption, it is possible to design a series of logical propositions that lead to contradiction (the interior angles are not equal to 90°). Since we know that each interior angle of a rectangle equals 90°, the original assumption must be false and, therefore, the opposite sides are parallel. Such ability to suspend the obvious temporarily in order to study the relationships involved marks the transition to Piaget's final stage of cognitive development.

Formal Operational Stage

The **formal operational** stage of development generally begins in early adolescence and continues through adulthood. Formal reasoning is characterized by the ability to carry out mental activity using imagined actions and symbols, divorced from their physical representation. Individuals at this stage are able to control variables systematically, test hypotheses, and generalize results to future occurrences. Proportional reasoning (e.g., computing distance using a map scale) also develops during this period. Piaget's final stage of intellectual development characterizes adult reasoning and mature problem solving ability.

Mechanism of Development

Piaget also described a mechanism to explain how individuals progress through these four stages of cognitive development. It is a two-step process in which new experiences are taken in, or **assimilated** into, what is already known in order to make the new experience understandable. Past understandings are subsequently changed by, or **accommodated** to, these new experiences. Piaget contended that, as learners assimilate input from the environment, the new information is not just stored like socks in a drawer but is integrated by the individual with what is already known (Piaget, 1960, 1965). For example, when a new geometric shape such as the pentagon is introduced, a list of unique features is not simply memorized but must be integrated with what is already known about triangles, squares, circles, and other figures.

Basically, Piaget proposed a very conservative system of learning. He argued that all new knowledge can be understood only in terms of what is already known. When it becomes increasingly difficult to fit in new experiences with what is already known, contradictions occur. For example, preoperational children experience no contradiction when responding illogically on a conservation of number task. However, once children are able to act on objects mentally and undo, or reverse, actions in their minds, the same task generates a contradictory effect. Piaget referred to this situation as a state of **disequilibrium**. To return to the more comfortable state of equilibrium, the mental structure involved must be modified to take into account, or accommodate, this new condition. Piaget called these mental structures that coordinate and organize assimilated knowledge **schema**.

Thousands of organizing structures, or schema, are developed over one's lifetime. Piaget argued that it is through the interactive process of assimilation and accommodation that individuals strive to maintain an integrated state of equilibrium and progress through the stages of development.

Piaget considered an individual's motivation for engaging in this developmental process to be internally driven. Although severely restricted social interaction or inadequate experience with objects and actions can retard the progress of intellectual development, he contended, attempts to accelerate development by external means are not necessary. That is, the physical world presents the necessary lessons, and the child is internally motivated to maintain a state of equilibrium (i.e., to learn) through the process of assimilation and accommodation. Given a reasonably complex environment, children will normally develop these mental actions, or **operations**, much in the way they develop language. Piaget cautioned parents and teachers to avoid the potential damage that can be caused by ignoring these internal actions, which are required for understanding. His theory implies that teaching in a way that encourages students to memorize information without rational understanding may retard, rather than accelerate, learning.

Though his theory of development has strongly influenced research on mathematics learning, Piaget himself cautioned educators not to over-interpret the significance of his work when applying it to classroom instruction. Several researchers have questioned Piaget's claims regarding internal motivation of cognitive development. Through direct instructional intervention and a careful selection of assessment tasks, children have demonstrated performance on conservation and classification tasks that can not be satisfactorily accounted for by Piaget's theory.

Brainard (1978) found that, through careful training, students who initially failed Piaget's conservation of number task were able to exhibit conservation behaviors sooner than predicted by Piaget's theory. Gelman and Gallistel (1978), by introducing a task more relevant to the child's experience, claimed to show conservation and rational counting behaviors in three- to-five-year-old children that also run counter to Piaget's prediction. In addition, cross-cultural research has cast doubts on the reliability of Piagetian tasks in accurately predicting the level of intellectual development in non-Western populations (Berry & Dasen, 1973; Cole & Scribner, 1974; Gay & Cole, 1967; Lancy, 1983; Serpell, 1976; Souviney, 1983). Despite the potential limitations of Piaget's theory, his developmental views continue to influence mathematics instruction throughout the world.

LEV VYGOTSKY

The Soviet psychologist Lev Vygotsky (1896–1934) criticized Piaget's emphasis on the individual's internally motivated interactions as the primary factor of cognitive development. In contrast to Piaget, Vygotsky (1978) emphasized the social basis of intelligence. He believed that mental operations are initiated through active social interaction with more competent peers and adults. These operations are subsequently internalized as necessary by the individual. For example, Vygotsky claimed that a child learning to speak requires social interaction. Every normal child that grows up around a spoken language becomes a competent language user around age three.

This ambitious undertaking is generally accomplished with little or no direct instruction. Depending on factors both internal and external to the child, however, the range of language competence varies considerably for any given age. If for some reason every child needed to develop language more rapidly, a social environment could be arranged to increase the amount of guided oral language interaction between novices and more accomplished peers and adults. Such a procedure would improve the rate of learning for some children, while others, due to developmental constraints, would achieve less. The issue is not whether a child can learn to speak (nearly everyone does) but the role of social resources in the process (Cole & Scribner, 1974). Parallel arguments can be made for the development of proficiency in other symbolic activities such as mathematics, art, and music.

Development and Learning

For the purposes of this discussion, **learning** is defined as the process of attaining the mental operations that vary from culture to culture according to local social needs (i.e., estimation of large quantities or techniques for calculating sums). **Development** is defined as the achievement of mental abilities that are universal among all normal human beings (i.e., language or conservation of number). Vygotsky and Piaget agreed that, since learning is more than simple imitation, a certain amount of preparedness must be present before an individual can be expected to internalize a new concept. Likewise, each theorist stressed active involvement on the part of the learner as fundamental to the learning process. However, whereas Piaget was concerned with the *development* of universal mental activities apparent in the *indi-*

vidual, which are less likely to be affected by direct instruction, Vygotsky's broader view included the consequences of social interaction on learning and development. For example, an objective test may indicate that two children are at the same level of mastery on addition exercises involving whole numbers. They may, however, differ greatly in their ability to engage, even with the aid of an expert, in similar but more complex tasks like solving problems *using* addition. This implies that the students are not at identical levels of achievement and may require different instruction to advance.

Vygotsky argued that instruction is most efficient when students cooperatively engage in activities within a supportive learning environment and receive an appropriate amount of guidance from the teacher. Cooperative learning activities, group problem solving, and cross-age tutoring are effective ways to structure lessons that increase the quantity and quality of social interaction and support within the classroom. The external support provided by the teacher, peers, and objects in the environment can be systematically withdrawn over time as individuals learn to complete a task on their own.

A child's level of development suggests a range of tasks that can be effectively addressed using such a system of dynamic supports. The lower limit is marked by concepts and skills already mastered. The upper limit is marked by those tasks the child would be able to complete only when provided with step-by-step instructions or expert advice. This range of potentially beneficial learning activities defines what Vygotsky termed the child's **zone of proximal development**.

The principles of dynamic support and the zone of proximal development recognize the active role in learning supplied by more accomplished peers and adults. Tasks children complete when working together are reasonable indicators of what they are likely to be able to do on their own in the near future. In Vygotsky's view, the role of the teacher is to organize interaction carefully in order to assist children in completing learning tasks and systematically withdraw these supports as students move to higher confidence levels in their unique zone of proximal development (Heap, 1986).

A way to structure lessons so that support can vary according to student needs employs the judicious use of perceptually salient manipulative materials. By embodying a new concept in several contexts, the learner can explore the regularities among them, perform actions that become internalized as mental sequences, and discover the underlying pattern relating the representations. For example, when learning the concept of *threeness*, a child can try out several groupings of objects in various locations and seek feedback on what counts as a group of three and what does not.

☐ Do all the objects have to be identical?
☐ Will it matter if the objects are arranged differently?
☐ Will the value change over time?

These questions and others must be resolved by the learner before the concept of *threeness* takes on the accepted meaning. In the next section, the instructional benefits of expressing a concept in a number of perceptually unique yet functionally equivalent ways will be explored.

JEROME BRUNER

Jerome Bruner (1966) proposed that new concepts be introduced to students using three levels of representation:

1. *Enactive* (concrete)
2. *Iconic* (graphic)
3. *Symbolic* (abstract)

Each level embodies a different amount of environmental support for the learner. At the **enactive** level, the teacher attempts to insure that a maximum of visual and tactile cues will be provided by the physical environment. In other words, the instructional environment and manipulative materials must provide a high level of **perceptual salience**. The amount of perceptual support is reduced at the **iconic** and **symbolic** levels, thereby requiring greater mental activity and coordination on the part of the individual. Ideally, as students progress through the levels of representation they abstract the similarities among the multiple contexts.

ZOLTAN DIENES

Zoltan Dienes (1964) elaborated on Bruner's approach by proposing a **multiple embodiment** theory of instruction specific to mathematics learning. Dienes felt that concept development resulted from the abstraction of features common to a set of carefully selected *enactive* experiences. For example, Dienes proposed that the concept of *place value* emerges from engaging in systematic counting experiences using various bases. It is because of this process of systematically increasing mental activity and level of abstraction that Bruner's levels of representation model and Dienes's multiple-embodiment theory of instruction coordinate well with Vygotsky's principles of zone of proximal development and dynamic support.

Various types of instructional materials and methods are available that provide a range of dynamic support for mathematics instruction. **Manipulative models** such as base-10 blocks, the abacus, and symbolic place-value tables allow children to add, subtract, multiply, and divide numbers before they would be able to complete meaningful symbolic computations on their own. As discussed in Chapter 5, the microcomputer holds perhaps the greatest promise for providing dynamic instructional support in a regular classroom setting. Lesson organization techniques and instructional materials that embody the principles of dynamic support will be introduced in every chapter of this book.

Multiple Representation of Concepts

Just as the notion of happiness can be expressed in a child's laughter, an ice cream advertisement, or a yarn spun by Mark Twain, mathematics concepts can be represented in multiple ways. A concept can be displayed using objects such as propor-

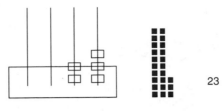

23

FIGURE 2.2

tional base-10 blocks, which offer a high degree of perceptual salience, an abacus, or only symbols. Success in answering place-value questions varies according to the type of representation employed. For example, young children often select **3** when asked whether the **2** or the **3** represent the larger value in the number **23**. The number 23 is displayed in Figure 2.2 using proportional base-10 blocks, using an abacus, and as a symbol. Each of these representations offers perceptual cues to help students determine which digit, the **2** or the **3**, represents the greater amount in the number **23**.

While the abacus offers a concrete representation of digit size and position, the base-10 blocks more clearly display the proportional relationship between the place values. The symbolic representation (i.e., 23) places even greater responsibility on the learner to maintain the proper relationship among the digits. While manipulative materials offer proven advantages for initial concept instruction, distinct disadvantages exist for long-term use. Bulky materials are inconvenient for prolonged use as a computation tool. While the abacus is used widely in parts of Asia, using base-10 blocks to compute your change at a grocery store would be inconvenient indeed. Such concrete displays also become unwieldy when numbers get large (i.e., greater than 1000) or small (i.e., less than $\frac{1}{100}$).

This limitation points out one of the reasons mathematics is such a powerful means of expression. Mathematics allows us to define operations with abstract *symbols* that would be inconvenient or impossible to construct in the real world. Symbols enable us to make assertions about the world that may not be concretely verifiable. We often carry out computations on large numbers that we have every reason to believe exist, but that we have not personally represented in any concrete manner. Have you ever really counted to 1000? How big a box would be needed to contain 100,000 marbles? Our economy is measured in trillions of dollars, but it would take a lifetime for one person to count even a billion one-dollar bills. One of the important goals of mathematics instruction is to have students understand the importance of symbols and how to effectively employ them when solving problems.

PIERRE VAN HIELE

Pierre van Hiele (1986) proposed the following multilevel model for thinking and learning about mathematics, particularly geometry concepts:

1. *Recognition*
2. *Description*
3. *Classification*
4. *Deductive analysis*

Children are normally expected to attain level three by the end of the sixth grade. Level four is generally attained by secondary students who demonstrate the ability to use deductive techniques to prove theorems. His proposal for levels of learning differs from Piaget's in that van Hiele was primarily concerned with how to stimulate the transition from one level to the next (i.e., teaching). Also, like Vygotsky, van Hiele's model views language and culture as more important to development and learning than does Piaget's model. Though van Hiele did not dispute the possibility of higher levels of thought, his work stresses the importance to educators of fully exploiting the potential of the first four levels. Further discussion of van Hiele levels of thinking is presented in Chapter 6.

DAVID AUSUBEL

David Ausubel (1968) advocated the widespread application of a Socratically inspired verbal form of instruction. An important feature of Ausubel's technique involves presenting, prior to instruction, one or more explicit statements about the learning objective, which he calls **advance organizers**. Following this statement of purpose, the teacher initiates an interaction involving carefully chosen questions and restatements of student responses to guide progress toward the desired goal. The lesson itself may involve the use of demonstrations, laboratory materials, visuals, or other instructional aids. Ausubel argued that advance organizers serve to keep the learner on task throughout the lesson. The teacher uses personal knowledge of the content to carry an appropriate amount of the argument as the students progress toward understanding. Though such verbal instruction must accommodate each learner's established language abilities, these procedures can be useful when employed by teachers who possess an excellent understanding of the subject matter and demonstrate effective discussion skills. However, verbal discussions alone are generally not sufficient to foster the mental actions necessary for rational understanding of mathematical operations.

ROBERT GAGNE

Robert Gagne (1965) argued for the systematic application of **task analysis** to aid the design of instructional materials and teaching procedures. The recommended process involves determining the sequence of subgoals associated with a complex task. Initially, a terminal goal such as the *addition of whole numbers* is specified. Skills necessary to attain this goal are classified as **intact skills** (those already known) and **enabling skills** (those that need to be learned). Finally, intact skills are arranged in a logical order so that enabling skills can be learned and the goal attained. Gagne concluded that it is through the coordination of these intermediate enabling skills that successful learning of complex tasks takes place.

Those critical of Gagne's work point out that it is this very act of coordination that novices find so difficult to achieve. Even if all the subgoals are met, there is no assurance that the learner will be able to complete the whole task independently. Effective teaching supports this coordination effort through the judicious organization of the learning environment and systematic social interaction.

While Vygotsky called for a careful structuring of the teaching/learning process, not just any structure will do. Breaking complex goals into smaller components may give only the illusion that learning is taking place. To exhibit competence, the novice must be able to coordinate these skills accurately under varying conditions. Using task analysis to develop an instructional hierarchy can give an adequate description of the whole task to an individual who is already competent at completing the task. Such a hierarchy might be an invaluable aid to teachers trying to unravel their own prior understanding of a concept or skill. Vygotsky would argue, however, that a novice is more likely to learn to coordinate the subgoals of a complex task if an expert provides a system of dynamic support that gradually transfers responsibility for coordinating task components to the learner.

The teacher plays a key role in this enterprise. It is the teacher who must understand the learning process, select instructional materials, organize a supportive classroom environment, and provide students with support at the appropriate teaching moment (Driscoll, 1980). The task is complex. However, a teacher who is familiar with the implications of research on mathematics learning can draw upon a variety of instructional aids and techniques to satisfy the wide range of learning requirements found in the typical elementary classroom.

LESSON PLANNING

When new concepts are being introduced, concrete representations offering maximum perceptual salience can be employed. A careful transition to symbolic representations takes place as students develop facility with the concept. To help children make this important transition, materials and activities must be selected that encourage the student to take on gradually greater responsibility for key elements in the task. This technique is an example of how the environment can be structured to provide dynamic support for learning based on Vygotsky's proposal for instruction.

Developmental Activity Sequences

Developmental activity sequences (DAS) can be assembled from a wide array of available instructional materials (Baratta-Lorton, 1976; Souviney, Keyser, & Sarver, 1978). The objective of this approach is to effect an orderly transition from the perceptually salient representations used to introduce a new concept to more versatile procedures employing abstract symbols. While practice is important throughout

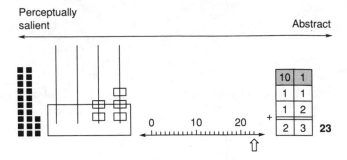

FIGURE 2.3

the sequence of activities, drill occurs primarily at the symbolic level. Figure 2.3 illustrates a sample continuum of instructional materials that might be employed in the development of whole number and place-value concepts.

In addition to the level of representation, it is also useful to consider a second factor when planning a sequence of lessons. Several types of learning tasks can be described at each level of representation. As shown in Figure 2.4, these include **free exploration**, **teacher-assigned exercises**, and **independent problem solving** (Wirtz, 1974). Developmental activity sequences can be assembled beginning with concrete free-exploration experiences to familiarize students with the instructional materials followed by concrete, graphic, and symbolic teacher-defined tasks. The

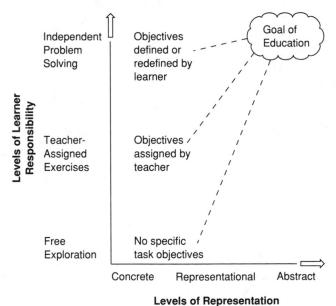

FIGURE 2.4

goal of instruction is to proceed gradually along both dimensions toward student-defined problem solving at the symbolic level.

Free exploration activities have no specific, externally imposed objective. A certain amount of messing around, or exploring relationships among objects and symbols, is an important feature of a successful sequence of lessons. Children should become familiar with instructional materials and secure in their use before they are expected to use them to carry out assigned tasks. Such exploration can be carried out at each level of representation, if necessary.

Tasks with teacher-assigned objectives are the most frequent type of lesson activity found in elementary classrooms. Textbook assignments and teacher talk are composed primarily of such exercises. These exercises can be *convergent* in nature (e.g., How much is $4+5$?), or *divergent* (e.g., What numbers have a sum of 9?).

A student-defined or independent problem-solving task is a situation recognized by the learner as posing a potential problem that he or she may wish to solve. Divergent type questions can encourage independent investigations. For example, when investigating an arithmetic problem such as determining sets of numbers that have the sum 9, children might go beyond considering only number pairs $(6+3)$ and explore larger sets of addends $(5+1+2+1)$ or the use of signed numbers $(^{+}12+{}^{-}3)$, might question if $(7+2)$ is the same as $(2+7)$, argue if $(9+0)$ is an acceptable solution, or find the total number of addend pairs for each sum 0 through 18.

Divergent thinking is considered by many creative individuals in our society to be an important cognitive ability. Practice with independent problem solving must begin early. By carefully arranging the types of instructional materials and social groupings in the classroom, the mathematics teacher can design zones of proximal development that give children practice engaging in independent and group problem-solving experiences.

Lesson Plan Outline

Well-planned, engaging lessons improve the quality of instruction and minimize classroom discipline problems. Though the task may at first seem overwhelming, long-term and weekly planning should become an integral part of the teacher's day. District and state curriculum guides, textbook scope and sequence charts, supplementary resource books, and personal experience all contribute to mapping out general plans for the school year. It is good practice to have children review previously learned material early in the year to gauge appropriate beginning levels of instruction. General, long-range objectives are then prepared for the entire class. This list will likely be altered from time to time as student needs and abilities unfold throughout the year.

Daily lesson plans should be prepared in advance. The beginning teacher generally finds it useful to describe each lesson in detail. This effort encourages the consideration of possible points of confusion and facilitates smooth transitions between lesson components. With experience, teachers become more efficient and generally require less time for planning and preparation. A number of lesson plan formats are available. Most formats can be adapted to address the three types of mathematics objectives: *concepts, facts and skills,* and *problem solving.* The following example is a lesson plan outline based on the **clinical teaching model** of instruction (Hunter, 1984).

Lesson Plan Outline

1. OBJECTIVE: The lesson objective should answer the questions:
   a. Learner will be able to _____
   b. Learner will demonstrate understanding by _____

2. MATERIALS: List all materials needed by the teacher and students.

3. ANTICIPATORY SET: Introduce the lesson by stating the objective and what the students will be able to do as a result. In the lesson plan, describe an opening that will motivate the students and demonstrate the relevance of the lesson.

4. INSTRUCTION: In this section of the lesson plan, describe the lesson content and sequence and list key questions to propel the learning forward. Include a list of new terms to be introduced, describe concepts to be modeled by the teacher, and include a pretest if appropriate.

5. GUIDED PRACTICE: Describe an observable manipulative experience, small group activity, or in-class seat work that requires the active participation of every student. As a result of observing guided practice, the teacher should be better able to determine future individual and group instructional needs.

6. CLOSURE: Include a brief, whole-group concluding activity that summarizes the learning and reviews student progress toward attaining the lesson objective. List key review questions and include a posttest as appropriate.

7. INDEPENDENT PRACTICE: Describe additional unguided practice such as appropriate homework or seat work assignment. Make sure these assignments are collected and/or reviewed on a timely basis.

For the following lesson plan, it is assumed the children know the values of coins, have had preliminary experiences with rational counting, and know the number symbols.

Sample Grade One Lesson Plan on Regrouping

1. OBJECTIVE: Learner will be able to regroup from the ones to tens column using base-10 materials as demonstrated by successfully completing five concrete examples and recording the results using paper and pencil.

2. MATERIALS: For each child: set of 25 unit cubes and five 10-rods, one foam-rubber die, sheet of construction paper, paper and pencil, textbook (p. 53). Overhead projector, transparency film, and overhead marker, 20 pennies and two dimes, set of base-10 materials, and a place-value mat (construction paper with heavy line dividing it in half).

3. ANTICIPATORY SET: On the overhead, show 5 pennies and ask the class to count them. Ask the class to count as the teacher adds one penny at a time until 10 pennies are showing. Ask if there is another way to show 10 pennies

(one dime). Exchange 10 pennies for one dime and continue placing pennies until 20 cents are showing. Ask if there is another way to show 20 cents (two dimes). Remove coins and repeat with base-10 blocks (unit-cubes and 10-rods). Then tell students they are going to learn how to *regroup* 10 ones into 1 ten just like changing pennies for dimes.

4. INSTRUCTION: Hand out cups of base-10 materials, sheet of construction paper, dice, and paper and pencil to each child. Hold up place-value mat and have each child draw a line dividing the construction paper in half. On the overhead, show similar place-value chart on transparency and have the class follow teacher actions to play Bankers Game. Teacher rolls die, reads value on top of die and places that many unit cubes in ones column (right side of mat). Children match action on their own mats. Process continues until 10 (or more) unit cubes are in the ones column. Ask the class if there is another way to show the cubes (exchange 10 one cubes for one 10-rod). Continue until reaching two 10-rods (20). Write the final value on the board (answers will vary from 20 to 25) and have the children do the same on a separate paper (show place-value headings and vertical line separating ones and tens column).

5. GUIDED PRACTICE: Have the children repeat the process and raise their hand when they reach two 10-rods. Walk around the room and assist with problems. When everyone reaches 20, increase the goal to three 10-rods, and if time allows, four 10-rods and five 10-rods.

6. CLOSURE: To check student understanding, ask one child to come to the overhead to regroup a pile of 13 unit cubes (one 10-rod and three unit-cubes). Write 13 on the board and have each child show the value using base-10 materials, draw a sketch of one 10-rod and three unit-cubes on their papers, and write the symbol. Write 18 and 25 as well and observe at least one solution for each child.

7. INDEPENDENT PRACTICE: Assign 10 exercises writing values for sets of objects (p. 53 in textbook) for seat work, or homework if not finished (allow use of base-10 materials as necessary).

SUMMARY

Chapter 2 introduces the developmental theories of learning proposed by Jean Piaget and Lev Vygotsky. These theories attempt to account for changes in cognitive performance experienced by children as they grow older. Both theories claim that cognitive development is the result of complex interactions between the learner and the environment. Piaget attributed growth through the sensorimotor, preoperational, concrete operational, and formal operational stages primarily to factors internal to the child, while Vygotsky proposed social interaction involving a process called the zone of proximal development as the fundamental motivation for learning. Jerome Bruner and Zoltan Dienes each proposed theories of instruction that exploit the developmental approach to learning. Bruner's model of instruction involves the presentation of mathematics concepts at three levels of abstraction: enactive (concrete), iconic (graphic), and symbolic (abstract). Dienes proposed embedding key mathematical ideas within sequences of carefully selected concrete activities, enabling learners to identify common features among the representations and thereby abstract the underlying concepts. The chapter proposes an instructional model called developmental activity sequences as a general procedure for lesson organization, and introduces a lesson plan outline.

COURSE ACTIVITIES

1. Describe the principles involved in the multiple representation of mathematics concepts. How can the principles be applied to learning? Select a mathematics concept and develop a lesson sequence that uses multiple representations as an instructional technique. Describe how this procedure could be used to address individual learning needs.

2. Review the conservation and classification tasks used by Piaget to assess cognitive development in children (see Labinowicz, 1980, for specific tasks). Select one task (e.g., conservation of number) and practice the procedures with a partner. If possible, administer the task to a preschool, first, and fourth grade student. Make an audio recording of the interactions and write a summary of each child's performance. Compare your results with those of your peers. Do your results agree? Describe the possible implications for instruction.

3. Review the theories of cognitive development and learning proposed by Piaget and Vygotsky. Describe the basic mechanism each proposed to explain how development takes place. List several similarities and differences. Describe possible implications for teaching associated with the two approaches.

4. Choose a mathematics concept with which you are reasonably familiar. Write a list of subskills and previously learned concepts you feel are required in order for you to define this concept. Try to develop as complete a list as possible. Do you feel your list of subskills and concepts is sufficient to teach this concept to a novice? Explain your reasoning.

5. Write a lesson plan to introduce the concept you identified in activity #4. How could this lesson be used as part of a developmental activity sequence based on Bruner's levels-of-representation model of instruction? How does Vygotsky's notion of the zone of proximal development apply to the lesson? List ways that the classroom environment could be arranged to provide dynamic support for learning.

6. Read one of the *Arithmetic Teacher* articles listed in the reference section. Write a brief report summarizing the main ideas of the article and describe how the recommendations for instruction might apply to your own mathematics teaching.

REFERENCES AND READINGS

Ausubel, D. (1968). Facilitating meaningful verbal learning in the classroom. *Arithmetic Teacher, 15*(6), 126–132.

Baratta-Lorton, M. (1976). *Mathematics their way*. Menlo Park, CA: Addison-Wesley.

Berry, J., & Dasen, P. (1973). *Culture and cognition: Readings in cross cultural psychology.* New York: Harper & Row.

Brainard, C. (1978). *Piaget's theory of intelligence.* Englewood Cliffs, NJ: Prentice-Hall.

Bruner, J. (1966). *Toward a theory of instruction.* New York: W. W. Norton.

Burns, M. (1977). *I am not a short adult: Getting good at being a kid.* Boston, MA: Little, Brown.

Cole, M., & Scribner, S. (1974). *Culture and thought.* New York: John Wiley.

Copeland, R. (1979). *How children learn mathematics: Teaching implications of Piaget's research.* New York: Macmillan.

Dienes, Z. (1964). *Mathematics in the primary school.* New York: Macmillan.

Driscoll, M. (1980). *Research within reach: Elementary school mathematics.* St. Louis, MO: CAMREL.

Flavell, J. (1963). *The developmental psychology of Jean Piaget.* Princeton, NJ: D. Van Nostrand.

Gagne, R. (1965). *Conditions of learning.* New York: Holt, Rinehart & Winston.

Gay, J., & Cole, M. (1967). *The new mathematics and an old culture.* New York: Holt, Rinehart & Winston.

Gelman, R., & Gallistel, C. (1978). *The child's understanding of number.* Cambridge, Harvard University Press.

Heap, J. (1986). *Collaborative practices during computer writing in a first grade classroom.* Paper presented at the American Educational Research Association Conference, San Francisco, CA.

Heddens, J. (1986). Bridging the gap between the concrete and the abstract. *Arithmetic Teacher, 33*(6), 14–17.

Hunter, M. (1984). Knowing, teaching and supervising. In P. Hosford (Ed.), *Using what we know about teaching.* (pp. 169–192) Alexandria, VA: Association for Supervision and Curriculum Development.

Labinowicz, E., (1980). *The Piaget primer.* Menlo Park, CA: Addison-Wesley.

Lancy, D. (1983). *Cross-cultural studies in cognition and mathematics.* New York: Academic Press.

Lancy, D., Souviney, R., & Kada, V. (1981). Intra-cultural variation of cognitive development: Conservation of length among Imbonggu. *International Journal for Behavioral Development, 4,* 445–468.

Piaget, J. (1960). *The psychology of intelligence.* Patterson, NJ: Littlefield, Adams.

Piaget, J. (1965). *The child's conception of number.* New York: W. W. Norton.

Serpell, R. (1976). *Culture's influence on behavior.* London: Methuen.

Souviney, R. (1980). Cognitive competence and mathematical development. *Journal for Research in Mathematics Education, 11*(3), 215–224.

Souviney, R. (1983). Mathematics achievement, language and cognition: Classroom practices in Papua New Guinea. *Educational Studies in Mathematics, 14,* 183–212.

Souviney, R., Keyser, T., & Sarver, A. (1978). *Math-matters: Developing computational skills with developmental activity sequences.* Glenview, IL: Scott, Foresman.

Suydam, M. (Ed). (1978). *Developing computational skills.* Reston, VA: National Council of Teachers of Mathematics.

Van Hiele, P. (1986). *Structure and insight: A theory of mathematics education.* New York: Academic Press.

Vygotsky, L. (1978). *Mind in society.* Cambridge: Harvard University Press.

Wirtz, R. (1974). *Mathematics for everyone.* Washington, DC: Curriculum Development Associates.

3

Teaching Mathematics Facts and Skills

When we try to pick out anything by itself, we find it hitched to everything else in the universe.

—John Muir

Upon completing Chapter 3, the reader will be able to

1. Describe two theories of rote and rational skill learning.
2. Describe two ways to categorize knowledge and give mathematical examples for each.
3. Describe the three components of a performance objective and write examples for a mathematics concept and a skill.
4. Describe four types of lesson organization and give examples of each.
5. Describe three types of affective assessment and give applications for each.
6. Describe four types of cognitive assessment and give applications for each.
7. Identify eight types of special-needs students and specify several ways to accommodate individual differences in the classroom.
8. List the five components of an individualized education program for special-needs students.
9. Write a sample lesson plan for a mathematics skill.

Learning mathematics concepts and learning mathematics skills are not separate processes. New concepts provide opportunities to develop new skills, and these skills in turn provide tools for extending conceptual knowledge. For example, mastery of the concept of place value allows the development of efficient techniques to add larger values than is possible using counting numbers or tally marks. Once proficiency in addition is established, other procedures can be introduced, such as multidigit subtraction and multiplication.

Both concept and skill development can be aided by the appropriate use of motivating problem situations. A new concept can be introduced to illuminate a relationship implied within a problem statement. For example, the concept of *average* can be introduced as a way to compare the relative running speeds of two classes.

Problems can also stimulate the practice of desired skills. For example, the following problem provides practice adding columns of one- and two-digit numbers. Assign the values one cent to twenty-six cents in sequence to each letter in the alphabet. Students can compute the value of their names and other words by calculating the sum of the values assigned to each letter (e.g., cat $\rightarrow$ 3¢ + 1¢ + 20¢ = 24¢). See if your name is worth exactly $1.00. A class can generate a list of $1.00 names and other words. Are there words worth more than $5.00? Problems like this integrate mathematics with other subject areas and help overcome the difficult challenge of making mundane practice intrinsically motivating.

ROTE AND RATIONAL LEARNING

Following in the development of the number concept, children are often introduced to an **algorithm**—a series of systematic procedures that can be invoked to solve routine computations. Algorithms are an important component of the basic mathematics skills. This chapter focuses on techniques for teaching the multiplication table, definitions of mathematical terms, computation algorithms, and other skills and facts.

Drill has traditionally played a central role in mathematics skill learning. In the 1920s, the elementary mathematics curriculum was strongly influenced by the work of psychologist Edward Thorndike. Thorndike (1922) argued that complex skills, such as the addition algorithm, consist of a sequence of smaller mental habits he called **bonds**. He proposed that educators fully define arithmetic skills by carefully identifying all the bonds necessary to carry out an operation. Appropriate instruction would then drill each bond in the hierarchy until it was mastered by the student. When the entire sequence was committed to memory, the skill would be considered automated, or mastered.

William Brownell (1935) disagreed with this argument and reported evidence that children carrying out arithmetic operations used an array of **rational** procedures other than the direct recall of bonds. He observed that children often used patterns involving grouping by tens, counting on fingers, or generalizing from known combinations to generate basic addition facts. Recent work by Rogoff and Lave (1983) and others suggests that adults also apply invented procedures when using arithmetic in their work and elsewhere.

Brownell expressed concern that premature drill would only give the *appearance* of competence. He argued that, although students might score well on timed exercises, the real test was whether they had sufficient understanding to apply the skill to problem situations (Ashlock & Washbon, 1978).

Although Thorndike believed that drill procedures would eventually allow nearly perfect computational accuracy, no large-scale study has confirmed this prediction. The results of several computer-aided instruction (CAI) projects carried out over the past two decades have indicated that once average accuracy reaches about 90 percent, additional practice has little value (Jacobsen, 1975; Judd & Glaser, 1969; Suppes & Morningstar, 1972).

Teaching Implications

The *Mathematics Framework for California Public Schools* (1985) summarizes the characteristics of teaching for understanding versus rote instruction using the contrasting statements listed in Table 3.1.

The results of research on rote drill versus rational learning support the position taken by the authors of the *Framework* that new skills should be introduced using

TABLE 3.1 Characteristics of Rational and Rote Learning

Teaching for Understanding	Teaching Rules and Procedures
Emphasizes understanding	Emphasizes recall
Teaches a few generalizations	Teaches many rules
Develops conceptual schemas or interrelated concepts	Develops fixed or specific processes or skills
Identifies global relationships	Identifies sequential steps
Is adaptable to new tasks or situations (broad application)	Is used for specific tasks or situations (limited context)
Takes longer to learn but is retained more easily	Is learned more quickly but is quickly forgotten
Is difficult to teach	Is easy to teach
Is difficult to test	Is easy to test

objects, or **models**, that make the underlying concepts more accessible to students. Once a rational basis has been established, a moderate amount of sharply focused drill is effective. Algorithms and number facts should be practiced briefly but frequently until automated. Subsequent practice is best presented in problem-solving contexts or mixed with other types of exercises.

ORGANIZING LEARNING OBJECTIVES

All ideas are not the same size. For example, once the multiplication table is memorized, recalling the answer for $7 \times 9 = ?$ is fundamentally different from knowing how to find the number of square-foot tiles needed to cover the floor of a seven-by nine-foot room. The answer in both cases is the same number, but the type of understanding necessary to solve each differs.

The first exercise requires that, when given the symbolic cues (7, $\times$, 9, $=$, $?$), a specific value be recalled from memory. If the multiplication table has been memorized, no understanding of what causes 7×9 to equal **63** is required. Either the correct value is recalled or it is not. One makes the mental association between 7×9 and 63. Employing a concrete model for multiplication allows students who cannot directly recall the product to reconstruct the solution using a sketch or an array of blocks.

The second case does not specifically state that the product of the room's dimensions gives the number of tiles required. An understanding of the underlying concept must be brought to bear on the problem situation in order to know the correct operation to employ. Prior experience working with rectangular arrays of objects and the multiplication concept offers a clue to solving this problem.

Benjamin Bloom (1965) and Robert Gagne (1970) described two distinct views of the various categories of knowledge. Each described a continuum beginning at the lowest level with stimulus-response associative bonds and, at the highest level, with problem-solving applications. Table 3.2 lists the main features of each categorization with illustrative examples. The purpose of the **performance terms** listed in column 4 is discussed in the following section on objectives.

It is helpful to keep these categories of knowledge in mind when planning lessons that require different instructional approaches.

TABLE 3.2 Categories of Knowledge

Gagne's Hierarchy	Bloom's Taxonomy	Mathematical Examples	Performance Terms
Associative: Ability to recall facts, procedures, and skills.	Knowledge: Ability to recall facts, procedures, and skills.	$6 \times 8 = 48$ $1, 2, 3, 4, \ldots ?$ $4 \times 6 = ?$	Lists, recalls, recites, remembers, names, selects, states.
Concept: Ability to recognize underlying patterns and relations among objects and events.	Comprehension: Ability to understand underlying concepts or principles involved in an idea or procedure.	$\bullet\;\bullet\;\bullet \rightarrow 3$ $14 < 21$	Distinguishes, estimates, explains, interprets, justifies, summarizes.
Principle: Ability to coordinate two or more concepts into a unified idea or procedure.		$28 \times 53 = 53 \times ?$ $\bullet\;\bullet\;\bullet\;\bullet$ $\bullet\;\bullet\;\bullet\;\bullet \rightarrow 3 \times 4$ $\bullet\;\bullet\;\bullet\;\bullet$	
Problem Solving: Ability to coordinate known skills, concepts, and principles and bring them to bear on the solution of new problems.	Application: Ability to use knowledge in new ways.	Find the area of a 3-meter by 4-meter floor.	Applies, constructs, demonstrates, discovers, modifies, predicts.
	Analysis: Ability to break knowledge into component parts.	Find the next value in the number pattern: $0, 1, 4, 9, 16, ?$	Deduces, diagrams, discriminates, identifies, illustrates, outlines.
	Synthesis: Ability to assemble known ideas into a new whole.	Collect data and make a graph to determine the average life of a pencil used in school.	Combines, compares, compiles, designs, induces, integrates, rearranges, reorganizes.
	Evaluation: Ability to critique the value of products and ideas.	In the number 23, which digit represents the larger value, 2 or 3? Justify your answer.	Compares, concludes, contrasts, criticizes, judges, justifies, interprets, proves.

1. **Associative**, or rote, learning requires considerable initial drill and periodic mixed practice in problem-solving contexts. Skills learned by rote are gradually forgotten when unused.
2. Learning **concepts** and **principles**—the simultaneous application of two or more concepts—requires active participation in carefully constructed sequences of lessons involving multiple representations. The recognition of patterns relating the different contexts leads to conceptual understanding.
3. Learning to apply skills and concepts to **problem-solving** situations develops through a gradual increase in student responsibility for key cognitive components of the task. Students learn to apply problem-solving strategies, such as guess-and-test, making a systematic list, and constructing a table or graph, to a wide range of nonroutine problems (see Chapter 4 for examples). Either lower level skills must be automated through appropriate practice, or children should have access to reference tables or computing devices.

Role of Performance Objectives

School districts commonly create lists of objectives to describe the required content of their mathematics curricula. **Performance**, or behavioral, objectives are intended to assist teachers with both the organization of lessons and the evaluation of student progress. A typical performance objective specifies:

1. A **description** of the task or competency to be displayed by the learner, specified by a performance term (such as *write*, *list*, *recite*, or *solve*).
2. The **condition**, or situation, under which learners will display their performance (work independently, use a table or calculator, on a paper-and-pencil test, and so on).
3. The **criterion**, or level of competence, to be displayed (e.g., with 80 percent accuracy, without error, will correctly solve four out of five problems).

Performance objectives are precisely written statements that help to facilitate lesson planning and the evaluation of achievement. Specifying a precise performance term is an important feature of objectives. Mager (1975) recommended that teachers employ performance specifications that are based on unambiguous, observable behaviors assumed to be indicators of desired mental abilities. Table 3.2 lists sample performance terms for each knowledge category. Table 3.3 lists ambiguous performance terms to avoid when writing objectives and more precise verbs that facilitate the evaluation of student progress.

Instructional objectives are used to describe more general areas of content. These objectives include only a description of the task without specified conditions or performance outcomes. Such statements are generally easier to read, though they may require more explicit specification to be useful for evaluation. The same learning task can be described as a general instructional objective or a specific performance objective.

1. *Instructional objective*: The student will learn to measure area in square centimeters.

TABLE 3.3 Performance Terms

Ambiguous Verbs	Precise Verbs
to appreciate	to build
to believe	to compare
to develop a feeling for	to construct
to grasp the significance of	to contrast
to have faith in	to design
to internalize	to draw
to know	to identify
to understand	to recite
	to solve
	to sort
	to write

2. *Performance objective*: Given five rectangular figures, the student will correctly compute the area in square centimeters of at least four figures by measuring their length and width with a ruler and computing the product using paper and pencil or a calculator.

Lists of instructional objectives are included in Appendix I. Performance objectives can be developed from these lists by specifying conditions and performance criteria.

A Word of Caution

You must be cautious about relying too heavily on performance objectives for daily planning. Although it is relatively easy to specify conditions and performance criteria for associative, or recall, tasks, it is difficult to make the same specifications for concept and principle learning. Any single performance criterion is but one observable indicator for a complex cognitive process. Slavish adherence to a single criterion may encourage a teacher to focus on superficial behaviors that give only the appearance of understanding. Performance objectives appropriate for such higher level cognitive tasks as problem solving and composition are exceedingly difficult to specify. A dependency on precisely stated performance objectives may also foster in the novice teacher the misimpression that teaching is an exact science.

Engaging All the Senses

Three of the senses—**hearing, sight,** and **touch**—are primary components of communication employed in the teaching/learning process. Each sense can be exploited under a range of classroom conditions to teach mathematics concepts, skills, and problem solving.

Effective teaching often engages more than one sense at a time. For example, when first introducing the triangle, a lesson might have the students handle several triangular objects, classify pictures of shapes used in building houses and bridges, engage in a verbal, twenty-questions-type activity, and as a follow-up experience, construct a triangle using bits of clay and toothpicks. The remaining senses—taste and smell—could be employed as well, in shaping, baking, and eating triangular cookies.

During the learning process, individuals frequently favor one sense over another. This preference is referred to as a student's **learning mode**, or style. Research to date on adapting instruction to individual learning styles has been largely inconclusive. It seems prudent, however, to organize lessons so that auditory learners have plenty to listen to and discuss, visual learners have ample graphic displays and demonstrations to study, and kinesthetic learners have concrete objects and models to move about in the environment. It is unlikely that any individual rigidly employs a single learning style for all new concepts encountered. A sound philosophy of instruction that embraces the use of multisensory materials and varied lesson organization seems advisable.

Lesson Organization

One way to classify lessons is according to the type of instructional input (auditory, visual, or kinesthetic) and the level of teacher and student interaction. Using these criteria, elementary teachers commonly employ four fundamental instructional techniques.

- [ ] *Lecture*—auditory and low-interaction
- [ ] *Discussion*—auditory and high-interaction
- [ ] *Demonstration*—visual/auditory and low-interaction
- [ ] *Experimentation*—kinesthetic/visual/auditory and high-interaction

Lectures are characterized by considerable teacher talk and relatively passive students. These lessons are deductive in nature, requiring students to generalize their own conclusions from stated principles and identify specific applications. Little interaction is required as the teacher gives instructions, explains procedures, and summarizes content.

Discussions engage students in verbal activity. Effective questioning techniques can help shift to students more responsibility for carrying on the interaction. Research indicates that most teachers' questions can be answered with a one-word utterance or short phrase. Such questions involve the recall of specific facts and generally require little analysis on the part of the student. With practice, **convergent** questions such as, How many sides does a triangle have? can be rephrased in a more **divergent** manner: What are the differences between a triangle and a square? Divergent questions increase the level of thinking required to formulate answers. A teacher who adopts this style of questioning places greater responsibility on students to carry on the discussion. A three- to five-second pause after a teacher asks a question, called the **wait time**, or response latency, has also been shown to encourage more thoughtful answers and a wider distribution of responses among children in the class (Rowe, 1974). The process is also likely to stimulate the production of thoughtful new questions.

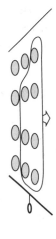

FIGURE 3.1

Demonstrations also place the teacher in the more active role. For example, working a subtraction exercise on the chalkboard coordinates a visual display with an auditory description of the steps involved. Student participation in demonstrations is also effective. For example, after the teacher has introduced subtraction, children can take turns demonstrating the procedure by moving counters on an overhead projector (see Figure 3.1).

Experiments involving concrete objects and symbolic representations require students working alone or in groups to engage in purposeful activity as they develop mathematical ideas. Activities that involve comparing objects, collecting and organizing information, and looking for patterns are often employed for initial concept development. This experience leads to more effective practice of skills and problem-solving applications.

ASSESSING LEARNING NEEDS AND ACHIEVEMENT

To be effective, you must develop expertise in assessing each student's strengths and weaknesses in both the **affective** and **cognitive** domains.

Affective Development

Affective development refers to one's attitudes, beliefs, and values about ideas, objects, and events in the world. Success in learning anything (including mathematics) depends to a large extent on one's attitude and motivation. Though research on the affective component of mathematics instruction is scarce (McLeod, 1986), it seems prudent to assume that the learning environment established by the teacher can significantly affect the way children view mathematics. Because children are particularly perceptive of others' fears and apprehensions, it is especially important for the teacher to present a positive attitude toward mathematics learning.

Cooperative Behavior

One of the main goals of this book is to engage you in a set of experiences that lead to a broader understanding of mathematics and mathematics education. To achieve this goal, manipulative instructional aids are frequently employed to motivate the development of concepts. Course activities for small groups are also suggested to encourage the cooperative use of social resources among the course participants.

It is likewise effective for children to work together in small groups on many learning activities. Children working on a single task in groups of two to four can often accomplish more than can students working alone. Groups should be encouraged to use each other as resources to answer questions and resolve problems. One useful rule for group work is to allow questions of the teacher only if the entire group can agree on the question. This gives children much-needed practice in developing organizational skills and cooperative behavior. Using this method, it is often productive to group together children with different skills and interests and to vary the group membership periodically.

Competitive Behavior

Another motivating force that pervades our society is competition. Most game structures use the dynamic interplay between cooperative and competitive behaviors as a basis for maintaining participant motivation. Team-based activities such as soccer and bridge require cooperation among team members while maintaining a competitive posture toward opponents. One can also compete with oneself in an attempt to improve on past performance.

Helping children redefine goals associated with competitive activities encourages self-improvement. Even team sports can be redefined to serve individual growth needs by subjugating the traditional goal of winning the game to the self-improvement goal of playing a better game. Children can thus predict their own level of performance and evaluate their own progress.

This process can be initiated in the classroom by encouraging cooperation among members of the class to complete a task that might otherwise be impractical. Individual skills improve as well as group skills. A popular, class-size problem involves adding, subtracting, multiplying, and/or dividing a small set of standard values (e.g., 1, 2, 3, 4, 5, 6; or 4, 4, 4, 4) to create a list of number sentences that gives each of the solutions, 1 through 100. For example, using four 4s, one can create equations that equal 1, 2, 3, and so on.

$$(4 \div 4) \times (4 \div 4) = 1$$
$$(4 \div 4) + (4 \div 4) = 2$$

$$((4 \times 4) - 4) \div 4 = 3$$
etc.

This class-size problem can also be assigned as homework to involve the entire family in a mathematical problem-solving experience. Though it is possible to introduce team competition (e.g., see which of two classrooms can find all 100 number sentences first), it is not necessary. When only a few students feel like winners, the debilitating effect of losing on the rest of the children reduces the overall effectiveness of competition as a motivating force. The intent is to harness the compelling features of competition without suffering potentially undesirable social side effects.

Affective Assessment

Often, the difficulties students experience in learning mathematics can be traced to lack of confidence. A history of incomplete instruction and of negative attitudes previously expressed by adults and peers can have a cumulative, disabling effect on learner motivation (Riel, 1982).

Attitude scales offer a systematic means to gain insight into the individual attitudes of an entire class. Such surveys ask children to indicate how strongly they agree (or disagree) with a given statement or how they feel about a particular mathematics exercise. To improve achievement, the teacher may wish to focus greater attention on planning engaging lessons on topics that students seem to dislike. The mathematics attitude scale designed by Lewis Aiken (1972) for grades 5–6 and older children is shown in the shaded box (p. 46). Two instruments adapted for grades 1–2 and 3–4 are shown in Figure 3.2. These instruments are scored by assigning 0 to the neutral feelings, $^{+}1$ (and $^{+}2$ for the Aiken scale) to the positive feelings, and $^{-}1$ (and $^{-}2$ for the Aiken scale) to the negative feelings. Totaling the responses gives a relative positive/neutral/negative attitude score for each child.

Sentence completion forms can also be used to assess student attitudes. Though the results are more difficult to quantify than attitude scales, the free response to items can give an accurate picture of student preferences and fears, especially those of younger children. The following are examples of sentence completion items:

1. When I am in math class, I feel. . .
2. My favorite part of math class is when. . .
3. What I just hate about math is. . .
4. Compared to other subjects, math is. . .
5. I want to learn math because. . .

Understanding more about the attitudes and interests of students can give the teacher insight into appropriate motivation, communication style, student grouping, and lesson planning.

Observation checklists can be used to gather information on individual students for instructional planning. The teacher selects a normal school day and keeps track of the amount of time a student devotes to optional mathematics activities and

Mathematics Attitude Scale

Directions: Please write your name in the upper right-hand corner. Each of the statements on this opinionnaire expresses a feeling or attitude toward mathematics. You are to indicate, on a five-point scale, the extent of agreement between the attitude expressed in each statement and your own personal feeling. The five points are: Strongly Disagree (SD), Disagree (D), Undecided (U), Agree (A), Strongly Agree (SA). Draw a circle around the letter or letters giving the best indication of how closely you agree or disagree with the attitude expressed in each statement.

1. I am always under a terrible strain in mathematics class.   SD D U A SA

2. I do not like mathematics, and it scares me to have to take it.   SD D U A SA

3. Mathematics is very interesting to me, and I enjoy arithmetic and mathematics courses.   SD D U A SA

4. Mathematics is fascinating and fun.   SD D U A SA

5. Mathematics makes me feel secure, and at the same time it is stimulating.   SD D U A SA

6. My mind goes blank and I am unable to think clearly when working mathematics.   SD D U A SA

7. I feel a sense of insecurity when attempting mathematics.   SD D U A SA

8. Mathematics makes me feel uncomfortable, restless, irritable, and impatient.   SD D U A SA

9. The feeling that I have toward mathematics is a good feeling.   SD D U A SA

10. Mathematics makes me feel as though I'm lost in a jungle of numbers and can't find my way out.   SD D U A SA

11. Mathematics is something that I enjoy a great deal.   SD D U A SA

12. When I hear the word mathematics, I have a feeling of dislike.   SD D U A SA

13. I approach mathematics with a feeling of hesitation, resulting from fear of not being able to do mathematics.   SD D U A SA

14. I really like mathematics.   SD D U A SA

15. Mathematics is a course in school that I have always enjoyed studying.   SD D U A SA

16. It makes me nervous to even think about having to do a mathematics problem.   SD D U A SA

17. I have never liked mathematics, and it is my most dreaded subject.   SD D U A SA

18. I am happier in a mathematics class than in any other class.   SD D U A SA

19. I feel at ease in mathematics, and I like it very much.   SD D U A SA

20. I feel a definite positive reaction toward mathematics; it's enjoyable.   SD D U A SA

Note: From "Research on Attitudes Toward Mathematics" by L. Aiken, 1972, *Arithmetic Teacher*, 19(3), pp. 230–31. Copyright 1972 by the National Council of Teachers of Mathematics. Reprinted by permission.

a. Grades 1–2

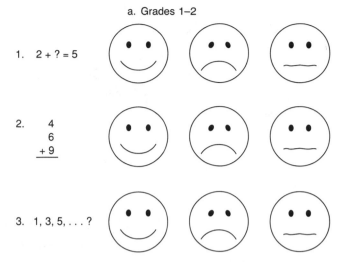

1.   2 + ? = 5

2.   4
        6
      + 9

3.   1, 3, 5, . . . ?

b. Grades 3–4

	Mark Your Response		
	Always	Sometimes	Never
1. I always say math is my favorite subject.	____	____	____
2. I think math is easy.	____	____	____
3. I feel sick when I take a math test.	____	____	____
4. Even when I try I get wrong answers.	____	____	____
5. I use math when I'm not in school.	____	____	____
6. I like doing hard math problems.	____	____	____
7. My parents help me with my math.	____	____	____
8. My parents tell me how they use math.	____	____	____
9. I think only really smart people like math.	____	____	____
10. I wish math was easy like last year.	____	____	____

FIGURE 3.2   Attitude Surveys

games. A record is also maintained of the components of the mathematics lessons in which the child actively participates. Actual classroom behavior and level of voluntary participation are most likely to reflect the child's interest in mathematics. It may be helpful to assign a classroom aide to assist with the observations. The results of such observations can help the teacher reorganize the environment to exploit the interests of the child in the service of learning mathematics. For example, if a child voluntarily joins small group math games, perhaps part of each math lesson can be planned to include small group work.

Cognitive Assessment

The development of intellectual, or cognitive, skills is a primary function of schooling. Table 3.4 lists several types of assessment available to assist the teacher in planning appropriate instruction to meet individual needs.

Developmental assessment provides information on the general cognitive potential of children. Success at conservation (number, length, area, volume, mass, time) and classification (logical classes, class intersection, class inclusion) tasks gives insight into the types of instructional aids, lesson organization, and student groupings likely to foster effective learning. Similarly, tasks such as controlling variables

TABLE 3.4 Types of Cognitive Assessment

Type of Instrument	What It Measures
1. Developmental assessment	Developmental level
a. Conservation tasks (individual)	
b. Classification tasks (individual)	
c. Controlling variables tasks (individual)	
d. Ratio and proportion tasks (individual)	
2. Intelligence quotient (IQ) tests	General intellectual ability
a. California Short Form Test of Mental Maturity (group)	
b. Stanford-Binet (individual)	
c. Wechsler Intelligence Scale for Children (individual)	
3. Norm-referenced achievement tests	Achievement relative to general population
a. Comprehensive Tests of Basic Skills (group)	
b. Iowa Test of Basic Skills (group)	
c. Metropolitan Achievement Tests (group)	
d. Stanford Achievement Tests (group)	
4. Criterion-referenced tests	Achievement relative to stated performance objectives
a. Diagnostic, placement, and mastery tests that accompany most textbooks (group)	
b. Peabody Individual Achievement Test (individual)	
c. Key Math Diagnostic Arithmetic Test (individual)	
d. Stanford Diagnostic Mathematics Test (group)	
e. Teacher-made diagnostic tests	

and applying proportional reasoning are important abilities for upper-grade students. Since developmental tasks are administered individually, they are not frequently employed as part of regular assessment procedures. For children displaying particular difficulty learning mathematics (or particular prowess), a careful analysis of conservation, classification, and formal operational abilities can provide you with additional insight into potential learning problems and suggest corrective action.

IQ tests have been used for over fifty years as a means of screening individuals for school and job opportunities. In recent years, these tests have come under increasing criticism. Opponents contend that an individual's IQ score is not fixed at birth but can be increased through careful training and that the tests are culturally biased. As a result, using IQ tests to screen students has distorted the distribution of culturally different and low socioeconomic students in gifted and other special education programs and, consequently, in the job market as well. In response to Arthur Jensen's claim (1969) that education and career decisions should be made based in part on IQ results, Benjamin Bloom observed:

> The psychologist may wish to speculate about how to improve the genetic pool—the educator cannot and should not. It is through the environment that [she or] he must fashion the educational process. If heredity imposes limits—so be it. The educator must work with what is left, whether it be 20% of the variance or 50%. (1969, p. 50)

Norm-referenced achievement tests are designed to indicate individual levels of performance compared with those of a large population of similar students. Because one achievement test cannot include items that address every important concept and skill taught at the elementary level, a limited number of items must be selected. It is possible for two properly designed tests to sample entirely different subsets of mathematics concepts and be equally **valid** (indicate overall mathematics achievement) and **reliable** (produce consistent scores among individuals possessing equivalent mathematics knowledge).

Results of norm-referenced achievement tests may be divided into subscales that indicate performance in computation, conceptual understanding, and mathematics applications. Performance can be reported using grade-level equivalent scores or percentile rankings. A grade-level score of 4.3 means that the student is performing at the grade 4, month 3 level. A result showing a student at the 75th percentile means that 75 percent of the children in the same grade scored the same or less. A child at the 50th percentile is performing at the expected grade level.

Criterion-referenced tests assess knowledge of specific objectives. The results of a well-prepared criterion-referenced test should indicate which concepts and skills a student has mastered and which require reteaching. Generally, only previously taught concepts and skills are included on such tests. Most textbook series include a set of criterion-referenced tests designed to help teachers assign students to appropriate groups (placement tests), assess performance relative to specific lesson objectives (achievement tests), and determine specific areas of student weakness to assist in corrective instruction (diagnostic tests).

Another type of criterion-referenced test is the **teacher-made test**. Teachers sometimes find that tests supplied with textbooks or other commercial instruments are unsuitable for their class. Teachers can easily design their own criterion-referenced tests by following these four steps:

1. Identify the *concept* or skill to be assessed.
2. Specify a limited number of *objectives.*
3. Design test *items* for each objective.
4. Evaluate item *validity* for future use.

A diagnostic test assesses in depth a limited number of objectives associated with a particular mathematics concept or skill. At the kindergarten—grade 1 level, assessment using three to five diagnostic tasks is likely to require a one-to-one session between the child and the teacher or aide. For older students, paper-and-pencil tests of ten to twenty items are more efficient.

Table 3.5 lists objectives and associated diagnostic tasks that assess counting subskills of kindergarten children, and Table 3.6 lists sample objectives and diagnostic items that assess multiplication skills of grade 4 students.

Once designed, the diagnostic test should then be administered to the intended group of children. It is important to evaluate carefully the results to see if any item is confusing or inappropriate. Discard any items that do not seem to assess the objective as planned. Retest these objectives with new diagnostic items. For example,

TABLE 3.5 Counting Objectives and Diagnostic Tasks

Objective	Diagnostic Task
Given a pile of 9 proportional rods, the child will place them in order from shortest to longest.	Give the child 9 different Cuisenaire rods in a cup. Pick any 3 rods and place them in order from shortest to longest. Replace the rods in the cup and ask the child to repeat the process with all 9 rods. If the child fails, repeat with smaller sets of rods.
Given numeral cards 1–9, the child will match the cards in order with a set of 9 seriated rods.	With the 9 rods in order, give the child a set of numeral cards 1–9. Ask the child to find the card showing the numeral 1. Tell the child to put the 1 card next to the shortest rod. Tell the child to continue placing the cards next to each rod in order until all the cards are placed. If the child fails, repeat with smaller sets.
Shown 9 identical objects in a line, the child will place 9 additional objects in one-to-one correspondence.	Place 9 identical blocks in a line. Give the child a supply of plastic chips in a cup. Ask the child to match each block with a chip. If the child fails, repeat with smaller sets.
Given a set of picture cards showing 1–9 objects and numeral cards 1–9, the child will place the correct numeral card on each picture card.	Place 9 picture cards in a line without regard to order. Mix the numeral cards and ask the child to match each numeral card with the correct picture card. If the child fails, place picture cards in order and repeat task or use smaller set.

TABLE 3.6 Multiplication Objectives and Diagnostic Items

Objective	Diagnostic Task	
Given a 1-digit times 2-digit multiplication exercise involving no regrouping, the child will compute the product using paper and pencil.	$\begin{array}{r} 24 \\ \times\ 2 \\ \hline \end{array}$	
Given a 1-digit times 2-digit multiplication exercise requiring regrouping, the child will compute the product using base-10 materials and record the results using symbols.	$\begin{array}{r} 25 \\ \times\ 3 \\ \hline \end{array}$	25 25 25
Given a 2-digit times 2-digit multiplication exercise, the child will compute the product using paper and pencil.	$\begin{array}{r} 22 \\ \times 16 \\ \hline \end{array}$	
Given a 2-digit times 2-digit multiplication exercise with a zero in the ones place, the child will compute the product using pencil and paper.	$\begin{array}{r} 50 \\ \times 23 \\ \hline \end{array}$	75

a subtraction item designed to assess regrouping might involve a zero in the minuend (e.g., $304 - 128 = ?$). Although this item does require regrouping, students who *are* able to regroup for subtraction across one place value may still miss this item, as it requires regrouping across two place values before subtracting in the ones column. With experience, designing diagnostic tests becomes second nature for most teachers.

A Final Observation

It is important to note that criterion-referenced tests are constructed from lists of performance objectives. The content assessed for a given test is, therefore, limited to mathematics content that can be expressed in performance terms. As observed earlier, while diagnostic items for facts and skills are relatively easy to write, concept and problem-solving items are more difficult. Teachers should be aware of this problem and avoid overemphasizing the role of diagnostic tests when planning mathematics lessons. Criterion-referenced test results, integrated with classroom observations and information on student attitudes, can enhance the teacher's ability to analyze student needs and plan effective instruction.

ACCOMMODATING INDIVIDUAL DIFFERENCES

Children who exhibit particular competence or protracted difficulty in learning mathematics often show marked improvement in performance and behavior if the teacher modifies the classroom environment and curriculum to better account for their specific needs. Several types of special-needs learners have been identified.

Students classified as **gifted and talented** include those who exhibit a high level of intellectual achievement, talent in the creative and performing arts, and/or leadership ability. Children in this category possess a high level of general ability, demonstrate above-average persistence in completing tasks, and show particular creativity in one or more domains (Johnson, 1983). The criteria for identifying gifted and talented children continue to be controversial. Each state has its own means of identification based on a combination of standard IQ scores, academic achievement, and teacher judgment. In some regions, only students who achieve IQ scores of 130 and above are included in special instructional programs. Historically, reliance on this single measure of performance has limited access to special programs for gifted children from different cultures and from low socioeconomic backgrounds. Also, due to peer pressure, students embarrassed by their own talents may attempt to conceal their abilities from their classmates, parents, and teachers. The teacher plays a key role in identifying such children, especially those from non-middle-class backgrounds.

Children who show the potential to excel in mathematics should be given additional opportunities to engage in problem solving and enrichment activities. Although gifted children can often finish more work in a given amount of time than normal students, teachers should resist the urge simply to assign additional exercises to gifted students similar to those given to the rest of the class. Increased performance expectations for gifted students should include measures of divergent thinking, flexibility, originality, and elaboration. Examples of appropriate activities that should receive additional emphasis when planning for gifted mathematics students are given in the problem-solving and special-needs sections in chapters 6 through 14.

It is also useful for high-achieving students to spend a limited amount of time tutoring other children who need additional help. Explaining concepts and procedures to others is an excellent way to extend what is already known and to practice interactional skills. Care must be taken, however, to limit the amount of time taken from the gifted student's own learning activities to tutor others during the school day.

The Bureau of Education for the Handicapped (BEH) estimates that about 12 percent of children eighteen years or younger are physically, emotionally, or intellectually impaired. Public Law 94–142: Education for All Handicapped Children Act of 1975 requires local districts to provide for the needs of special students (except gifted and talented) in the **least restrictive educational environment**. It also provides for due process and guaranteed access by parents to all test results and student files. This law has had the effect of including many children in the regular classroom who otherwise would be segregated into special classes or schools.

Many special-needs children experience protracted difficulty in learning mathematics. With such students, the teacher should avoid using the same methods employed by previous teachers to simply reteach concepts and skills. A consistent pattern of failure will almost surely lead to frustration, intimidation, and reduced motivation. Inappropriate previous instruction, whether at school or at home, will likely continue to affect attitudes and behavior in school unless a knowledgeable teacher intervenes with alternative methods. The use of manipulative models and other specialized techniques for teaching the special-needs child will be presented in chapters 6 through 14.

Approximately 3 percent of the school-aged population are classified as **learning disabled**. These students exhibit normal intelligence; perform at least one year below grade-level norms in reading, mathematics, or other subjects; may display perceptual, memory, or orientation impairments; and may be easily distracted or hyperactive. The classification is not intended to include children with learning problems primarily associated with sensory or physical impairments, mental retardation, cultural differences, or other environmental factors. An additional 2 percent of the school population are classified as **educable retarded** (IQ performance range of 50–75) and about 5 percent are **hearing, speech, visually,** or **physically impaired**.

The classification **behavior disordered** includes those children whose inability to learn cannot be explained by intellectual, sensory, or health factors (Bower, 1969). All children exhibit behavior problems from time to time as a natural part of the maturation process. If these occurrences are mislabeled by the school and become self-fulfilling, irrevocable damage may be done to students' school careers. While surveys show that classroom teachers report as many as 20 percent of their students as having behavior problems, the severely behavior disordered are estimated at only 1 percent of the school-aged population (Kelly, Bullock, & Dykes, 1977). This disparity may partially explain the disproportionate number of boys, culturally different, and low socioeconomic children classified as behavior disordered in the public schools today.

Table 3.7 summarizes the identifying characteristics of special-needs students and lists several teaching suggestions for each category (Bley & Thornton, 1981; Glennon, 1981; Hallahan & Kauffman, 1978; Mercer & Mercer, 1985; Thornton, Tucker, Dossey, & Bazik, 1983).

Individualized Education Programs (IEPs)

Public Law 94–142 requires that an Individualized Educational Program (IEP) be developed annually for each special-needs student. The intent of the law is to require districts to make provisions to include, or **mainstream**, each child in the least restrictive educational environment subject to the limitations imposed by individual handicaps (Fennell, 1984). The IEP is a detailed program of instruction written for an individual child. The child's parent, the teacher, a district representative, and, as appropriate, the child must be involved in writing and approving an IEP. Each IEP must contain a statement of

1. The student's current level of performance.
2. Short-term objectives and annual goals.
3. The planned instructional schedule, including the percentage of time spent in various environments and the personnel involved.
4. Special equipment or instructional aids required.
5. Performance objectives with evaluation criteria.

As the mainstreaming requirements of PL 94–142 increase the proportion of special-needs children in the regular classroom, it becomes more important for teachers to be familiar with instructional procedures appropriate for each type of special student. Teaching special-needs children has much in common with quality

TABLE 3.7 Teaching Suggestions for Special-Needs Children

Identifying Characteristics	Teaching Suggestions
1. Gifted and talented	
Verbal fluency Superior problem-solving ability Uncommon creativity Mature humor Persistent, self-actuated behavior Willing to take intellectual risks Flexible thinker	Attempt to identify children who show particular ability or creativity. Be particularly observant of culturally different and low socioeconomic children who may mask abilities due to peer pressure. Provide challenging assignments that tap problem-solving and organizational skills. Maintain high performance expectations. Utilize community resources, such as a local computer company, to expand interests and problem-solving opportunities. Work with parents and auxiliary school personnel to identify special equipment and resources available to assist with lesson planning. Encourage a limited amount of cross-age and peer tutoring using high achievers as tutors.
2. Hearing impaired	
Minor to total loss of hearing Speech may be affected if loss occurred before onset of language	Maintain eye contact when speaking. Minimize background noise and distraction. Provide systematic evaluation and immediate feedback of performance. Use visuals and the overhead projector. Never talk with back to the class. Write key points and vocabulary on the board. Try to stand still when giving directions at a distance of about 6 feet from the child. Speak naturally without exaggerated movements.
3. Speech impaired	
Slurred or halting speech Reluctance to talk in class	Encourage verbalization. Avoid interrupting incorrect verbalization. Provide corrective feedback in a supportive manner. Work with speech therapist by providing a list of current vocabulary.
4. Visually impaired	
Minor to total loss of sight Reading ability may be affected if loss occurred prior to schooling	Place students in appropriately lighted areas and encourage them to use their available sight. Treat blind and partially sighted students, as much as possible, like other students. Maintain high achievement expectations. Foster independence by minimizing dependence on sighted students. Maintain a consistent organization of desks, tables, and so on, and when changes occur, help the child establish new mobility patterns. Use raised-letter or braille labels to adapt instructional materials (e.g., rulers, dice, cupboards). Encourage positive interaction among all children in the class.

TABLE 3.7 *continued*

Identifying Characteristics	Teaching Suggestions
5. Physically impaired	
Health or nervous system disorders Physical handicaps	Become informed about specific physical limitations. Organize physical arrangements to accommodate needs (e.g., ramps, grab rails in bathroom) and provide special equipment (e.g., typewriter or word processor, special desk). Maintain high achievement expectations. Encourage positive interaction among all children in the class.
6. Educable retarded	
Poor language skills Very low general academic ability Limited attention span Limited problem solving ability	Carefully sequence instruction, making small but consistent steps. Provide concrete materials that offer optimal perceptual salience. Provide frequent opportunities for review using games and activities. Encourage verbalization of learned processes and skills. Provide systematic evaluation and immediate feedback on *each* learning activity. Maintain high expectations for performance of basic skills required for everyday life.
7. Learning disabled	

Perceptual deficits
  Seeing part/whole relationships
  Hearing patterns when counting
  Distinguishing between symbols (e.g., 2 and 5)
  Distinguishing between spoken numbers (e.g., 15 and 50)
  Reversing digits and place values
  Following written and oral directions
Memory deficits
  Copying assignments and taking dictation
  Retaining basic number facts and algorithms
Integrative deficits
  Solving word problems
  Counting-on in sequence
  Oral drills
	Carefully assess current performance and commence instruction at appropriate level.
	Minimize negative criticism of student's work by praising correct elements no matter how insignificant.
	Provide systematic evaluation and immediate feedback of performance.
	Carefully select instructional materials to provide appropriate sensory input for concept development (e.g., base-10 blocks, squared paper place-value tables).
	Provide a private study area for easily distracted students.
	Utilize available resource personnel.
	Encourage positive interaction among all children in the class.

8. Behavior disordered	
Easily distracted Physically active Excessive mood swings Impaired social skills Medically diagnosed autistic and schizophrenic children	Discuss child's history, behavior, and classroom management and teaching techniques used previously by support personnel (special education teacher and psychologist). Establish a consistent daily instructional schedule: a jointly constructed learning contract is often useful in keeping student on task. Establish clear limits of acceptable behavior and consequences, and be calm and consistent in applying them. Consistently identify the behavior as being inappropriate, not the child. Establish a time-out area for the child to go when behavior is disruptive. Maintain realistic expectations for behavior and academic performance.

instruction for all children. If the teacher plans intellectually stimulating lessons that approach the subject matter at an appropriate level of abstraction, many potential instructional problems and behavior-related disruptions can be avoided (Baroody, 1986). Children excited about learning are far more likely to channel their energies in productive ways, making school a healthier place for everyone.

THE TEACHER'S ROLE

The classroom teacher plays a central role in helping children meet individual learning objectives in school. All facets of a child's life contribute to school success. Enlisting support from the home and community is essential, as well as establishing a classroom environment that encourages children to take intellectual risks. Little real intellectual growth is likely unless existing knowledge and beliefs are freely explored and challenged.

It is essential that teachers of mathematics bring to bear on the instructional task a broad range of skills and knowledge. Although it is necessary to possess a complete understanding of the art and science of mathematics, this knowledge alone will not guarantee that an individual will become a master teacher. Along with content knowledge, the teacher must understand expected cognitive and affective developmental patterns; must be expert at organizing and managing complex instructional materials and events; must communicate effectively with and enlist the active participation of parents, colleagues, and other members of the community; must draw upon support personnel for assistance with special-needs children; and, perhaps most importantly, must engage in an ongoing program of self-renewal and personal development to keep professional skills up-to-date and to maintain respect for the learner and the teaching/learning process.

Assigning Homework

Actively involving family members in the learning process generally improves a student's school performance. However, homework assignments that require technical assistance not readily available in the home should be avoided. Homework activities are most appropriate if designed to take advantage of special features of the home environment that are unavailable at school. For example, children can collect data on the number of pets on their street, which can then be graphed and analyzed the following day in school. Parents should be encouraged to include their children in ongoing cooking and building activities to provide practical measurement experiences. Often parents can read to or be read to by their children and provide the one-on-one practice drills that are difficult to arrange in school.

Care should be taken to ensure that parents are aware of the intended learning objective of a homework assignment and that they have reasonable performance expectations. A note to parents describing the assignment and giving examples of what should be expected on the part of the student can make homework more successful. Children can also incorporate learning activities into common events at home. For example, children can engage in commercial drills, seeing how many multiplication flash cards they can answer correctly during television commercials. The creative teacher invents homework assignments that support classroom instruction yet take advantage of the special features of the out-of-school environment.

Lesson Planning

Advanced planning for lessons helps the beginning teacher think through the sequence of teaching actions, materials needed, and organizational logistics before standing in front of the class. The following sample is an adaptation of the clinical teaching model lesson outline presented in chapter 2. This lesson presupposes that the students have already mastered the concept of multiplication.

Sample Grade 4 Lesson Plan for Multiplication Facts 0–5

1. OBJECTIVE: Learner will be able to recall the multiplication facts for all combinations of factors 0 through 5 on a paper-and-pencil quiz (lesson will be repeated using other sets of factors).

2. MATERIALS: Each child needs: 5 × 5 bingo card with randomly placed products to 25, supply of plastic chips. Set of open number sentence strips on transparency film showing all factor combinations 0–5, sample bingo card on transparency, overhead projector, basic fact quiz.

3. ANTICIPATORY SET: On the overhead, show bingo card and one fact strip. Ask the class, Who knows the product? If answer appears on the card, read the number sentence and the product, then place a chip over the answer. If not, repeat until a correct product appears on the card. Explain that the class is going to play multiplication fact bingo to practice the basic facts and will have a quiz at the end of the period to check which facts each student has memorized.

4. INSTRUCTION: Teacher leads the first game.

5. GUIDED PRACTICE: The winner (the first student to get five correct products in a vertical, horizontal, or diagonal row) becomes the leader. The leader picks the person who is to say the product aloud. *Alternate version*: Silent bingo— no number sentences or products are spoken during the game. Leader writes down each number sentence selected. Winner must read off all five winning products and associated number sentences. Leader checks for accuracy and corrects mistakes. The game can be played so that winners get credit if they give any correct number sentence (e.g., 1 × 4 or 2 × 2), or the exact number sentence may be required.

6. CLOSURE: To check facts memorized by each student, give written quiz of 36 randomly selected basic multiplication facts 0–5.

7. INDEPENDENT PRACTICE: Assign students to play multiplication fact bingo in small groups, or if time is short, with family for homework (students will need to construct sets of bingo cards and appropriate number sentence flash cards to take home).

SUMMARY

Chapter 3 presents techniques for teaching mathematics facts and skills. The behaviorist theory of mathematics learning proposed by Edward Thorndike argues that learning arithmetic skills is the result of accumulating a hierarchical set of mental habits called bonds. William Brownell countered this claim by demonstrating the existence of rational procedures other than direct recall that were employed by children as they carried out arithmetic operations. Brownell's proposal for rational learning remains a central feature of quality mathematics instruction today.

A mathematics curriculum is specified using performance objectives or more general instructional objectives. Performance objectives include a description of the learning task, the students' working conditions, and the specification of criteria for success. Instructional objectives provide a description of the learning task without identifying the working conditions or success criteria. Objectives can be more easily specified for the knowledge and comprehension levels of Bloom's categories of knowledge than for higher level problem-solving tasks.

Several types of instruments are available to assess student affective and cognitive development. Each type of assessment measures different student attitudes and abilities and has specific applications in the classroom. Two frequently used instruments are the sentence completion form for affective assessment and the teacher-made criterion-referenced test for cognitive assessment. Individually administered developmental tasks can also be useful for assessing mathematics readiness of young children and special-needs students.

Public Law 94–142 ensures greater access to regular classroom instruction for many special-needs children. The teacher must provide instruction for such main-streamed students using a five-part Individualized Education Program (IEP). Several categories of special-needs students are recognized in the public schools, and specialized instructional techniques may be required to ensure adequate mathematics achievement for all children.

COURSE ACTIVITIES

1. Select a skill such as writing the counting numbers in order or measuring the length of objects using a ruler. Develop an intrinsically motivating problem situation that could be used to provide practice in the selected skill. Give the problem to your peers and, if possible, an appropriate group of students to evaluate its effectiveness in providing the intended practice experience.

2. Select an interesting problem from a resource book of elementary mathematics problems (see the reading list below and in chapter 4 for examples). List the mathematics concepts that you feel should be introduced to aid in the solution. Describe how you could integrate these concepts into a discussion of the problem situation.

3. List the categories of knowledge as defined by Benjamin Bloom. Write ten examples of your own mathematics knowledge on separate file cards and work in small groups to classify them into the appropriate categories. Describe how this classification of knowledge might be useful in organizing instruction.

4. Describe the three components of a performance objective. Select an instructional objective from the appendix and rewrite it in performance terms. List several of the advantages and disadvantages of using performance objectives for instructional planning.

5. Using the examples given in this chapter, develop and administer a mathematics attitude sentence completion form to several of your peers or, if possible, a class of elementary students. Summarize the results in a table and discuss the attitude patterns you discover. If you surveyed elementary students, discuss with the classroom teacher how your results compare with the teacher's assessment of students' attitudes. How could you use the results to improve mathematics instruction?

6. List the four types of cognitive assessment instruments discussed in this chapter and the purpose of each. Identify which types of tests are currently used in schools in your area. How are the results used? What instructional decisions are made based on these results? Using specific examples from your own experience, discuss the uses and abuses of tests with your peers.

7. Visit a special education teacher in your area. Discuss the range of services available to special-needs students and their teachers. If possible, review an individual education program and discuss the IEP development process with a knowledgeable teacher. Compare it with the list of the specific components that must be included in an IEP.

8. Using the outline given in chapter 2, design a lesson plan to teach a mathematics skill such as the addition algorithm for two-digit numbers. Share your lesson plan with your peers and discuss each component. Focus particularly on the Anticipatory Set used to motivate the skill.

9. Read one of the *Arithmetic Teacher* articles listed in the reference section. Write a brief report summarizing the main ideas of the article and describe how the recommendations for instruction might apply to your own mathematics teaching.

REFERENCES AND READINGS

Aiken, L. (1972). Research on attitudes toward mathematics. *Arithmetic Teacher, 19*, 229–234.

Ashlock, R., & Washbon, C. (1978). Games: Practice activities for the basic facts. In M. Suydam (Ed.), *Developing computational skills, 1978 Yearbook of the National Council of Teachers of Mathematics*. Reston, VA: National Council of Teachers of Mathematics, 39–50.

Baroody, A. (1986). The value of informal approaches to mathematics instruction and remediation. *Arithmetic Teacher, 33*(5), 14–18.

Bley, N., & Thornton, C. (1981). *Teaching mathematics to the learning disabled*. Rockville, MD: Aspen Systems.

Bloom, B. (1969). Replies to Dr. Jensen's article. *ERIC/ECE Newsletter*. ERIC Clearinghouse on Early Childhood Education, Urbana, IL: University of Illinois, 3, 50.

Bloom, B., Englehart, N., Furst, E., Hill, W., & Krathwohl, D. (1965). *Taxonomy of educational objectives—The classification of educational goals, Handbook I: Cognitive domain*. New York: David McKay.

Bower, E. (1969). *Early identification of emotionally handicapped children in school* (2nd ed.). Springfield, IL: Charles C. Thomas.

Brownell, W. (1935). Psychological considerations in the learning and the teaching of arithmetic. In *The teaching of arithmetic: The tenth yearbook of the National Council of Teachers of Mathematics*. New York: Columbia University, Teachers College.

Bushaw, D., Bell, M., Pollak, H., Thompson, M., & Usiskin, Z. (1980). *A sourcebook of applications of school mathematics*. Reston, VA: National Council of Teachers of Mathematics.

Charles, R. (1983). Teaching: Evaluation and problem solving. *Arithmetic Teacher, 30*(5), 6–7, 54.

Fennell, F. (1984). Mainstreaming in the mathematics classroom. *Arithmetic Teacher, 32*(3), 22–27.

Gagne, R. (1970). *The conditions of learning*. New York: Holt, Rinehart & Winston.

Glennon, V. (1981). *The mathematical education of exceptional children and youth, an interdisciplinary approach*. Reston, VA: National Council of Teachers of Mathematics.

Greenes, C., Spungin, R., & Dombrowski, J. (1977). *Problem-Mathics: Mathematical challenge problems with solution strategies*. Palo Alto, CA: Creative Publications.

Hallahan, D., & Kauffman, J. (1978). *Exceptional children*. Englewood Cliffs, NJ: Prentice-Hall.

Hodges, H. (1983). Learning styles: Rx for mathophobia. *Arithmetic Teacher, 30*(7), 17–20.

Jacobsen, E. (1975). *The effect of different modes of practice on number facts and computational abilities*. Unpublished manuscript, University of Pittsburgh, Learning Research and Development Center.

Jensen, A. (1969). How much can we boost IQ and scholastic achievement. *Harvard Educational Review, 39*, 1–123.

Johnson, M. (1983). Identifying and teaching mathematically gifted elementary school children. *Arithmetic Teacher, 30*(5), 25–26, 55–56.

Judd, W., & Glaser, R. (1969). Response latency as a function of training method, information level, acquisition, and over learning. *Journal of Educational Psychology Monograph, 60*(4), 1–30.

Kelly, T., Bullock, L., & Dykes, M. (1977). Behavioral disorders: Teachers' perceptions. *Exceptional Children, 43*(5), 316–318.

Krulik, S., & Reys, R. (Eds.). (1980). *Problem solving in school mathematics: 1980 Yearbook*. Reston, VA: National Council of Teachers of Mathematics.

Mager, R. (1975). *Preparing instructional objectives* (2nd ed.). Belmont, CA: Pitman Learning.

Mathematics framework for California public schools. (1985). Sacramento: California State Department of Education.

McLeod, D. (1986, May). *Affective influences on mathematical problem solving*. Paper presented at the Conference on Affective Influences on Mathematical Problem Solving, San Diego, CA.

Mercer, C., & Mercer, A. (1985). *Teaching students with learning problems*. Columbus, OH: Merrill.

Riel, M. (1982). *Computer problem solving strategies and social skills of linguistically impaired and normal children*. Unpublished doctoral dissertation, University of California, Irvine.

Rogoff, B., & Lave, J. (1983). *Everyday cognition: Its development in social context*. Cambridge: Harvard University Press.

Rowe, M. (1974). Relation of wait time and rewards to the development of language, logic and fate control: Part II—Rewards. *Journal of Research in Science Teaching, 11*(4), 291–308.

Sears, C. (1986). Mathematics for the learning disabled child in the regular classroom. *Arithmetic Teacher, 33*(5), 5–11.

Suppes, R., & Morningstar, M. (1972). *Computer-assisted instruction at Stanford, 1966–1968: Data, models, and evaluation of arithmetic programs*. New York: Academic Press.

Thorndike, E. (1922). *The psychology of arithmetic*. New York: Macmillan.

Thornton, C., Tucker, B., Dossey, J., & Bazik, E. (1983). *Teaching mathematics to children with special needs*. Menlo Park, CA: Addison-Wesley.

4

Teaching Mathematical Problem Solving

A child who expects things to "make sense" looks for the sense in things...

—Mary Baratta-Lorton

Upon completing Chapter 4, the reader will be able to

1. Describe the important features of routine and nonroutine mathematics problems and give an example of each.
2. Describe five techniques for teaching children how to approach story problems.
3. Describe the four steps of a basic problem-solving plan and give an example of a model solution.
4. Give examples of several strategies for solving nonroutine problems.
5. Describe three techniques that enable special-needs students to engage in problem-solving experiences.
6. Write a first draft of a philosophy of teaching.

Children are generally curious to discover how things work. Mathematical concepts and skills can help them make accurate of predictions about how things work and assist with the solutions of problems. The following story illustrates the different ways a problem might be addressed by three people of various backgrounds.

An engineer, a mathematician, and an anthropologist were taking basic survival training, and their instructor was giving a session on survival in the desert. He maintained that a person stranded in the desert—alone, without water, but within sight of an oasis—should walk half the distance to the oasis, then rest, then walk half the remaining distance, rest again, and continue this procedure until the oasis is reached.

The mathematician immediately responded that following this procedure would always result in dying of thirst. She claimed that if, after each rest, the thirsty traveler walked only half the remaining distance to the oasis, there would be an infinite number of rest stops, and the person would never arrive! One solution might be to walk halfway plus one step before each rest.

The engineer agreed with the logic of the mathematician but recognized that real-world problems often do not exactly follow their underlying mathematical models. In this case, he was sure that, due to round-off error, he could get close enough to the oasis using the recommended procedure to satisfy his thirst.

The anthropologist questioned the very basis for the recommended survival procedure. Observing that animals and people who live in the desert almost always travel during the night, he pointed out that, following the local custom, the half-way rule could not work, since distance from the oasis would be difficult to judge without sunlight. He suggested that the traveler should rest during the day and walk as far as possible during the night, using the stars to guide the way.

Each of these solutions is correct when guided by the assumptions of the solver. Mathematical models can often motivate critical insights necessary to solve theoretical problems. To be useful in the real world, however, the solution must be empirically tested by conducting an experiment. Finally, the solution should be evaluated against competing solutions to assess its relative merits and social acceptability.

Even simple strategy games can take on new meaning when sequences of moves are systematically analyzed. For example, let's explore the game of multi-pile NIM. In this game, fifteen counters are positioned in three rows of seven, five and three counters, respectively (see Figure 4.1).

Rules : Two players take turns removing one or more counters from any row. Each player may remove counters from one row only on each turn but may switch to another row on the next turn. The player who removes the last of the fifteen counters wins.

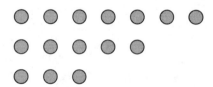

FIGURE 4.1   7/5/3 NIM Game

NIM playing skill improves with experience as players learn to recognize winning patterns. For example, observe that a player who leaves an opponent faced with two equal rows can always win. Regardless of how many counters are removed from either of the two equal rows by Player 2, Player 1 can make the rows equal again by removing the same number of counters from the other row. Eventually, Player 1 will be left with the last counter to remove (see Figure 4.2).

By working backward from states of **parity** (demonstrated by the paired rows in the example), it is possible to establish a series of moves that guarantees the first player will win regardless of the countermeasures employed by Player 2. As we

FIGURE 4.2   Winning Strategy: Leave Two Equal Rows

will see later in this text, such explorations not only lead to winning strategies but may also encourage students to discover generalized solutions for a whole family of related problems.

> Play 7/5/3 NIM several times with a partner and try to work out a winning strategy that assures victory for Player 1. Use the technique of working backward from various winning states for Player 1. Keeping careful records of each move will help you unravel this problem. (Hint: one solution involves the powers of 2: 1, 2, 4, 8). As an extended activity, add one more counter as a fourth row. How would this starting display affect your winning strategy?

CHARACTERISTICS OF GOOD PROBLEM SOLVERS

Recent research indicates that students who are successful at solving problems tend to focus on the problem's underlying mathematical structure. Novice problem solvers often attend to irrelevant, surface details of the problems, ignoring important structural relationships (Krutetskii, 1976; Silver, 1979).

Good problem solvers also seem to structure mathematical concepts and rules into related chunks that make these problem components easier to recall at appropriate times (Chi, Feltovich, & Glaser, 1981). Conversely, novice problem solvers tend to memorize mathematical knowledge as unrelated rules, making it far more difficult for them to perceive critical relationships that may play an important role in solving a problem. Such rote learning also makes it more difficult to recognize inconsistencies in a set of procedures (Brown & Burton, 1978).

Good problem solvers demonstrate the ability to identify pertinent information and to ignore irrelevant or superficial problem features. This ability enables the capable problem solver to view a specific problem as one example of a whole class of similar problems, a critical attribute that seems to separate good problem solvers from their less capable peers (Carpenter, 1985).

In this chapter, we will review specific strategies for solving mathematical problems. The techniques employed encourage exploration of the underlying mathematical structure associated with each problem.

ROUTINE AND NONROUTINE PROBLEMS

For simplicity, we will divide mathematics problems into two characteristic types: routine **story problems** and nonroutine **process problems**.

Story Problems

The word, or *story*, problems that appear at the ends of textbook chapters are designed to provide follow-up practice in applying previously learned concepts and skills. Story problems are also employed on a daily basis to practice reading skills and to motivate new mathematical ideas and procedures.

To solve story problems, students first read and evaluate the problem statement, and then select an appropriate operation and apply it to the values given (LeBlanc, 1982). As students gain experience, multistep exercises are introduced.

One-Step Story Problem

Alan and Nick each bought the same stamp for their collection. Alan paid $14. Nick paid $1 more. How much did Nick pay for his stamp?

Multistep Story Problem

Alan bought two stamps from Nick, one for $10 and another for $14. Later, he sold the first stamp for $12 and the second for $16. How much profit did Alan make on the sale?

These two examples are fairly typical of traditional story problems found in elementary mathematics texts. Notice that all the values included in each problem must be employed to find a solution. In everyday problem solving, it is almost always necessary to select among a range of values present in the problem situation in order to derive a satisfactory solution. In fact, it is often the process of systematically identifying and rejecting extraneous information that makes real-world problems difficult to solve (e.g., selecting the best mortgage for a house, or picking a stock to buy by reviewing corporate financial statements).

Other Types of Story Problems.   Problems with extraneous or insufficient information are beginning to appear in textbooks; they give students practice in selecting critical values and identifying missing information required for a solution.

Extraneous Information Problem

Alan bought three stamps from Nick for $10, $12, and $15. He sold the first two stamps for $14 each. How much profit did Alan make on the sale?

Missing Information Problem

Alan bought two stamps from Nick. He sold the first for $12 and the second for $14. How much profit did Alan make on the sale?

To introduce such story problems, have students identify the extraneous or missing information in a set of specially prepared story problem cards. It is also advisable to include one or two story problems containing only the required values. Additional techniques for introducing story problems include:

1. Headline problems
Write a **headline** summary for a problem. Students supply the story to go with the headline.
Example of headline:
Nick Has 10 More Stamps Than Alan

	Possible Story: Nick has 75 stamps. Alan has 65 stamps. How many more stamps does Nick have?
2. No-value problems	Present story problems with **no values** specified. Students are asked to describe (orally or in writing) the solution procedure. Students can supply their own values if they wish. Example: Alan and Nick each have a large stamp collection. How many stamps do they have altogether?
3. Fill-in value problems	Write story problems in which the values must be filled in by the students before solving them. Example: Nick has _____ stamps in his collection. He sold _____ stamps to Alan. How many stamps does Nick have left?

Story problem instructional activities are included in several of the following chapters. While practice solving routine story problems is important at all grade levels, children should also learn strategies for solving nonroutine process problems.

Process Problems

Process problems differ from story problems in that they cannot be solved immediately by selecting and applying one or more appropriate operations. Process problems can also be used daily to motivate new skills and concepts or as follow-up activities to apply previously learned mathematical ideas and procedures. Solving process problems requires flexible thinking and good organizational skills. Success often relies on an ability to exploit the underlying mathematical structure of a problem using a general solution strategy (such as working backward in the 7/5/3 NIM game). Examples of several general solution strategies are presented in the next section.

The following example offers insight into the different *quality* of mathematical thinking required to solve process problems successfully:

> Alan bought a stamp for $10 and sold it to Nick for $12. Later, he bought it back from Nick for $14 and resold it to another collector for $16. Did Alan make a profit on the transaction? If so, how much?

To solve this problem, one must go beyond the misleading surface features presented in the problem statement and question the underlying assumptions. In Alan's case, at least two seemingly logical explanations give rise to different amounts of profit:

> *Explanation 1: Alan bought the stamp for $10 and sold it to Nick for $12, giving him a $2 profit. Alan bought the stamp back from Nick for $14 at a*

$2 loss. Finally, Alan sold the stamp for $16, which generated a $2 profit. Alan's total profit was ($2 − $2) + $2 = $2.

Explanation 2: Assume that Alan made two separate transactions. He bought a stamp for $10 and sold it for $12, giving him a $2 profit. He purchased a stamp for $14 and sold it for $16, giving a $2 profit. Alan's total profit was $2 + $2 = $4.

Think about this problem carefully. Do both solutions seem reasonable? Check with other students to see if they agree with you. Would the solution be different if Alan repurchased the stamp for $14 from someone other than Nick? Try drawing a picture representing the transactions to justify your answer.

When Alan computes his profit for the Internal Revenue Service, he cannot report either a $2 or $4 profit for these transactions. Something must be wrong with the assumptions underlying one of these solutions.

In Explanation 1, an assumption is made that the difference between the first sale ($12) and the repurchase price ($14) counts as a loss. This would be true only if the stamp was never resold. Profit is computed by subtracting the most recent purchase from the subsequent sale price. Profit has nothing to do with an object's prior price history. For example, imagine that the price of the stamp jumped to $1000 after Alan sold it to Nick. If Alan bought it back from Nick for $1000 and sold it for $1002, he would still make a profit of $2 on this transaction plus $2 from the original purchase at $10 and sale to Nick for $12, giving a total profit of $4. Based on the solution procedure suggested in Explanation 1, Alan would have sustained a loss of $1000 (i.e., ($2 − $1000) + $2 = ⁻$1000).

This extreme example clearly shows the faulty logic employed in Explanation 1. In Explanation 2, the solver correctly assumes no connection between the two transactions (Alan could have purchased two different stamps).

Each of the following chapters includes activities that offer you and your students an opportunity to solve mathematical problems. The process involves actively constructing models, analyzing relationships, risking conjectures, and justifying conclusions. Problem solving is integral to the enterprise of learning mathematics concepts and skills and is itself one of the most important goals of mathematics education.

EMPLOYING PROBLEM-SOLVING STRATEGIES

A set of classroom-tested problem-solving strategies to help children learn to solve process problems has been identified by teachers and researchers (Charles & Lester, 1982; Souviney, 1981; Suydam, 1980). These strategies can be successfully applied to a wide range of problems; alone or in combination, they can help children uncover the mathematical structure underlying novel problem situations. Some strategies may already be familiar and others may be new. Examples of several strategies are described in "Step 2: Select a Strategy."

Organizing for Problem Solving

Careful planning is required to integrate effectively the use of problem-solving strategies into mathematics instruction. Problem solving rarely involves an uninterrupted line of reasoning from the problem statement to the solution. Children often make false starts and encounter blind alleys. Students need to feel comfortable while trying out untested ideas. A teacher can encourage students to invest the time and energy necessary to solve nonroutine problems by

1. Providing an environment that encourages students to take intellectual risks and explore untested alternatives.
2. Valuing all answers as potentially useful.
3. Praising persistence and unexpected solutions.
4. Evaluating the quality of students' problem-solving efforts.
5. Allowing group work during problem-solving sessions.
6. Encouraging students to work on problems at home with their families.
7. Establishing a systematic schedule for integrating problem-solving sessions into each strand of the mathematics curriculum.
8. Introducing a systematic problem-solving plan.
9. Asking appropriate questions during each problem-solving session.

A FOUR-STEP PROBLEM-SOLVING PLAN

George Polya (1957) described a general problem-solving plan consisting of four interrelated steps:

☐ Understanding the problem
☐ Selecting a strategy
☐ Carrying out the strategy
☐ Evaluating the results

This four-step plan, summarized in the boxed insert, has been successfully adapted by teachers and researchers for use in elementary school instruction (Charles & Lester, 1982; Souviney, 1981). It is particularly helpful in solving nonroutine mathematics problems. This problem-solving plan closely parallels the **writing process** method recommended for teaching written expression in schools (Mehan, et al., 1986). Each technique specifies a preparatory stage to encourage the student to gain a broader understanding of the problem goals and conditions prior to commencing the task. This stage is called **prewriting** when creating a composition and **understanding the problem** when solving problems.

Four-Step Problem-Solving Plan

Step 1: **Understand the Problem**
- ☐ Relate given facts and conditions to the problem goal
- ☐ Coordinate current problem with previously solved problems

Step 2: **Select a Strategy**
- ☐ Guess-and-test
- ☐ Substitute simple values
- ☐ Divide problem into subtasks
- ☐ Conduct an experiment
- ☐ Design a model
- ☐ Draw a sketch
- ☐ Make a systematic list
- ☐ Make a table
- ☐ Construct a graph
- ☐ Reduce to a simpler case
- ☐ Search for a pattern
- ☐ Construct a general rule (function)
- ☐ Work backward
- ☐ Add something to the problem situation

Step 3: **Carry Out the Strategy**
- ☐ Persistently follow through with the solution strategy
- ☐ Maintain accurate records of the data collected
- ☐ Relate progress to Steps 1 and 2

Step 4: **Evaluate the Results**
- ☐ If a solution is uncovered, refine the result and relate to other problems
- ☐ If not, reevaluate understanding and seek a new solution strategy

---

Step 1: Understand the Problem

The first step in solving a problem is to understand the information given in the problem statement and the intended goal. The following techniques have proven to be useful in helping students understand the problem situation and select a solution strategy. Students can

1. Restate the problem in their own words.
2. Use materials or sketches to model the problem situation.
3. Make a list of all given facts.
4. Make a list of the stated conditions and restrictions.
5. Write the stated goal in their own words.
6. Establish whether an exact or approximate answer is desired.
7. Make a list of related information.
8. List unstated, or implicit, conditions.
9. Compare the current problem to problems solved previously.
10. Work with a partner or small group to discuss the problem.

Questions carefully posed by the teacher can stimulate student thinking during the preliminary phase in the problem-solving process. This is particularly true with young students and students who prefer verbalizing and role-playing problem situations.

Questions should promote critical thinking on the part of the students. Novice problem solvers should be encouraged to make conjectures, evaluate the ideas of others, and alter their positions in light of new evidence. The attempt to manufacture convincing arguments encourages students to look beyond the surface features stated in a problem statement and to explore the underlying mathematical structure and other associated conditions. It is rarely helpful for the teacher to provide answers to problems at this point. It is far better for students to examine their own conjectures and discuss them with their peers. Appropriate questions might include

1. What do you know about the problem?
2. Is it like any other problem you have solved?
3. What are some reasonable answers? Does the answer have to be exact? Could there be more than one correct answer?
4. Does your answer make sense? Can you convince me?
5. What can you do to find out?
6. What is wrong with that answer?

It is equally important for teachers to view the occurrence of incorrect answers and flawed reasoning as opportunities for learning. It is through the process of recognizing and attempting to repair inconsistencies in logic that students learn to question their current understanding of the problem situation.

The success of this approach depends to a great extent on how effectively the teacher can establish an atmosphere of acceptance for solving problems. Teachers should try to accept all answers as appropriate for discussion. The student's persistence and the *quality* of an argument should be praised more than accuracy and speed in achieving a solution.

Novices often find it helpful to work in pairs or in groups of four. For a given problem, students should also discuss the range of appropriate answers. For example, when asking students the population of the United States, we would generally be satisfied with an answer rounded to the nearest million persons; for a smaller country such as Belize, however, we might expect a value rounded to the nearest hundred thousand persons. Questioning of this type helps students gain experience dealing with the complexity of mathematical thinking and nonroutine problem solving. It is especially important for novice problem solvers to take time to explore and discuss the special features of each problem situation before continuing with the solution.

Step 2: Select a Strategy

Once the problem situation is well understood, a systematic approach, or solution strategy, can be selected. Often, novice problem solvers move to this step in the plan prematurely, before enough effort has been spent on understanding the problem itself. This frequently results in a failure to identify key structural features of the problem that lead to the selection of an effective solution strategy.

Encourage students to select a strategy based on past experience with related problems or on insight gathered from exploring the structure of the problem statement itself. The teacher can play a critical role at this stage by offering hints to assist the process of inspiration. One of the creative challenges for teachers is recognizing when to provide guidance and when to allow the struggle to continue unassisted.

One way of viewing this subtle process is to consider that anyone can solve any problem if given sufficient expert assistance. The role of the teacher is to broaden the student's range of independent problem-solving abilities until the student requires little or no external assistance (see the discussion of zone of proximal development in chapter 2). These problem-solving abilities can often be applied to only a narrow class of problems. With sufficient experience, however, it may be possible to transfer skills from one problem-solving domain to another. For example, an accomplished engineer might learn to be an architect more easily than someone without a similar problem-solving background.

The teacher's task becomes one of recognizing the **teaching moment** and providing a level of guidance appropriate to the ability and experience of the learner. Polya (1957) suggested that questions and hints on selecting a solution strategy should be offered in such a manner that they could have occurred to the students themselves. Questions of this sort might include:

1. What is the goal? What are you trying to find?
2. What information is given? What do you know?
3. What special conditions or restrictions apply?
4. Can you remember a problem with a similar unknown or goal? Could you use its solution to help in any way?
5. Can you restate the problem in your own words?
6. What are some possible and reasonable answers?
7. If you can't solve this problem, can you first solve a similar one?
8. Did you use all the given information?
9. Is there unstated information that could be useful?
10. How do the stated conditions restrict the solution?

Only as a last resort should the teacher offer specific suggestions, for example, "Perhaps you should average these six amounts." Remember that revealing answers to problems prematurely is unlikely to help students develop the ability to select efficient problem-solving strategies.

When mounting an assault on a novel problem, students have found one or more strategies from the following arsenal to be effective. Many process problems succumb to more than one technique, though whole classes of problems may be solved through the judicious application of one favorite strategy. These strategies are explained in more detail as they are employed in solving process-type problems in "Sample Process Problems and Solutions."

Guess-and-Test. This familiar trial-and-error technique can be used to solve a wide range of problems. Guess-and-test can be thought of as a preliminary, informal experiment motivated by intuition. The solver makes an educated guess and then tests to see if the answer solves the problem. If it does not, the guess is altered, based on the result of the test, and is tested again until a satisfactory answer is obtained. Its simplicity makes the guess-and-test strategy well-suited for use by elementary students.

What it gains in simplicity, however, it loses in efficiency. Generally, the guess-and-test method is more time-consuming than other strategies. It may, however, provide critical initial insight into a problem situation.

Substitute Simple Values.   Substituting less complex numbers for messy decimal and fractional values can reduce students' stress when solving problems. This technique often enables children to focus on the underlying structure of the problem. Once a solution path has been determined using the simpler substituted values, the original numbers can be reintroduced and the steps repeated.

Divide Problem into Subtasks.   Insight can be gained into some complex problems by separating them into more manageable components. After two or more component problems are carefully specified, each can be solved in turn and the results combined to solve the original problem.

Conduct an Experiment.   Many real-world problems in business, science, and engineering are solved by conducting carefully designed experiments. Children can also design and conduct experiments as a practical strategy for solving mathematics problems. Experiments require designing a physical representation of a problem situation. Children must learn to be systematic in carrying out an experiment. Sketches, lists, tables, and graphs can be used to organize data gathered from experiments. Once the results are organized, students can search for patterns that will help reveal the solution.

Design a Model.   Sometimes it is impossible or dangerous to carry out an experiment by using the objects and reenacting events described in the problem situation. Designing a model that embodies the essential features of the problem situation may help lead to a solution.

Draw a Sketch.   Drawing a sketch or diagram of a problem situation may help students visualize a solution.

Make a Systematic List.   The solutions to many problems result from a careful listing of all possible outcomes.

Make a Table.   Organizing data into a table can simplify the presentation of information and lead to the discovery of patterns and other clues to a solution.

Construct a Graph.   Graphing information can result in a visual display that uncovers underlying relationships that might otherwise go unnoticed.

Reduce to a Simpler Case.   When a problem requires a long series of actions, it is often helpful to look at what happens in the first few steps of the process. Patterns discovered in these early stages can then be used to predict what will happen as the process continues.

Search for a Pattern.   Looking for numerical and geometric patterns often provides clues to the structural relationships in a problem situation.

Construct a General Rule (Function).   By writing a formula, or function, that describes the pattern underlying a problem situation, a solution may be generalized to a whole class of related problems.

Work Backward.   This strategy can be useful when the solution to a problem requires describing the systematic steps to achieve a known goal, or end state. To employ this strategy, one starts with the known end state and works backward to determine the sequence of steps to reach the beginning state. Carrying out these steps in reverse will give the desired solution.

Add Something to the Problem Situation.   Sometimes adding new elements to the problem situation (e.g., drawing a diagonal in a rectangle) aids in the solution.

Step 3: Carry Out the Strategy

Once the facts, conditions, and goal of a problem are understood, and a solution strategy has been selected, the next step is to apply the chosen strategy persistently. It should be noted that, in practice, Steps 1, 2, and 3 are *not* independent activities. For example, in trying to understand a problem, a person may find it useful to apply a favored strategy immediately to gather further information. This insight may lead to the selection of a more effective strategy.

Carrying out a solution requires persistence. The student's task is to determine whether or not the chosen strategy generates meaningful clues toward unraveling the problem. These clues may take the form of patterns that relate the problem to a previously solved example. When carrying out a strategy, children should be encouraged to

1. Keep accurate records (tables, sketches, lists, graphs).
2. Stick to their chosen strategy until evidence suggests specific changes.
3. Carefully monitor their thinking during each step in the solution.
4. Put the problem aside for a day or so if no progress is being made and then try it again.
5. Select another strategy based on insight gained from their initial attempt.

Step 4: Evaluate the Results

Once a problem is solved (or attempted without success), children should review the solution process for two reasons. First, it provides an opportunity for learners to evaluate and refine their results. Second, it brings the process of solution into sharper focus. Children are better able to describe their intuitive processes immediately following a successful problem-solving experience. Students can evaluate the validity of their results, and, more importantly, they can be encouraged to compare the current solution process with those of previously solved problems.

Once a successful solution has been found, students should

1. Describe orally or in writing the steps taken and the strategies employed in solving the problem.
2. Discuss the form of the answer.
3. Compare the problem and solution to previously solved problems.
4. Solve extensions of the problem using the same strategy.
5. Evaluate alternative solutions reported by peers.

SAMPLE PROCESS PROBLEMS AND SOLUTIONS

Each of the following nonroutine process problems is solved using the four-step plan and one or more of the fourteen solution strategies. Though each solution is described in the four-step sequence, note that in actual solutions, the steps are likely to be far more recursive (i.e., when carrying out the strategy, new insights may lead to a more effective solution strategy).

Problem 1

Understand the Problem

A teacher assigned homework on two facing pages. The sum of the page numbers equaled 65. What were the homework page numbers?

Look at any book. Notice that facing pages are always displayed with an even page number on the left and the consecutive odd page number on the right. Therefore, the homework pages numbers must be consecutive.

Select a Strategy.   *Guess-and-test*

Carry Out the Strategy.   Compute the sum of any two facing page numbers (e.g., 24 + 25 = 49). Since the answer is too small, make the second guess larger (e.g., 40 + 41 = 81). Since this result is too large, the answer must lie between these two guesses. Additional carefully chosen guesses will narrow the range of possible solutions, eventually yielding the answer (32 + 33 = 65).

Evaluate the Results.   The homework is on pages 32 and 33. Notice that the sum of any two consecutive numbers is always odd. Why do you think this is true? If it is true, then the sum of facing page numbers must be odd as well. Could the teacher assign facing pages with a sum of 64? How about 63? Find all the possible solutions of this problem for hundred-page book.

Problem 2

Understand the Problem

Table 4.1 shows the expected miles per gallon of gasoline for a car with and without the air conditioner running. How much extra will it cost to run the air conditioner for a 4-hour, 45-minute trip at 55 miles per hour if gasoline costs $1.09 per gallon?

We need to determine the cost of the gasoline required to make the trip with the air conditioner turned off and with it turned on. The trip is about five hours long, and the gasoline costs a little over $1.00 per gallon.

Select a Strategy.   *Substitute simple values*

Carry Out the Strategy.   Let's assume the trip was only one hour long and that gasoline cost exactly $1.00 per gallon. In one hour, at 55 miles per hour, the car would travel 55 miles. At this speed, the car gets 29 miles per gallon without air conditioning. Therefore, the car burns one gallon of gasoline each 29 miles. A 55-mile trip would require about two gallons of gasoline (55 ÷ 29). A 5-hour trip would require about 10 gallons of gasoline (275 ÷ 29). The cost of the gasoline would be about $10.00 (10 × $1.00). Using the same procedure, we can now substitute the actual values.

$$4 \text{ hours, } 45 \text{ minutes} = 4.75 \text{ hours}$$
$$4.75 \times 55 = 261.25 \text{ miles}$$
$$261.25 \div 29 = 9.01 \text{ gallons (without air conditioning)}$$
$$261.25 \div 26 = 10.05 \text{ gallons (with air conditioning)}$$
$$9.01 \times \$1.09 = \$9.82 \text{ (without air conditioning)}$$
$$10.05 \times \$1.09 = \$10.96 \text{ (with air conditioning)}$$

Evaluate the Results.   The extra cost of running the air conditioner for the entire trip was $1.14. It costs about $0.24 per hour to run the air conditioner at 55 mph in this car ($1.14 ÷ 4.75). Using the mileage chart, explore how the air conditioning costs change as a result of increased or decreased speed.

TABLE 4.1 Miles Per Gallon (mpg)

	40 Mph	45 Mph	50 Mph	55 Mph	60 Mph
Without air conditioning	34 mpg	33 mpg	31 mpg	29 mpg	26 mpg
With air conditioning	32 mpg	31 mpg	28 mpg	26 mpg	22 mpg

Problem 3

Understand the Problem

A town decided to paint house numbers on the mailbox in front of each home to improve mail delivery. The houses were numbered in sequence (1, 2, 3, 4, . . .). It cost $1 to paint each digit on the mailboxes. If the Town Council needed to budget exactly $600 for the job, how many houses were there in the town?

The cost for each mailbox is related to the number of digits in the house number. The house numbers must be one, two, or three digits long, because if there were any four-digit house numbers, the cost would exceed $600 even if each house cost only $1.00.

Select a Strategy.   *Divide problem into subtasks*

Carry Out the Strategy.   First, compute the cost of the single-digit house numbers 1 through 9 ($1 × 9 = $9). Next, compute the cost for double-digit house numbers 10 through 99 ($2 × 90 = $180), giving a total cost of $189 for the first 99 homes. For the houses numbered 100 through 199, the cost is $300 ($3 × 100 = $300) giving a total cost of $489 for houses 1 through 199. The remaining $111 ($600 − $489 = $111) is enough to paint 37 additional triple-digit numbers (111 ÷ 3 = 37). Organizing the information as in Table 4.2 can help to solve this problem.

TABLE 4.2

House Numbers	Cost for Painting	Number of Houses
1–9	$9	9
10–99	$180	90
100–199	$300	100
200–236	$111	37
Total	$600	236

Evaluate the Results.   There must be 236 houses in the town if exactly $600 was budgeted to paint the house numbers. How many houses would there be in another city if the mayor had to budget $2893 to paint the house numbers?

Problem 4

Understand the Problem

Using only a 1-pound, 3-pound, and 9-pound weight, how many different quantities can be weighed on a pan balance?

TABLE 4.3

Left Pan	Right Pan
1-ounce weight	1 ounce of rice
3-ounce weight	2 ounces of rice + 1-ounce weight
3-ounce weight	3 ounces of rice
1-ounce + 3-ounce weights	4 ounces of rice
9-ounce weight	5 ounces of rice + 1-oz. + 3-oz. weights
9-ounce weight	6 ounces of rice + 3-ounce weight
9-ounce + 1-ounce weights	7 ounces of rice + 3 ounce weight
9-ounce weight	8 ounces of rice + 1 ounce weight
9-ounce weight	9 ounces of rice
9-ounce + 1-ounce weights	10 ounces of rice
9-ounce + 3-ounce weights	11 ounces of rice + 1-ounce weight
9-ounce + 3-ounce weights	12 ounces of rice
9-oz. + 3-oz. + 1-oz. weights	13 ounces of rice

A pan balance allows the user to weigh objects by comparing them with standard weights. Weights can be placed on the balance singly or in groups. Weights can be placed on both pans if necessary. Pound weights are heavy, so let's substitute 1-ounce, 3-ounce, and 9-ounce weights so we can use a simple classroom balance.

Select a Strategy.   *Conduct an experiment*

Carry Out the Strategy.   Using a pan balance, a supply of rice, and one each of the three weights (1-ounce, 3-ounce, and 9-ounce), systematically weigh different amounts of rice.

List each of your measurements as shown in Table 4.3.

Evaluate the Results.   All quantities of rice from 1 to 13 ounces can be accurately measured using only these three weights and a pan balance. Using 1-, 3-, and 9-pound weights would give similar results. The problem could also have been solved by adding and subtracting combinations of the values 1, 3, and 9 (e.g., 9 + 3 = 11 + 1). How many quantities of rice could be measured if we included a 27-pound (or ounce) weight in the collection? Notice that this set of weights contains all powers-of-3 units. Do powers of 2 or powers of 4 have the same property?

Problem 5

Understand the Problem

The boss of a small business noticed that the telephone line was busy about half the time when she called her office while away on business. She wondered how the situation would change if she had a second line installed for incoming calls.

TABLE 4.4

Head/Head	Head/Tail	Tail/Tail
25	51	24

If incoming calls have a 50–50 chance of getting through with one line available, what will happen if two lines are available? Assuming that the same number of people are trying to call the business, how many callers on the average will hear a busy signal if a second line is installed? It is too expensive to put in a second line to test its effectiveness.

Select a Strategy. *Design a model*

Carry Out the Strategy. Let *heads* on a coin represent a successful caller and *tails* represent a caller receiving a busy signal. Flipping one coin 100 times models 100 attempted calls to the business with only one telephone line. There should be about 50 heads and 50 tails in 100 flips. Installing a second telephone line can be modeled by flipping two coins. Receiving a *head* on either coin indicates a successful call and *two tails* represent a busy signal. The actual results of 100 flips of two coins are shown in Table 4.4.

Evaluate the Results. On the average, tail/tail appears about 25 percent of the time. This indicates that installing a second line would reduce the chances of getting a busy signal to about 25 out of each 100 calls. Notice that there are about twice as many instances of head/tail than either head/head or tail/tail. To understand why, use a penny and nickel instead of two identical coins. It is easy to see that the combination of a head on the penny and a tail on the nickel is a different event from that of a tail on the penny and a head on the nickel. A similar principle applies in the case of the two telephone lines. With two incoming phone lines, there are three conditions that will allow the call to go through (Line 1 not busy and Line 2 busy, Line 2 not busy and Line 1 busy, or Lines 1 and 2 not busy). There is only one condition that will cause a busy signal (Lines 1 and 2 both busy). Three out of the four conditions will allow a successful call.

Problem 6

Understand the Problem

If the shadow of a 6-foot-tall man is 4-feet long, how tall is a tree that casts a 10-foot shadow?

At any spot on earth at a specific time during the day, the length of a shadow is proportional to the height of the object casting the shadow. In this case, we need to find the proportional relationship between the man's shadow length and his height in order to predict the height of the tree based on its shadow length.

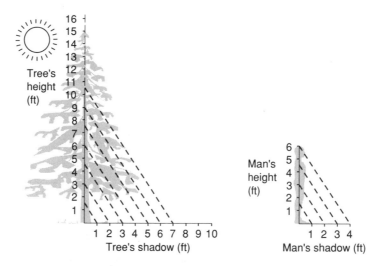

FIGURE 4.3

Select a Strategy. *Draw a sketch*

Carry Out the Strategy. First, make a scale drawing of the situation using squared graph paper. Label the three known lengths and use a dotted line to represent the unknown tree height. Connect the top of the man's head to the end of his shadow to complete outline of the triangular shadow region. Draw parallel lines connecting each 1-foot marker along the man's shadow to points along the vertical axis representing the man's height. Notice that 1 foot of the shadow's length is created by 1.5 feet of the man's height (see Figure 4.3). Follow a similar procedure and connect the 1-foot shadow markers to the 1.5-foot markers on the vertical line representing the tree's height. Continue drawing parallel lines until you reach the 10-foot marker on the tree shadow axis. This line crosses the tree height axis at 15 feet.

Evaluate the Results. The tree's height is 15 feet. Every foot of shadow at this particular time of day is the result of 1.5 feet of object height. This relationship can be stated as the ratio of 1.5 to 1. The relationship between object height and shadow length can also be expressed as a ratio using only whole numbers (a ratio of 3 to 2). Why are these ratios equivalent? An easier way to measure the height of tall objects is to wait until the man's shadow equals his height. At this moment, the height of all objects in the immediate area should equal their shadow lengths (have a ratio of 1 to 1).

Problem 7

Understand the Problem

How many ways are there to make change for a quarter?

Using pennies, nickels, dimes, and a quarter, we need to find all the combinations of one or more coins that equal twenty-five cents. As many of each coin as necessary can be used.

Select a Strategy. *Make a systematic list*

Carry Out the Strategy. Since it is hard to keep track of each combination of coins mentally, we need to list each combination systematically so we will know when we have found them all. The following listing begins with the coin of largest value, the quarter, and works toward the coin of least value. Once all the combinations are listed for each set of larger coins, one of these coins is removed and the process repeated.

Systematic List of Coins

	Quarters	Dimes	Nickels	Pennies
1.	1	0	0	0
2.	0	2	1	0
3.	0	2	0	5
4.	0	1	3	0
5.	0	1	2	5
6.	0	1	1	10
7.	0	1	0	15
8.	0	0	5	0
9.	0	0	4	5
10.	0	0	3	10
11.	0	0	2	15
12.	0	0	1	20
13.	0	0	0	25

Evaluate the Results. There are thirteen ways to make change for a quarter. Can this technique be used to find the number of ways to make change for $1.00? Look for patterns in the systematic list to make the job easier. Perhaps the problem could be divided into smaller components, each of which could be solved by a different person and the results combined.

Problem 8

Understand the Problem

If ten persons attend a meeting, how many handshakes will be required for each person to greet every other person exactly once?

We need to determine how many total handshakes are necessary for each person to shake hands exactly once with every other person at the meeting. For example,

TABLE 4.5

Person No.	1	2	3	4	5	6	7	8	9	10
Handshakes	9	8	7	6	5	4	3	2	1	0

if Alma shakes hands with Luis, then Luis does not need to shake hands with Alma again. To make it easier to keep track of the pairings, assume that only two people shake hands at one time.

Select a Strategy.   *Make a table*

Carry Out the Strategy.   Construct a table that lists the number of handshakes required for each person if the greetings are initiated by one person at a time. The first person must shake hands with nine others. Once these greetings are accomplished, Person 1 can sit down at the conference table. The next person then shakes hands with the eight remaining persons. Table 4.5 shows the results of this process.

Evaluate the Results.   A total of forty-five handshakes is required for ten persons to greet each other exactly once. Similar tables can be constructed to calculate the number of handshakes needed for larger meetings. For example, if twenty persons attended the meeting, 190 handshakes would be required. How many handshakes would be needed if the custom was for three persons to shake hands at one time instead of two persons?

Problem 9

Understand the Problem

A turtle and a rabbit ran a 1000-meter race. Because in previous races, the turtle ran 10 meters each minute, and the rabbit ran 100 meters each minute, the turtle was given a 500-meter head start. Who won the race? Was it a fair race?

The turtle started the race at the 500-meter mark, and the rabbit started at the 0-meter mark. Assume the turtle ran a constant 10 meters per minute and the rabbit ran 100 meters per minute. Either the turtle reached the finish line first or the rabbit passed the turtle somewhere between the 500-meter mark and the finish line.

Select a Strategy.   *Construct a graph*

Carry Out the Strategy.   Construct a graph on squared paper as shown in Figure 4.4. Plot the race track distance along the vertical axis and the race time along the horizontal axis. The rabbit started at the 0 mark and ran 100 meters each minute. Follow the rabbit's progress by plotting a point that represents the time it takes to

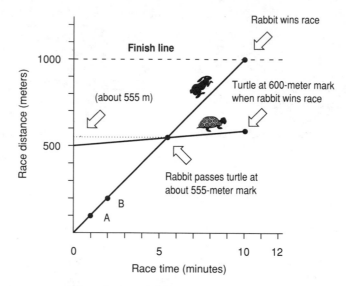

FIGURE 4.4   Who Wins the Race?

cover each 100 meters (point A represents 1 minute for 100 meters; point B represents 2 minutes for 200 meters; and so on). Connecting these points with a line gives a graph of the rabbit's speed. A similar procedure is followed for the turtle, who started at the 500-meter mark. The graph of the turtle's speed is not as steep as the graph of the rabbit's speed.

Evaluate the Results.   The rabbit won the race. The rabbit passed the turtle at about the 555-meter mark (see Figure 4.4). The turtle was at the 600-meter mark when the rabbit crossed the finish line. The race took ten minutes. If the turtle continued the race, it would finish forty minutes after the rabbit. Can this result be shown on the graph? The race did not seem to be fair, based on the expected performance of both contestants and the starting positions. Where would the turtle have to start in order to have a sporting chance to win?

Problem 10

Understand the Problem

How many blocks are needed to build a staircase twenty steps high following the pattern shown in Figure 4.5?

The staircase is constructed by stacking identical cubes into columns, each column one cube higher than the previous column. Assume the staircase is only one cube thick.

Select a Strategy.   *Reduce to a simpler case*

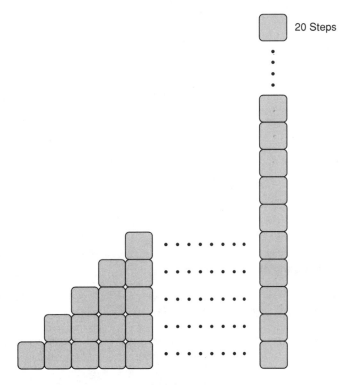

FIGURE 4.5   Twenty-Step Staircase

Carry Out the Strategy.   Instead of trying to count the blocks needed for a twenty-step staircase, let's look at smaller staircases and see if we can find a relationship between the number of steps and the number of blocks needed for construction.

> 1 step needs 1 block
> 2 steps need 3 blocks
> 3 steps need 6 blocks
> 4 steps need 10 blocks
> 5 steps need 15 blocks

Notice that the total number of blocks required to construct the four-step staircase equals the number of blocks needed to construct the three-step staircase (six blocks) plus the number of additional blocks needed to construct the fourth step (four blocks). Continuing the list to the nineteenth step (190 blocks) allows us to compute the number of blocks needed to build the twenty-step staircase.

Evaluate the Results.   A total of 210 blocks is required to build a twenty-step staircase (190 + 20). The number of blocks needed to build larger staircases can be calculated by continuing the above list.

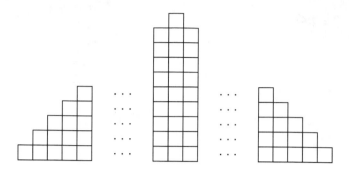

FIGURE 4.6 Ten-Step Double Staircase

Problem 11

Understand the Problem

How many blocks are needed to construct a double staircase ten steps high following the pattern shown in Figure 4.6?

A double staircase has the same number of ascending and descending steps. The height of the staircase is the number of steps required to reach the top.

Select a Strategy.   *Search for a pattern*

Carry Out the Strategy.   Let's look at simple cases of this construction problem and search for a pattern that will help us count the number of blocks required. A one-step double staircase requires only one block. A two-step double staircase requires four blocks. Table 4.6 shows the number of blocks required to build one- through five-step double staircases.

To find the next value in the table, let's look at how the values in the second column increase. In Table 4.7, the difference between each adjacent pair of values in column two seems to follow the odd-number pattern (3, 5, 7, 9, . . .). This pattern

TABLE 4.6

Double Staircase Height	Blocks Required
1	1
2	4
3	9
4	16
5	25
6	?

TABLE 4.7

Double Staircase Height	Blocks Required	Difference
1	1	
2	4	3
3	9	5
4	16	7
5	25	9
6	36	11 (?)
7	?	
8	?	
9	?	
10	?	

predicts that the next difference will be 11. Counting the blocks in a six-step double staircase verifies this prediction.

Therefore, the number of blocks required for each successive step in a double staircase seems to grow according to this odd-number pattern. Continuing the table enables us to calculate the number of blocks required for any double staircase.

Evaluate the Results. To calculate the number of blocks needed to construct a ten-step double staircase, we must continue the pattern to find the number needed for nine-steps (81 blocks) and add the ninth value in the odd-number series (19). Therefore, 100 blocks are needed to construct a ten-step double staircase (81 + 19). The solution to this problem is related to the previous staircase problem. Notice that a double staircase is the sum of two sequential single staircases (e.g., a four-step plus five-step single staircase make a five-step double staircase). Check to see if this prediction holds for larger staircases.

Problem 12

Understand the Problem

Same double staircase problem as above.

Using a pattern requires that all previous steps be computed in sequence to solve a larger problem (i.e., you can calculate the answer to the twenty-step problem *only* if you know the solution to the nineteen-step problem). Can we find a rule that directly relates the staircase height to the number of blocks required for construction?

Select a Strategy *Construct a general rule (function)*

Carry Out the Strategy. Table 4.8 lists the number of blocks needed to construct a one- through ten-step double staircase. Instead of looking for a pattern that

TABLE 4.8

Double Staircase Height	Blocks Required
1	1
2	4
3	9
4	16
5	25
6	36
7	49
8	64
9	81
10	100

relates the values down the second column, let's find a direct relationship between each staircase height value in column one and its corresponding block count in column two.

To help visualize this relationship, look at the shape resulting from reorganizing a double staircase that is shown in Figure 4.7.

The number of blocks required can be computed by multiplying the staircase height by itself (5 × 5 = 25 blocks). This general *rule*, or function, seems to work for all double staircases.

Evaluate the Results. The total number of blocks required to build a ten-step double staircase is 100 blocks (10 × 10 = 100). The numbers listed in the second column of Table 4.8 are the *square numbers*. A square number results from multiplying any whole number by itself (i.e., n × n = n^2). There is a significant advantage to finding a general rule that relates each value in the first column in a table to its corresponding value in the second column. For example, if you wish to find the number of blocks necessary for a 75-step double staircase, you can compute the result directly using the rule (75 × 75 = 5625). If you relied on a pattern to find the solu-

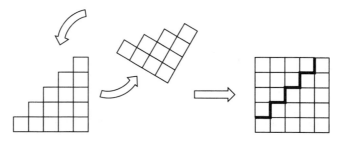

FIGURE 4.7 Double Staircase Rule

tion, you would first have to compute all the solutions through the seventy-fourth step!

Problem 13

Understand the Problem

Arrange a deck of cards numbered 1–10 so that while holding the deck facedown, you can deal out the cards in order from 1 to 10 by alternately slipping the top card under the deck and dealing the next card faceup on a table (see Figure 4.8).

Select a Strategy. *Work backward*

Carry Out the Strategy. To arrange the cards in the original deck, begin with the final goal (i.e., the end state is a line of cards on the table in order from 1 to 10). Deal the cards in *reverse* order. Working backward, pick up the 10-card and slip the bottom card to the top (because you are only holding one card, the deck does not change). Pick up the 9-card, place it facedown on top of the deck and slip the bottom card on top (the dealer now holds two cards with the 9-card on the bottom and the 10-card on top). Continue the process until all the cards are in the deck.

Evaluate the Results. Deal the cards using the newly configured deck and carefully follow the original instructions. The result should be a line of cards in sequence from 1 to 10. To extend the activity, slip two cards under the deck each time before dealing a card on the table. You can also write letters on the cards to spell out names and funny sayings.

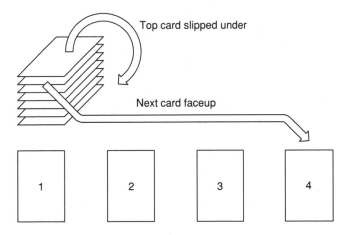

FIGURE 4.8 Dealing Cards 1–10

Problem 14

Understand the Problem

Suppose a bag contains a quantity of identical marbles. If only one marble at a time can be removed, observed, and then returned to the bag, how can the total number of marbles in the bag be accurately estimated?

Only one marble can be removed at a time, and then it is returned to the bag with the others. The marbles in the bag cannot be marked in any way.

Select a Strategy.   *Add something to the problem situation*

Carry Out the Strategy.   Guessing is the only available strategy unless we break the traditional mind-set and consider adding a new element to the problem situation. First, remove one marble and carefully observe its characteristics. Locate a supply of, say, twenty identical marbles and mark each with an *S* for *seed*. Thoroughly mix the original marble and the 20 new seed marbles with the other marbles in the bag. Draw one marble at a time, keeping track of the number of original and seed marbles. The following proportion can be computed to predict the number of original marbles in the bag.

$$\frac{\text{Number of seed marbles}}{\text{Number of seed marbles drawn}} = \frac{\text{Number of original marbles}}{\text{Number of original marbles drawn}}$$

If, in 100 trials, 40 seed marbles and 60 original marbles were drawn and replaced, the number of original marbles (x) in the bag can be computed as follows:

$$\frac{20}{40} = \frac{x}{60}$$

$$x = (20 \times 60) \div 40 = 30 \text{ original marbles}$$

Evaluate the Results.   Comparing the number of seed marbles drawn to the number actually put into the bag offers a way to estimate the number of original marbles in the bag. If 200 samples were drawn, would the prediction improve? Would putting more seed marbles into the bag give good predictions with fewer trials? Techniques like this are used by wildlife management scientists to estimate the population of birds and fish in the wild. Can you think of other possible applications of this technique?

PROBLEM-SOLVING LESSON PLAN

The following sample lesson plan is an adaptation of the clinical teaching model outline presented in chapter 2. This lesson presupposes that the students have already mastered necessary computation skills or have access to a calculator. Marilyn Burns (1987) provides an excellent series of problem-solving lesson plans for Grades 3–6 along with a detailed description of their implementation in actual classrooms.

Sample Grade 6 Lesson Plan for Introducing the Four-Step Problem-Solving Plan

1. **OBJECTIVE:** To introduce a four-step problem-solving plan for nonroutine process problems.

2. **MATERIALS:** A supply of twenty-five wooden blocks for each student; paper and pencil; overhead projector; blank transparency and overhead pen.

3. **ANTICIPATORY SET:** With the overhead projector turned off, build a five-step staircase of blocks (build the staircase flat on the surface of the overhead). A few inches to the right, build a ten-block stack (also flat on the surface of the overhead) and cover it with a piece of paper (leave enough space to insert Steps 6–9 later). Turn on the overhead projector and ask the students what the construction looks like. Ask them to duplicate the staircase on their desks and to count the number of blocks required. Then remove the paper covering the ten-stack and discuss how many blocks it would take to make a ten-step staircase.

4. **INSTRUCTION:** Have the students mark off four sections on a piece of paper by dividing it in half vertically and horizontally (show an example on a transparency). Number the sections 1, 2, 3, and 4. Ask the students to write a description of the problem in Section 1. Encourage them to use their own words. Have them include all the facts, unstated assumptions, and the problem goal. In Section 2, have them list ways they may approach the problem (draw a picture, use the blocks, make a table, and so on). After everyone has written something in Sections 1 and 2, have the students begin to solve the problem. They will quickly find out that they do not have enough blocks and will need to try some other strategy. Have them use Section 3 to show sketches, computations, and any notes about their progress.

5. **GUIDED PRACTICE:** Walk around the room observing, posing questions and making comments such as, Could drawing a sketch help? or, Maybe if you recorded your results in a table it would be easier to keep track of your answers. Some students may decide to pool their blocks and solve the problem by constructing the ten-step staircase.

6. **CLOSURE:** After about fifteen minutes, have individuals tell the whole group how they attempted the solution. Have children describe how they attempted to solve the problem in Section 4. Make sure they explain why they think their answer is correct.

7. **INDEPENDENT PRACTICE:** Assign the problem of counting the number of blocks needed to construct a twenty-step staircase. Ask if anyone can find a way to predict the number of blocks needed for even larger staircases (e.g., fifty steps or one hundred steps).

ADAPTING INSTRUCTION FOR CHILDREN WITH SPECIAL NEEDS

Developing problem-solving skills is important for all children. While decision making, language usage, visual memory, and logical reasoning (all of which contribute to effective problem solving) are areas of difficulty for learning-disabled students, several techniques are available to help special-needs children learn to solve problems (Thornton & Bley, 1982).

Children who have difficulty reading can more easily participate in solving word problems that are presented orally or on tape. They should also follow along using a printed statement of the problem. Participation can also be improved if problem situations are presented using pictures to help represent the problem. For example:

Mona's mother bought 6 prizes for the party. ● ● ● ● ● ●
She bought 4 more the next day. ● ● ● ●
How many ● did she buy altogether? _____

To help students understand what information is given in a written problem statement and what is needed to solve the problem, you can highlight key values and phrases using a colored marker. For example:

Jose has **17** peanuts, and Alice has **12**. (**bold** = red)
Altogether, there are _____ **peanuts**.

As students gain experience, they can use markers to highlight key information in problems themselves. Different colored markers can be used to highlight values and key phrases in two-step problems. For example:

One quarter equals 25 cents.   (**bold** = red)
A stack of *16 quarters is 1 inch tall*.   (*italics* = blue)
Joan is *50 inches tall*.
How much is a stack of quarters as tall as Joan worth?
1. *How many quarters equals Joan's height?* _____ quarters
2. How much is the stack worth? _____ cents _____ dollars

Students must first have a good understanding of the problem situation and the required mathematics operations. It is often helpful to review any arithmetic concepts involved in a problem before it is presented. For example, a preliminary worksheet can be assigned on which all the problems are worked, and the student must insert the correct operation.

$$
\begin{array}{ccc}
35 & 34 & 315 \\
\square\ 16 & \square\ \ 9 & \square\ 145 \\
\hline
19 & 306 & 460
\end{array}
$$

Learning-disabled students also need practice predicting estimated answers as a way to check if a subsequent solution is reasonable. Predicting a reasonable answer before attempting a solution also helps students visualize the problem situation. As a warm-up activity, prepare a set of task cards that list the possible estimated solutions for word problems. Have students select the most reasonable solution using only mental computation. For example:

Each Valentine card costs 22 cents.
Rob has 31 classmates.

How much will Rob spend to buy a card for everyone in his class?

1. More than $5.00.   2. Less than $5.00.

Students then solve each problem and check their solutions against the estimate. Children can make up their own task cards using problems from a textbook. Each task card should list one good solution estimate for the problem and one or more poor solution estimates. Have the class members exchange cards, determine the best estimated solution for each problem, and check their worked solutions against the estimates.

SUMMARY

Problem solving is one of the important goals of mathematics instruction. A mathematical problem is a situation in which the context is readily understandable but no solution is immediately apparent. Children at all levels can successfully engage in problem-solving experiences. Two types of problems were discussed: story problems and process problems. Story problems can be characterized as word problems that contain all the necessary information to effect a solution. Generally, the solver must identify one or more operations to carry out on the values given in the problem statement. Process problems are nonroutine situations that often require additional information or the reorganization of given facts in order to effect a solution. Strategies such as making a table, looking for patterns, and constructing general rules can be employed to solve process problems. Generally, process problems can be solved in more than one way and may have multiple answers. Polya's four-step problem-solving plan can be employed by elementary children to solve routine word problems and nonroutine process problems. Specialized techniques can help special-needs children learn to solve mathematics problems.

COURSE ACTIVITIES

1. Work with a partner to develop winning strategies for the 7/5/3 NIM game. Use suggestions presented in the text and work backward from various winning end states. Keep accurate records of each successful sequence of moves. Once a consistent winning strategy is found, solve these extensions.

a. The player removing the last counter loses.

b. Begin with four rows containing seven, five, three, and one counters, respectively. Play using goal of player re-moving last counter wins, then player re-moving last counter loses.

2. Using any basal mathematics textbook, rewrite a selection of ten story problems on five-by-seven-inch cards to include

a. Ten examples of extraneous information problems.

b. Ten examples of missing information problems.

c. Ten examples of no-value problems.

d. Ten examples of story problem headlines.

e. Ten examples of fill-in value problems.

Working in pairs, shuffle the fifty problem cards and practice identifying each type. Select ten cards at random and solve each problem using the four-step problem-solving plan described in the text. Provide missing information and values as appropriate. Make sure to describe your solution in writing. If possible, do this activity with a small group of elementary students in a classroom setting. Discuss the results with your peers.

3. Select one of the fourteen solved process problems described in this chapter. Carry out the suggested solution strategy, keeping a detailed record of each step. Write a convincing argument in defense of the solution. Present the solution to a small group of peers and discuss the appropriateness of each argument. Answer any questions raised in the "Evaluate the Results" section.

4. Select one of the unsolved extensions of a process problem presented without solution in this chapter (or see resource books listed in references). Review the procedures associated with the four-step problem-solving plan. Working in pairs, attempt to solve the problem by carefully following the steps in the plan. One partner should observe, make suggestions, and keep detailed notes as the other partner carries out the solution. Feel free to switch roles and talk about each step in the solution. Using your notes, discuss the results of your efforts (successful or otherwise) with your peers. Try to present a convincing argument to support your strategy and, if appropriate, the solution.

5. Using the clinical teaching outline given in chapter 2 (or other appropriate format), design a lesson plan to introduce the four-step problem-solving plan described in this chapter. Share your lesson plan with your peers and discuss each component. If possible, try introducing the lesson to a small group of children.

6. Using the ideas presented in chapters 1–4, write a one-page, personal philosophy of teaching. Your discussion should relate how you plan to organize classroom practice based on your current understanding of how children learn mathematics concepts, skills, and problem solving.

7. Read one of the *Arithmetic Teacher* articles listed in the reference section. Write a brief report summarizing the main ideas of the article and describe how the recommendations for instruction might apply to your own mathematics teaching.

REFERENCES AND READINGS

Barson, A. (1985). And the last one loses! *Arithmetic Teacher, 33*(1), 35–37.

Bell, E., & Bell, R. (1985). Writing and mathematical problems solving: Arguments in favor of synthesis. *School Science and Mathematics, 85*(3), 210–221.

Brown, J., & Burton, R. (1978). Diagnostic models for procedural bugs in basic mathematics skills. *Cognitive Science, 2*, 155–192.

Burns, M. (1977). The good time math event book. Palo Alto, CA: Creative Publications.

Burns, M. (1977). *I hate math*. New York: Little, Brown.

Burns, M. (1978). *The book of think*. New York: Little, Brown.

Burns, M. (1985). The role of questioning. *Arithmetic Teacher, 32*(6), 14–16.

Burns, M. (1987). *A collection of math lessons*. New Rochelle, NY: Cuisenaire Company of America.

Butts, T. (1980). Posing problems properly. In S. Krulik & R. Reys (Eds.), *Problem solving in school mathematics: 1980 Yearbook* (pp. 23–33). Reston, VA: National Council of Teachers of Mathematics.

Carpenter, T. (1985). Research on the role of structure in thinking. *Arithmetic Teacher, 32*(6), 58–60.

Charles, R. (1981). Get the most out of "word problems." *Arithmetic Teacher, 29*(3), 39–40.

Charles, R. (1983). Evaluation and problem solving. *Arithmetic Teacher, 30*(5), 6–7.

Charles, R., & Lester, F. (1982). *Teaching problem solving: What, why and how*. Palo Alto, CA: Dale Seymour.

Chi, M., Feltovich, P., & Glaser, R. (1981). Characterization and representation of physics problems by experts and novices. *Cognitive Science, 5*, 121–152.

Dudeney, E. (1966). *The Canterbury puzzles*. New York: Dover.

Erlwanger, S. (1975). Case studies of children's conceptions of mathematics (Part 1). *Journal of Children's Mathematical Behavior*, Summer, 157–283.

Fennell, F., & Ammon, R. (1985). Writing techniques for problem solvers. *Arithmetic Teacher, 33*(1), 24–25.

Gardner, M. (Ed.). (1952). *Mathematical puzzles of Sam Loyd*. New York: Dover.

Gardner, M. (1956). *Mathematics, magic and mystery*. New York: Dover.

Gardner, M. (1966). *Mathematical puzzles and diversions*. New York: Dover.

Gilbert-Macmillan, K., & Leitz, S. (1986). Cooperative small groups: A method for teaching problem solving. *Arithmetic Teacher, 33*(7), 9–11.

Ginsburg, H. (1983). *The development of mathematical thinking*. New York: Academic Press.

Greenes, C., Spungin, R., & Dombrowski, J. (1977). *Problem-mathics: Mathematical challenge problems with solution strategies*. Palo Alto, CA: Creative Publications.

Jacobs, H. (1970). *Mathematics: A human endeavor*. San Francisco: W. H. Freeman.

Krulik, S., & Reys, R. (Eds.). (1980). *Problem solving in school mathematics: 1980 Yearbook*. Reston, VA: National Council of Teachers of Mathematics.

Krutetskii, V. (1976). *The psychology of mathematics abilities in school children*. Chicago: University of Chicago Press.

LeBlanc, J. (1982). Teaching textbook story problems. *Arithmetic Teacher, 29*(6), 52–54.

Lesh, R., & Landan, M. (Eds.). (1983). *The acquisition of mathematical concepts and processes*. New York: Academic Press.

Mathematical thinking issue. (1985). *Arithmetic Teacher, 32*(6).

Mehan, H., Miller-Souviney, B., Riel, M., Souviney, R., Whooley, K., & Liner, B. (1986). *The write help*. Glenview, IL: Scott, Foresman.

Meiring, S. (1980). *Problem solving—A basic mathematics goal*. Columbus: Ohio Department of Education.

Moses, B. (1982). Individual differences in problem solving. *Arithmetic Teacher, 30*(4), 10–14.

Moyer, M., & Moyer, J. (1985). Ensuring that practice makes perfect: Implications for children with learning disabilities. *Arithmetic Teacher, 33*(1), 40–42.

Polya, G. (1957). *How to solve it*. Princeton: Princeton University Press.

Polya, G. (1962). *Mathematical discovery; On understanding, learning and teaching problem solving*. New York: John Wiley.

Problem solving issue, (1978). *Arithmetic Teacher. 25*(6).

Seymour, D., Holmberg, V., & Laycock, M. (1973). *Aftermath*. Palo Alto, CA: Creative Publications.

Seymour, D., & Shedd, M. (1973). *Finite differences: A problem solving technique*. Palo Alto, CA: Creative Publications.

Shoenfeld, A. (1985). *Mathematical problem solving*. Orlando: Academic Press.

Sharron, S. (Ed.). (1979). *Applications in school mathematics: 1979 Yearbook*. Reston, VA: National Council of Teachers of Mathematics.

Silver, E. (1979). Student perceptions of relatedness around mathematical verbal problems. *Journal for Research in Mathematic Education. 10*, 195–210.

Souviney, R. (1977). Recreational Mathematics. *Learning Magazine*, May, 55–56.

Souviney, R. (1979). Math problems—Life problems. *Teacher Magazine*, February, 49–51.

Souviney, R. (1981). *Solving problems kids care about*. Glenview, IL: Scott, Foresman.

Suydam, M. (1980). Untangling clues from research on problem solving. In *Problem solving in school mathematics: 1980 Yearbook.* (pp. 34–50) Reston, VA: National Council of Teachers of Mathematics.

Thornton, C., & Bley, S. (1982). Problem solving: Help in the right direction for LD students. *Arithmetic Teacher, 29*(6), 26–27, 38–41.

Wertheimer, M. (1959). *Productive thinking.* New York: Harper & Row.

Wirtz, R. (1976). *Banking on problem solving.* Washington, DC: Curriculum Development Associates.

5

Computers and Calculators in Mathematics Education

Problems that we must solve are going to become insoluble without computers. I do not fear computers. I fear the lack of them.

—Isaac Asimov

Upon completing Chapter 5, the reader will be able to

1. Briefly describe the history of the development of computers.
2. Identify the four basic components of a microcomputer and give an example of each.
3. List at least ten characteristics of good educational software.
4. Describe six types of instructional computer use and give an example of each.
5. Describe five elementary classroom applications for calculators.
6. Specify criteria for selecting classroom calculators.
7. Describe how calculators and computers can be used as aids in problem solving and give an example for each.

The simultaneous development in England, Germany, and the United States of the predecessor to the modern digital computer was a closely guarded military secret in the early 1940s. ENIAC, the first operational computer, was developed in 1945 at the University of Pennsylvania. After World War II, several companies continued to develop more powerful computing devices for military and commercial purposes. During the 1950s and 1960s, a number of projects explored the application of computers to the learning process, but due to the relative expense of computers at that time, none had a lasting effect on classroom practice.

The current revolution in educational computing is the result of dramatic improvements in our ability to miniaturize electronic components and produce them at low cost. This effort was begun as part of the space exploration effort and reached a watershed in 1971 with the development of Intel's microprocessor, the first computer on a chip. Since then, significant advances in microelectronics and information storage devices have made possible the manufacture of low-cost processing circuits to run clocks, automobiles, appliances, and inexpensive microcomputers that rival the capabilities of the largest mainframe computers of the previous generation.

By the mid-1980s, over one million computers were in use in American elementary and secondary schools, four times the number reported only two years earlier (Becker, 1985). This versatile new technology has the potential to influence our society as profoundly as did the Gutenberg printing press, which revolutionized communication over 400 years ago.

COMPUTERS AND LEARNING

Mathematics teachers today are finding it increasingly important to be knowledgeable about the potential uses and limitations of microcomputers as tools for learning. While it may not be essential for most teachers to know how to write a computer **program** (a list of special words and symbols that direct the operation of a computer), knowing how to operate a computer, integrate its use into classroom lessons, and effectively evaluate instructional software is now considered an essential skill for beginning teachers.

Microcomputer Components

Several versatile microcomputers are marketed for personal and classroom use. Prices range from under one hundred dollars for relatively simple devices to several thousand dollars for remarkable, multipurpose systems. Hardware selection depends primarily on the intended applications. In general, it is wiser to buy a computer that runs the software you wish to use than to base your selection exclusively on a computer's power, memory, or cost.

All microcomputers have the following **hardware** features in common:

☐ Input device—keyboard, mouse, turbo-ball, joy stick, graphics pad, touch-sensitive screen.
☐ Output device—monitor, printer, plotter, overhead-projection pad.
☐ Microprocessor and memory—central processing unit (CPU), random access memory (RAM), and read only memory (ROM).
☐ Data storage device—floppy disk, hard disk, CD-ROM, cassette tape.

> Invent a metaphor for each of the four categories of components (e.g., input device is like a pencil or telephone dial). Write a short story about how you might use the four components of a computer to help you plan a camping trip.

Computer Software

Thousands of instructional programs are now available for the microcomputers currently used in schools. Good software, like a good book, requires considerable creativity and skill to produce. Most high-quality commercial products are costly, though some excellent **public domain** programs (e.g., *FrEdWriter*, an easy-to-use classroom word processor) are free and are widely used. Like other instructional materials, software should be carefully evaluated prior to classroom use.

To help teachers select good quality software, several national and regional organizations evaluate software products and report the results in readily accessible form. A list of software review journals and educational computing magazines appears in the appendix of this text. Local organizations of computer-using educators may also offer up-to-date information on available software and provide access to a low-cost software exchange network.

It is always good practice to preview software before it is purchased to ensure that it is easy to operate and can be effectively integrated into ongoing lesson objectives. Since quality software is generally expensive, many companies offer liberal preview policies for educators. Some software publishers supply multiple-copy lab packs of a product at a substantially reduced per-disk cost. Make sure the publisher agrees in writing to replace any product at low cost if it is defective or becomes damaged. Ann Lathrop (1982) argues for rejecting any software that displays the following characteristics.

1. Emits audible response to student errors—no student should be forced to advertise mistakes to the whole class.
2. Rewards failure—some programs make it more fun to fail than to succeed.
3. Has sound that cannot be controlled—the teacher should be able to easily turn sound on and off.
4. Exhibits technical problems—software should be written so that it will not crash if the user accidentally types the wrong key; incorrect responses should lead to software-initiated help comments.
5. Has uncontrollable screen advance—advancing to the next page should be under user control, not automatically timed.
6. Lacks adequate instructions—all instructions necessary to run the program must be interactively displayed on the screen (in a continuously displayed instruction window, if possible).
7. Contains factual errors—information displayed must be accurate in content, spelling, and grammar.
8. Insults or makes sarcastic and derogatory remarks—a student's character should not be impugned.
9. Is poorly documented—demand a teachers' guide of the same quality as those that accompany textbooks or other teaching aids.
10. Is copy protected or no back-up copy is provided—publishers should recognize the unique vulnerability of magnetic disks and offer low-cost replacement.

Figure 5.1 is a software evaluation form published by the Teacher Education and Computer Center and California Library Media Consortium for Classroom Evaluation of Microcomputer Courseware (1986 revision).

TYPES OF EDUCATIONAL SOFTWARE

One convenient way to classify computer use is to specify how **initiative**, or control over the operation of the program, is distributed between the user and the computer

(Levin & Souviney, 1983). Computers were first used in classrooms to provide automated drills, called **computer-aided instruction** (CAI), and to provide an environment for students to learn how to write programs.

Computer-Aided Instruction (CAI)

When using CAI software, the user responds to questions or situations posed by the computer. The response, generally a value or multiple choice selection, is evaluated by the program, and the computer initiates the next appropriate display. This type of application places most of the initiative for sequencing, pacing, and content within the program itself. The user responds to predetermined screen displays controlled by the computer program. The following are examples of commercially available CAI software.

Addition Logician. Part of the *Mastering Math Series* (MECC), this program contains a set of four addition drills and one review program for the Grade 3 level. Randomly generated exercises are presented at an appropriate level of difficulty. After successfully completing five examples, the user is presented with an interesting game to play against the computer. Students are given a session score at the end of each lesson, and a record is kept for review by the teacher.

Challenge Math. This number operation practice software (Sunburst Communications) provides practice using basic whole number and decimal operations in an environment motivated by intruders from space, a dinosaur, and a mysterious mansion. The difficulty level can be set by the teacher.

Computer Programming

Programming a computer places most of the initiative with the user. A program is a set of instructions (words and characters) that tell the computer what to do. This list of instructions must be very detailed, since a computer always does exactly as directed—nothing more and nothing less. Programmers must initiate all the instructions to control the operation of the computer and the content of the screen displays. Some newer programming languages give useful error messages when the programmer has made a mistake, but in general, the programmer must exhibit a high level of initiative when writing a program.

Programming Languages. **BASIC** is a widely known, general purpose programming language available on all popular microcomputers. It is reasonably easy to learn, because most BASIC statements and commands are written in English. **Logo** was developed at MIT as a graphics-oriented programming environment for children (Papert, 1980). Using Logo commands, young programmers direct an electronic turtle to draw figures on the screen. Many other languages, such as Pascal, C, and Fortran, are available on microcomputers for use by secondary students and professional programmers.

A. Descriptive Information

Title _____ Version _____

Publisher _____ Computer (brand/model) _____

Appropriate subject area(s) _____

Appropriate grade levels: (circle) K 1 2 3 4 5 6 7 8 9 10 11 12 College Teacher-use

Type of program (check all that apply):

__ drill/practice	__ tutorial	__ simulation	__ educational game
__ teacher utility	__ demonstration	__ testing	__ problem solving
__ word processing	__ game	__ other: _____	

Describe the program (content, main objective, how students interact with program): _____
_____

Describe management system, if any, including scoring or performance reporting and the number of students/classes permitted by program: _____

B. Evaluative Information:
Student Response

Grade levels in which used: _____ Subject: _____

Behavior observed which indicates learning took place: _____

Other reactions of students: _____

Any problems experienced, special preparation of students required: _____

Evaluation Checklist

Yes	No	Not Applic.		**General Design:**
__	__	__	**1.**	Effective, appropriate use of the computer?
__	__	__	**2.**	Design based on appropriate instructional strategies?
__	__	__	**3.**	Follows sound instructional organization?
__	__	__	**4.**	Fits well into the curriculum?
__	__	__	**5.**	Free of programming errors, reliable in normal use?
__	__	__	**6.**	Publisher's objectives are stated and met?

General Design: __ EXCELLENT __ GOOD __ OK __ POOR __ NOT USEFUL

				Content:
__	__	__	**7.**	User may choose from levels of difficulty?
__	__	__	**8.**	Branches to easier or harder material in response to student performance?
__	__	__	**9.**	Factually correct?
__	__	__	**10.**	Punctuation, spelling, and grammar are correct?
__	__	__	**11.**	Free of excessive violence or competition?
__	__	__	**12.**	Free of stereotypes—race, gender, ethnic, age, handicapped?
__	__	__	**13.**	Interest, difficulty, typing, and vocabulary levels are appropriate and are commensurate with student skills?
__	__	__	**14.**	Program data, speed, word lists, etc., can be adapted by instructor?
__	__	__	**15.**	Responses to student errors are helpful?

FIGURE 5.1 Evaluation of Instructional Courseware Form

— — — **16.** Responses to student errors are non-judgmental?
— — — **17.** Responses to student success are positive and appropriate?

Content: __ EXCELLENT __ GOOD __ OK __ POOR __ NOT USEFUL

Ease of Use:

— — — **18.** Program may be operated easily and independently by student?
— — — **19.** Expected user responses for program operation are consistent?
— — — **20.** Instructions within program are clear, complete, concise?
— — — **21.** Instructions can be skipped or called to screen as needed?
— — — **22.** Instructions on how to end program, start over, are given?
— — — **23.** Menu allows user to access specific parts of program?
— — — **24.** Answers may be corrected by user before continuing with program?
— — — **25.** Pace and sequence can be controlled by user?
— — — **26.** Screens are neat, attractive, well-spaced?
— — — **27.** Well designed graphics are used appropriately?
— — — **28.** Sound is appropriate, may be turned off?
— — — **29.** Classroom management, if any, is easy to use?

Ease of Use: __ EXCELLENT __ GOOD __ OK __ POOR __ NOT USEFUL

Motivational Devices (Check all which are used):

____ color   ____ game format   ____ timing
____ scoring   ____ personalization   ____ graphics for instruction
____ sound   ____ random order   ____ other: _____
____ graphics for reward

Motivational Devices: __ EXCELLENT __ GOOD __ OK __ POOR __ NOT USEFUL

Documentation (Check all which are provided in the package):

____ instruction manual   ____ procedures for installation   ____ workbooks
____ instructions appear on ____ teacher's guide   ____ student worksheets
     the screen   ____ instructional objectives   ____ suggested activities
____ describes required   ____ tests   ____ other: _____
     hardware

Documentation: __ EXCELLENT __ GOOD __ OK __ POOR __ NOT USEFUL

Evaluation Comments

Describe any special strengths of the program: _____

Weaknesses/Concerns/Questions: _____

Compare with similar programs: _____

Overall Opinion

__ **EXCELLENT** __ **GOOD** __ **OK** __ **POOR** __ **NOT USEFUL**

Note: From *Educational Software Preview Guide* by the Teacher Education and Computer Center and California Library Media Consortium for Classroom Evaluation of Microcomputer Courseware, 1986. San Diego: San Diego County Office of Education. Copyright 1986 by California State Dept. of Education. Reprinted by permission.

FIGURE 5.1 *continued*

Low User Initiative		Mixed Initiative			High User Initiative
←————————————————————————————————————→					
Computer-aided instruction	Tutorials	Simulations	Tools	Authoring systems	Programming languages

FIGURE 5.2 Computer Use Continuum

For experienced programmers, programming can also be a powerful problem-solving tool. The ability to invent and adapt programs requires a clear understanding of the problem situation, logical thinking, and good organizational skills—all useful components of effective problem solving. Examples of instructional applications of BASIC and Logo programming for Grades 4–6 are presented later in this chapter.

A significant amount of time is needed for elementary students to learn enough about a programming language to make it a truly useful problem-solving tool. Due to limited access to computers and an already busy teaching schedule, elementary teachers focus less on programming than on other instructional applications of computers.

Software tools, or mixed initiative software, fall near the middle of the computer use continuum shown in Figure 5.2. Such software takes greater advantage of the interactive capabilities of the computer than CAI applications yet requires significantly less instructional time to learn than programming. **Tutorials**, **simulations**, **tools**, and **authoring systems** are finding widespread applications in elementary classrooms.

Tutorials

Unlike CAI software, which provides drill for previously learned skills, computer tutorials attempt to teach new concepts. Tutorials generally rely on multiple choice responses to computer displayed questions. Students are guided through the predetermined content based on the accuracy of their responses. More sophisticated tutorials identify students who exhibit a particular error pattern and divert them to another part of the program to review the relevant concept or skill. Then, when an appropriate level of performance is attained, the students are returned to the point where the instructional sequence was interrupted. Tutorials may also be used to reintroduce previously taught concepts. The following software is an example of a simple tutorial program.

Fraction Bars Computer Program. This set of seven disks provides practice solving fraction problems (Scott Resources). Three examples of each type of exercise are given at the beginning of each lesson as a tutorial. Next, randomly generated exercises are presented, first using graphic displays and then using abstract examples. Word problems, games, and a quiz are also included on each disk (for Grades 4–6).

Simulations

Activities that enable participants to experience key features of real-world situations that would otherwise be too inconvenient, expensive, or dangerous to recreate in the classroom are called simulations. As an interactive medium, microcomputers are particularly effective at supporting simulations.

Simulations developed especially for instruction have *embedded tasks* to guide the user through a series of carefully sequenced activities. Programs like the following present problems to solve, ask questions, offer suggestions, and provide various kinds of user support.

The Market Place.   This popular software gives elementary students practice in coordinating several key factors associated with running a successful bicycle shop, lemonade stand, or other business (MECC). Students determine the cost of various raw materials and the selling price of finished products. The program supplies random weather conditions that may affect sales, and the user can check profits over a period of several days. Variables can be altered to maximize profits (for Grades 4–6).

Microworlds.   A new type of simulation called a *microworld* is designed to alternate the responsibility for guiding progress through a task between the user and the program. Such mixed-initiative programs generally begin with software-controlled instructions or examples of the task. The system then passes responsibility to the user to practice required procedures and explore the well-defined objects and relationships contained within the microworld. Users then select among software-defined tasks and tutorials to extend their understanding, with the option to return to the microworld environment to test their knowledge and conduct experiments.

Microworlds are engaging for students and exploit the interactive capabilities of microcomputers more fully than do other types of software. These programs offer a well-defined environment of objects and relations that can be used to build machines and carry out experiments. A set of predefined tasks or problems is included to assist the user in refining patterns and understandings. The following program is an example of a microworld.

Rocky's Boots.   This program allows the user to build functioning machines from electronic components (The Learning Company). The user has a set of tools and electronic components that can be connected on the screen to make larger machines. The user can also select challenging problems that are solved by constructing various devices to sort colors and shapes (for Grades 4–6).

Tools

A microcomputer can be used as a tool to accomplish a wide range of tasks. Students and teachers can use a software tool to assist in:

☐ Writing
☐ Constructing graphs and charts

☐ Designing graphics and animation
☐ Composing music
☐ Managing information
☐ Telecommunicating

Word processors, **painting programs**, and **music synthesizers** extend the features of traditional tools. Software tools respond only to the user-initiated commands; they do not pose questions or offer suggestions. A writer can enter and easily rearrange the lines of a poem; an artist can design an animated cartoon and adjust elements in each frame; or an astronomer can identify a constellation as viewed from the South Pole or from the moon. Each of these functions can be activated by loading a different program from a disk into the computer memory. Such tools do not provide specific content instruction but offer support to carry out common tasks more easily (Levin & Souviney, 1983).

Writers may initially express ideas more freely, as using a word processor (e.g., *AppleWorks*, *FrEdWriter*, or *The Writer's Assistant*) makes it easier for them to revise and polish their work than is possible with traditional technology (Mehan et al., 1986). Individuals lacking the dexterity to play a musical instrument or draw using pen and paper may produce creative works with the aid of music or graphic design software that provides greater control over the composition and revision process.

Routine **telecommunication** access to massive data bases, such as stock market statistics and airline schedules, is now possible through the use of a modem, a device that connects a computer with a telephone line, and telecommunications software. Electronic message systems have also become inexpensive enough to compete with regular mail. At the elementary school level, *computer-pal* letters are routinely being exchanged between computers over telephone lines by children in classrooms separated by thousands of miles (Levin, Riel, Boruta, & Rowe, 1985).

Manipulating and processing numbers (telephone numbers, grades, money, dates) and text (addresses, surveys, bibliographies, historical anecdotes) is also greatly facilitated by software tools.

Data base management software allows the user to enter items of information about each member of a group or collection of objects. As a classroom demonstration, personal information about each class member could be entered (name, address, telephone number, birthday, place of birth, parent's name(s), number of pets, number of siblings). The data base could then be asked questions about the class. For example, the data base could be asked to systematically search for and print out a list of everybody with a birthday in October.

Spreadsheet software is like a two-dimensional calculator. While a calculator works with only one set of values at a time, spreadsheets allow the user to enter numbers and text in separate locations on the screen (called cells), do calculations on whole rows and columns of values at one time, and show the results on the screen. For example, the formulas for the area ($A = \pi r^2$) and circumference ($C = 2\pi r$) of a circle can easily be included within the same spreadsheet, so that, for any value r, the resulting area and circumference will be displayed. Classroom applications for these versatile tools are now being designed and tested by teachers at all grade levels. The following are examples of software tools that have been used successfully in elementary classrooms. Several examples of mathematics software are described in the section "Solving Problems with Computers" later in this chapter.

Botanical Gardens. This genetics laboratory tool suggests no specific tasks but allows students to systematically test variables that affect plant growth (Sunburst Communications). Several fictitious plant seeds are available for testing in the greenhouse. Users can create their own plants for exploration in the Genetics Lab and read about the plants in a Library. Results are graphically displayed on the screen but cannot easily be printed (for Grades 4–6).

Bank Street Laboratory. This set of laboratory instruments allows students to measure and compare temperature, light, and sound intensity (Holt, Rinehart & Winston). The software produces graphs of student experiments to aid in the display of results. Patterns and functions relating physical phenomena can be explored (for Grades 4–6).

FrEdWriter. This popular, free, public-domain word processor for the classroom has on-line help and a simple built-in authoring system to help teachers design on-screen prompts for computer-supported lessons (CUE Softswap/Steel Publishing) (for Grades 2–6).

Authoring Systems

Simplified programming languages specifically designed to help students and teachers develop their own educational software are called **authoring systems**. Authoring systems do not assume that the user has prior programming experience.

Writing ToolMaker. This authoring tool allows teachers (or students) to easily develop computer activities by displaying personalized questions, menus, or other on-screen prompts (InterLearn Inc.). Later, children can use the software tools created by the *ToolMaker* to enter responses to these prompts. Graphics can also be displayed. The tool creator can specify which text will be saved on disk for later revision using a word processor (for adults).

PILOT. The authoring system, *Programmed Instruction Learning Or Teaching* (PILOT), enables novice users to write software that presents a series of personalized questions with multiple-choice responses. Feedback for correct and incorrect answers can be supplied as the program is created (for adults).

Classroom Computer Use

Teachers have successfully used a single machine with a class of students by setting up a computer learning center within the classroom (Miller-Souviney, 1985). Pairs of children can be scheduled throughout the day to work on software selected to complement ongoing instruction. Effective uses of computers include extra drill, concept development, and problem-solving experiences. Teachers should avoid using computer games as, for example, a reward for good behavior, because it is an inefficient use of the machine. To ensure equal access for all children in the class, activities such as programming that demand considerable computer time may need to be scheduled on a voluntary basis during lunch or after school.

A computer can also be used to demonstrate a concept during a lesson. By connecting the computer to a large-screen television or overhead projector display pad, the teacher can display graphs or other computer-generated visuals. An advantage of a computer demonstration over traditional displays is that professional-looking images can be developed in **real time** using actual information provided by the class. For example, in a single period a class could construct, display, and print separate graphs that summarize student television viewing preferences, eating habits, and family membership. Subsequent discussion of these graphs is likely to be more meaningful because the class participated directly in their construction. When reviewing software for classroom use, it is a good idea to consider its possible demonstration applications as well as its instructional uses.

Some schools have established a central computer laboratory that can be shared by all classes. Teachers either bring their entire classes at a scheduled time or, if a computer resource teacher is available, periodically send smaller groups of children for instruction.

Whether computers are available in a laboratory or in individual classrooms, it is important for every student to have equal access to them. As computers become more widely available, the problems associated with equitable scheduling are likely to diminish. Facility with computers will become more and more a basic skill, and it will be increasingly important for teachers to develop effective techniques for integrating this technology into everyday classroom activities.

INSTRUCTIONAL USES OF CALCULATORS

An Agenda for Action (1980) recommended that teachers at all levels take full advantage of calculators and computers in mathematics instruction. Based on more recent research results, the National Council of Teachers of Mathematics (1987) issued a more detailed *Position Statement on Calculators in the Mathematics Classroom*. The authors recommended that calculators be integrated into the mathematics program at all grade levels in classwork, homework, and evaluation. Relatively free access to calculators in schools would greatly reduce the amount of time spent on computation drills, allowing increased emphasis on problem solving, reasoning skills, and mathematics applications (National Council of Teachers of Mathematics [NCTM], 1987). Specifically, NCTM recommends that all students use calculators to

☐ Concentrate on the problem-solving process rather than on the calculations associated with problems;

☐ Gain access to mathematics beyond the students' level of computational skills;

☐ Explore, develop, and reinforce concepts including estimation, computation, approximation, and properties;

☐ Experiment with mathematical ideas and discover patterns;
☐ Perform those tedious computations that arise when working with real data in problem-solving situations.

Several states have taken the position that calculators should be available for a range of instructional and evaluation applications for all children in Grades K–12 (California State Department of Education, 1985; Carter & Leinwand, 1987). In particular, calculators have been employed to improve students' understanding of basic concepts, provide additional skill practice, and support real problem solving in the classroom.

Learning to calculate accurately is an important part of growing up in our complex society. Many events in everyday life involve working with various kinds of numbers. Counting, making change, comparison shopping, figuring sales tax, filing income taxes, doing carpentry, and working in the garden all require an ability to manipulate numbers sensibly. It is not only necessary to estimate approximate answers and calculate accurately when required, but it is also important to have a good understanding of the underlying concepts in order to know when to apply the appropriate procedure. Getting the right answer does little good if you've solved the wrong problem!

When inexpensive calculators first became available in the mid-1970s, there was concern about their potential negative effects on the development of children's computational skills and mathematics learning. During the past decade, over two hundred studies of classroom calculator use have found virtually no measurable negative effects on mathematics learning (Driscoll, 1981; Hembree & Dessart, 1986; Suydam, 1979). In fact, classroom instruction research has shown the calculator to be a useful tool to support the teaching of mathematics concepts, skills, and problem solving.

Twenty-first century children will still need to learn the basic facts and computation algorithms. An understanding of number operations and facility in computing small numbers are fundamental to the efficient application of arithmetic to real-world problems. However, solving problems with fractions, decimals, or very large and small values can be facilitated by using a calculator. Students can focus more attention on understanding problems and determining appropriate solution procedures if a calculator is available to handle the routine calculations (Comstock & Demana, 1987). Students may also be more willing to risk a tentative solution if they are not confronted with several minutes of tedious arithmetic required to test their conjectures (for sample activities, see the section "Solving Problems with Calculators").

Calculator Applications

Many calculator activities and strategy games have been designed to give children practice discovering patterns, thinking logically, and checking results (Immerzeel & Ockenga, 1977). For example, calculators can be effectively employed to

1. Develop the concept of place value (e.g., enter the value 68,341 into a calculator; what number must be subtracted in order to leave a zero in the thousands place while all the other digits remain the same?).

2. Carry out tedious calculations or those that exceed students' current level of ability (e.g., a second grader might recognize the need to add the cost of ten items purchased at the store but need help with the computation).
3. Practice estimation skills (e.g., enter 142 ÷ 16, estimate the solution, then press = to check your accuracy).
4. Give instant feedback for basic fact drills (e.g., enter 3 + 4, think of the answer, and press = to check your solution). Note: Inexpensive, preprogrammed devices are available that embed this activity in a game format.
5. Enhance insight into why algorithms work (e.g., to demonstrate that 24 × 39 = 936, first calculate the subproducts 4 × 39 and 20 × 39, then add the results 156 + 780 = 936).
6. Introduce new concepts such as percent, negative numbers, and square root (e.g., entering the subtraction exercise 4 − 7 = displays the solution $^-$3).
7. Play number-sequence strategy games (e.g., for two players sharing a calculator, see who can get to 21 first by starting at 0 and alternately adding 1 or 2 to the displayed sum).

Selecting a Calculator

Calculators with a wide range of features are available for less than five dollars each. For Grades K–2, a minimum of three to four calculators should be available for independent learning center applications. For Grades 3–6, at least fifteen to twenty calculators should be available for pairs of students to use during problem-solving sessions. Encourage students to bring a calculator from home for their personal use in school and for homework.

When selecting a calculator for classroom use, consider the following criteria.

1. Good quality keys and a large, easy-to-read, eight- to ten-digit liquid crystal display.
2. A battery life of at least ten thousand hours, or it can be solar-powered.
3. Pressing any operation key (+, −, ×, ÷) when carrying out a chain of calculations should cause the pending operation to be executed (e.g., entering 2 + 3 + should display the value 5).
4. The availability of a *memory register* (memory $^+/_-$), which makes two-step computations easier because the result of an intermediate step can be stored and recalled later (e.g., the exercise, (3 + 4) × (5 + 6) = ? can be solved by pressing keys in the following sequence: 3 + 4 = (answer 7) Memory $^+/_-$ 5 + 6 = (answer 11) × Recall Memory = (answer 77)).
5. The availability of a *constant mode*, which means that the calculator will repeat an operation if the key is pressed two or more times (e.g., to show that (4 × 3) can be calculated by adding (4 + 4 + 4), press 4 + + + to display the answer 12).

Additional desirable functions include:

6. A *square root* key that causes the number in the display to be replaced with its square root (e.g., entering 25 and then pressing the square root key gives the answer 5).
7. A *change sign* key ($^+\!/\!_-$) that allows the user to change the sign of any number displayed or entered (e.g., entering the value 3 $^+\!/\!_-$ should display $^-$3).
8. A *reciprocal* ($\frac{1}{n}$) key that causes the number in the display to be replaced with its reciprocal (e.g., entering the value 4 and then pressing the reciprocal key gives the answer 0.25 or $\frac{1}{4}$).

SOLVING PROBLEMS WITH CALCULATORS

Electronic calculating devices are useful problem-solving tools. In the following example, children can use a calculator to complete routine calculations, freeing them to focus attention on the underlying structure of the problem.

Suppose you wish to construct a box from a 1 × 1 meter piece of cardboard. Boxes are formed by cutting identical squares from each corner and folding each side into position (see Figure 5.3).

As you vary the size of the identical squares cut from each corner, different boxes can be constructed. Which box will have the greatest volume? First, let's explore some questions students may ask as they try to understand this problem.

1. Is it obvious that the boxes constructed using the method described will, in fact, have different volumes?
2. How many different boxes can be constructed?

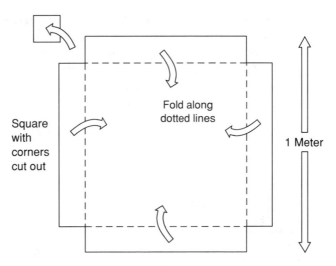

FIGURE 5.3

TABLE 5.1 Box Volume

Edge of Square Removed (cm)	Volume of Box (cm^3)
10	64,000
20	72,000
30	?
40	?
50	?

3. Is there a limit to how large the volume can get?

4. What is the largest square that can be removed from each corner and still make a box? The smallest?

5. How do you compute the volume of a box?

6. How accurate should the answer be?

To answer these questions, students could draw sketches and compute the volumes of the resulting boxes using a calculator or actually construct boxes using paper or cardboard until they are satisfied that the boxes do represent a range of volumes and that it is impossible to make them all. One possibility is to simplify the problem by allowing only squares with edges that are whole centimeters (1 cm, 2 cm, 3 cm, and so on) or multiples of 10 centimeters (10 cm, 20 cm, 30 cm, and so on) to be cut from the four corners. The resulting volumes could be computed using a calculator and compiled as in Table 5.1.

> Continue the solution and find the box with the greatest volume. Determine to the nearest centimeter the size of the square that, when removed from each corner, will result in the greatest volume. Use a calculator as required. Try to answer the six questions posed in the problem introduction.

SOLVING PROBLEMS WITH COMPUTERS

Problem-solving applications of computers can be divided into three categories.

☐ Problem-solving software
☐ Software tools
☐ Programming

Each application demands a different level of teacher expertise and computer use (Kantowski, 1983). Although it is preferable that teachers have facility with all three types of applications, with appropriate problem-solving and tool software even teachers with limited programming experience can effectively use computers in their mathematics classrooms.

Problem-Solving Software

A number of quality software products support the development of mathematical thinking and problem solving. The following examples present the user with a series of problems associated with a particular environment. Students are encouraged to use problem-solving strategies (guess-and-test, working backward, making a systematic list, constructing a table, and so on). Varying amounts of help are provided on the screen depending on the program's difficulty level and the user's experience. Problem-solving software can be incorporated into a learning center containing one or more computers or can be the focus of a teacher-directed activity for the whole class. Generally, teachers select software that aligns with the ongoing mathematics curriculum rather than adjust the instructional sequence to accommodate a particular software program.

Problem-Solving Strategies. This software is a good example of computer-supported problem-solving activities (MECC). The disk contains four programs, *Diagonals*, *Squares*, *Thinking With Ink*, and *Pooling Around*. The first two programs are tutorials that show applications of guess-and-test, systematic listing, and simplifying strategies. *Diagonals* asks students to find the number of diagonals in regular polygons. In *Squares*, the user must find the total number of squares in a five-by-five grid. *Thinking With Ink* is a game in which students must minimize the cost of painting a map where adjacent countries cannot be painted the same color. *Pooling Around* explores the number of times a pool ball hits the edge of the table and predicts in which pocket it will end. The software supports the problem solver by generating lists, tables, and graphic displays on request (see Figure 5.4). These programs take good advantage of the interactive graphic capabilities of the computer, redrawing the maps and pool tables while simultaneously updating systematic lists or tables (appropriate for Grades 4–6).

Gertrude's Secrets and Puzzles. These programs give students practice in classifying shapes, sizes, and colors (The Learning Company). Gertrude the Goose presents classification tasks that students carry out using interactive graphics. Students can design their own shapes in a shape editing room on the screen and use their objects when classifying (see Figure 5.5). Tasks encourage guess-and-test, systematic listing, and working backward strategies (*Gertrude's Secrets*—Grades K–3; *Gertrude's Puzzles*—Grades 3–6).

Software Tools

Computer software tools, originally created to improve productivity in business, have been adapted for use in schools. The software tool emulates traditional technology (a typewriter, pen and art board, or accounting pad) yet extends the capability of each.

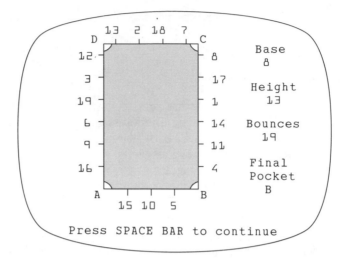

FIGURE 5.4 Pooling Around. *Note:* From "Pooling Around" [Computer program] by MECC, 1983. Copyright 1983 by MECC. Reprinted by permission.

Some systems now allow interaction between tools. For example, on more expensive machines, a column of values in a spreadsheet can be automatically displayed in graph form (e.g., *Excel* for the Macintosh). Soon, similar capabilities are likely to appear in software designed for school use.

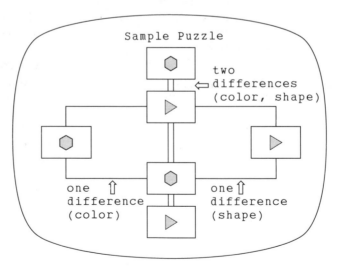

FIGURE 5.5 Gertrude's Puzzles. *Note:* From "Gertrude's Puzzles" [Computer program] by Learning Company, 1982. Copyright 1982 by Learning Company. Reprinted by permission.

MECC Graph. This program allows children to enter information, and the computer creates a nicely formatted graph that can then be printed (MECC). Like the calculator, a graph-generating tool gives students greater freedom to concentrate on the structure of problem situations.

For example, suppose a graph generator such as *MECC Graph* was available to help solve the box construction problem introduced in the previous section. The volume for each box could be quickly graphed. The graph in Figure 5.6 shows that the maximum volume occurs when the edge length of the corner squares is about 20 centimeters. If we calculate in 1-centimeter steps the volumes of the boxes that result from removing squares 15–25 centimeters on an edge, the solution will be more accurate (see Figure 5.7).

The answer (17 cm) is clearly apparent when displayed in graph form. Even greater accuracy can be achieved by computing volumes of boxes resulting from removing squares with edge lengths of between 16 and 18 centimeters in 1 millimeter steps (i.e., 160 mm, 161 mm, and so on). While it would be too time-consuming to have students draw many trial graphs displaying increasingly accurate representations of this problem, using a computer graph generator provides a handy alternative. Such graphs also serve as excellent aids when students attempt to communicate and justify their solutions as recommended in Step 4 of the problem-solving plan.

Primary level students can also use graphs to help them solve problems and explain their results. For example, suppose a class was asked to find the total number of pets owned by students in the room. Each student could fill out, in class or for homework, a pet checklist listing common household pets (cats, dogs, snakes, fish, etc.). Students could enter their results on a prepared bulletin board graph (see Figure 5. 8). The teacher (or a student) could then enter the information into a graph generator, print the graph, and quickly duplicate a copy for each student. The graph could then be used to answer a series of related questions.

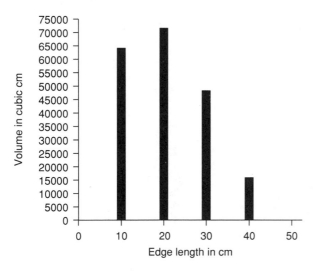

Edge length (cm)	Volume (cm^3)
0	0
10	64000
20	72000
30	48000
40	16000
50	0

FIGURE 5.6 Box Volume

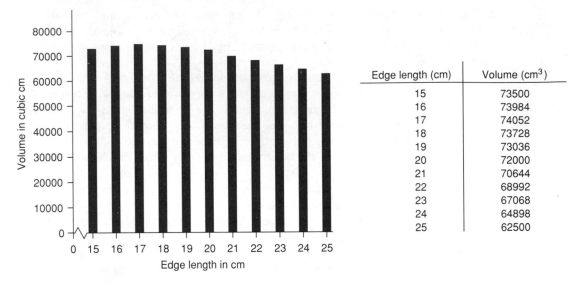

Edge length (cm)	Volume (cm^3)
15	73500
16	73984
17	74052
18	73728
19	73036
20	72000
21	70644
22	68992
23	67068
24	64898
25	62500

FIGURE 5.7 Box Volume (Close Up)

1. Who has the fewest pets?
2. Who has the most pets?
3. How many pets are there altogether?
4. What is the difference between the most and fewest pets?
5. How many students have no pet, one pet, two pets, and so on?

New graphs could be constructed to show the number of each type of pet. Questions could be posed based on the popularity of each type of pet (see Figure 5.9).

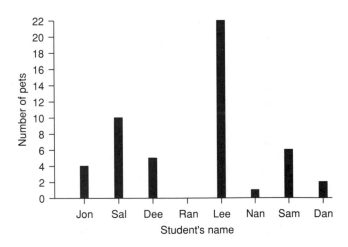

FIGURE 5.8 How Many Pets?

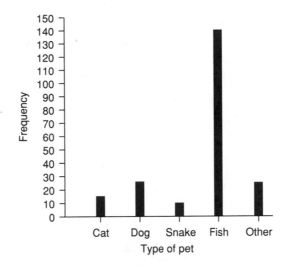

FIGURE 5.9 Which Pets Are Popular?

The Writer's Assistant. This word processor (Figure 5.10) is particularly useful for mathematics teachers because it is designed for ease of use and flexibility and can later be used to write and debug Pascal programs (Encyclopaedia Britannica Educational Corporation and InterLearn Inc.).

A word processor can be a useful tool to help students solve problems. Recent research indicates that students who write about their solutions to problems are better able to apply similar strategies to subsequent problems (Bell & Bell, 1985). Having students write their own mathematics story problems has also been recommended as an effective instructional technique (Fennell & Ammon, 1985; Graves, 1978).

The intrinsic motivation associated with using a word processor can encourage students to employ writing as a tool to analyze and justify their solutions. To be effective, students need to have regular access to a computer equipped with a word processor. A word processor allows users to enter, store, and print text. Later, the original text can be retrieved for editing and adding new ideas to the existing report. Over the course of a year, students can compile a notebook on disk, detailing each problem solution. These reports can be printed and illustrated. Of course, if no word processor is available, a problem-solving notebook can be hand written.

The use of other computer-supported tools should be explored as they become available in our rapidly changing software market. Promising recent developments include **idea processors**, such as *Calliope* (Innovision), that help users outline their ideas and plan solution strategies. Such simple data base managers and spreadsheet programs as *AppleWorks* are finding applications in elementary classrooms. These productivity tools can be employed to manage a myriad of details when solving problems and to encourage students to focus their attention on the structure of the problem itself.

New developments in software design are focusing on **domain-specific** tools. In the area of writing, software such as the *Interactive Writing Tools* (Encyclopaedia

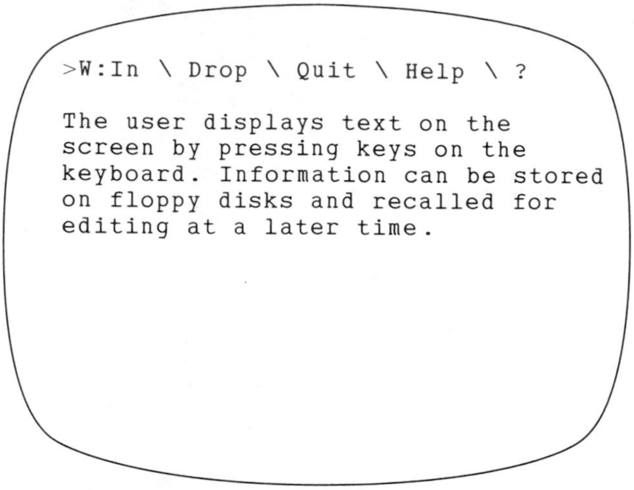

```
>W:In \ Drop \ Quit \ Help \ ?

The user displays text on the
screen by pressing keys on the
keyboard. Information can be stored
on floppy disks and recalled for
editing at a later time.
```

FIGURE 5.10 The Writer's Assistant. *Note:* From "The Writer's Assistant" [Computer program] by Encyclopedia Britannica Educational Corp. and InterLearn, Inc., 1983. Copyright 1987 by InterLearn, Inc. Reprinted by permission.

Britannica Educational Corporation) and *FrEdPrompts* (CUE Softswap/Steele Publishing) have been developed to assist in the composition of narratives, essays, poetry, letters, and other specific forms of writing (Mehan et al., 1986).

Like the calculator, tools of this type do not provide instruction. Students use these special-purpose environments to help solve problems posed by the teacher or those they ask themselves. The following are two examples of domain-specific software tools currently available for problem-solving instruction.

SemCalc. This semantic calculator employs **unit analysis** as a problem-solving strategy. It can be used effectively by children in Grades 4 and up (Sunburst Communications). For example:

> If it takes 3 hours to fly to your destination traveling 330 miles per hour, how long will it take flying 500 miles per hour?

To solve this problem, *SemCalc* asks the user to analyze the units in the problem (e.g., miles/hr × hr = miles). It then constructs a formula relating the units, allows the user to insert the values, and computes a trial solution (see Figure 5.11). The user must determine if the answer is reasonable (for Grades 5–6).

The Geometric preSupposer: Points and Lines. This tool allows children in Grades 5 and 6 to carry out geometric constructions involving points, segments, angles, triangles, quadrilaterals, polygons, and circles (Sunburst Software). It also allows the user to measure lengths, angles, areas, and the circumference of circles. Using basic constructions and measures, students can systematically test conjectures about geo-

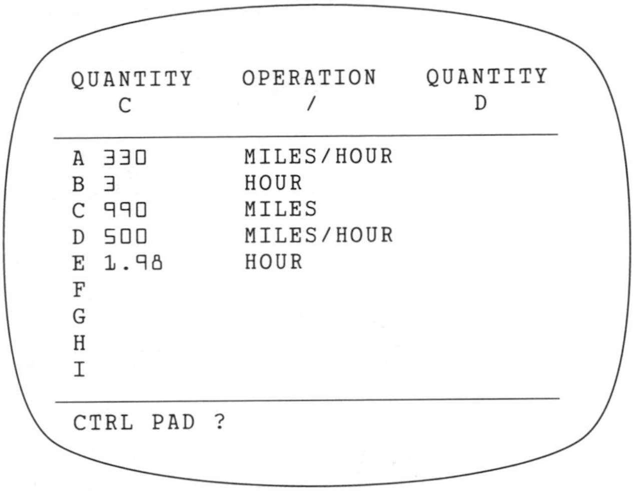

```
    QUANTITY      OPERATION    QUANTITY
        C             /            D

   A  330         MILES/HOUR
   B  3           HOUR
   C  990         MILES
   D  500         MILES/HOUR
   E  1.98        HOUR
   F
   G
   H
   I

   CTRL  PAD  ?
```

FIGURE 5.11 SemCalc. *Note:* From "SemCalc" [Computer Program] by Sunburst Communications, 1983. Copyright 1983 by Sunburst Communications. Reprinted by permission.

metric relationships and make generalizations across classes of figures. It calculates measures of segments and areas and computes ratios. The program poses no problems itself but assists students in carrying out repetitions of user-designed geometric constructions and measurements. For example, children can use the preSupposer to demonstrate that the area of each of the two triangles formed by the diagonal of a rectangle equals one half of the area of the original figure (see Figure 5.12) (for Grades 4–6).

Programming

Programming offers a third way in which computers can be used in the teaching of problem solving. Without a program, a computer is like an airplane without a pilot. A computer's potential to carry out tasks is activated only when it receives a detailed list of instructions provided by the program. Programs can be written by the user or by someone else and stored on a disk.

Many computer languages have been designed to simplify the task of writing programs. Two popular languages used at the elementary level are BASIC and Logo. The BASIC language is available for all popular microcomputers. It is relatively easy to learn how to write BASIC programs to manipulate numbers and text. Logo is a graphics-oriented language that allows even very young children to draw geometric figures on the monitor and save them on disk.

BASIC. The following examples are written in Applesoft BASIC for the Apple //e, //c, or //gs computer. However, each program will work with minor changes on other

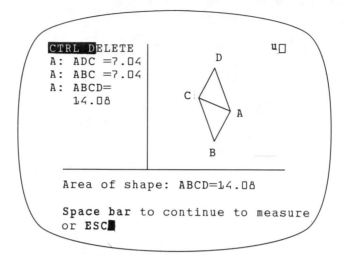

FIGURE 5.12 Geometric preSupposer. *Note:* From "The Geometric preSupposer" [Computer program] by Sunburst Communications, 1986. Copyright 1986 by Sunburst Communications. Reprinted by permission.

popular computers. See the user's manual supplied with the computer for specific details.

For novices, it is often a useful technique to begin with a completed program and encourage students to change specific values and to observe the effect on the output displayed on the screen. In the following BASIC program, notice that each line begins with a number. These are called **line numbers**. The computer reads the line with the smallest number and executes the instruction that follows. It then reads the next line, executes that instruction, and so on.

BASIC Count Program

```
1 LET COUNT = 1
2 PRINT COUNT
3 LET COUNT = COUNT + 1
4 IF COUNT < 25 THEN 2
5 END
```

This BASIC program will print the counting numbers 1 through 24. The first line instructs the Apple to set aside a location in memory named COUNT and to put the value 1 in it. The second line instructs the Apple to display the value in COUNT on the screen. The next line is the BASIC instruction for taking the current number in COUNT and replacing it with that value increased by 1. The fourth line asks the computer to make a decision: If the value in COUNT is smaller than 25, then jump back to line 2; when the value in COUNT reaches 25, the computer goes on to line 5 and stops. By carefully keeping track of the values in COUNT, it is possible to predict what numbers the computer will display on the screen.

Turn on the Apple Computer. Hold down the CTRL (Control) key and press RESET. You should see a blinking cursor. Type NEW and press RETURN to clear the memory. Enter the BASIC count program into the Apple. Type RUN and press RETURN. Make a record of the output. Next, retype Line 1 as follows and press RETURN (this will change only Line 1 and leave the other lines unchanged):

```
1 LET COUNT = 5
```

What numbers do you think will be displayed on the screen when you run this new program? Why? Type LIST and press RETURN. Notice Line 1 has been replaced as above. Now type RUN and RETURN. Make a record of the displayed numbers. Change the value 5 in Line 1 to 10, 17, 20, and then 25. Run the program and record the results after each change. Change the value in Line 1 back to 1. Try changing the value added to COUNT in Line 3 to 2, 3, and then 5. Run the program each time and record the results. Change the value 25 in Line 4 to 50, 100, and then 1000. Run the program each time and describe the results. Try to make the program count down instead of up.

Novices can explore simple BASIC programs and attempt to unravel the logic used by the programmer. The following examples introduce the BASIC statements FOR...NEXT and INPUT. Programs like these can be used to provide practice in logical thinking and looking for patterns. See the references at the end of the chapter for sources of additional programs.

BASIC Count 2

```
1 FOR COUNT = 1 TO 25 STEP 1
2 PRINT COUNT
3 NEXT COUNT
4 END
```

BASIC Count 3

```
1 INPUT X
2 FOR COUNT = 1 TO X STEP 1
3 PRINT COUNT
4 NEXT COUNT
5 END
```

The following BASIC program generates the table of volumes for the box construction problem presented earlier. Notice the inclusion of a REMember statement, which reminds the program author about the purpose of the program. Also, note that line numbers are multiples of ten to allow room for additional lines if the program needs to be altered later (EDGE = 100 is the length of edge of original cardboard square; HEIGHT = height of box; DEPTH = length of box; WIDTH = width of box; VOLUME = volume of box). (LENGTH cannot be used as a variable name since the LEN in LENGTH is a reserved programming word in Applesoft BASIC.)

Maximum Box Volumes

```
1 REM BOX VOLUME
10 LET EDGE = 100
20 PRINT "HEIGHT", "DEPTH+WIDTH", "VOLUME"
30 FOR HEIGHT = 1 TO EDGE/2 STEP 1
40 LET DEPTH = EDGE-(2*HEIGHT)
50 LET WIDTH = DEPTH
60 LET VOLUME = DEPTH*WIDTH*HEIGHT
70 PRINT HEIGHT,DEPTH,VOLUME
80 NEXT HEIGHT
90 END
```

Enter the program to compute volumes for the box construction problem. Press CTRL–S to stop the program to view the results on the screen. If a printer is available, type PR#1, press RETURN, and RUN the program to print the output. Type PR#0 and press RETURN to display the output on the screen once again. Compare the resulting table to your previous results. Change the value of EDGE from 100 centimeters to 1000 millimeters. What is the EDGE length, to the nearest millimeter, of the identical squares removed from each corner that results in the maximum box volume? Remember that the results will be expressed in millimeters and cubic millimeters. Use CTRL–S to stop the display in order to view the values on the screen (it takes a long time to print out the results on paper). Can you change line 30 in the program to display only the values in a narrow range around the correct answer?

Logo. In the late 1960s, the MIT Logo Group headed by Seymour Papert developed the Logo computer language as a Piaget-inspired learning environment for children. A key feature of this environment is a technological **turtle**, which draws graphics on the monitor under student control. Students can write Logo programs called **procedures** to guide the turtle's path. These procedures can be saved on disk for later use.

Logo has found widespread application in elementary schools. It is employed as an introduction to programming for young children and as a tool to aid the development of problem-solving skills. In particular, Logo is a convenient environment to practice the problem-solving strategies of subdividing problems into manageable units, searching for patterns, and evaluating solutions for errors in logic called **bugs**. The following activity uses the MIT (Terrapin/Krell) version of Logo run on an Apple //e, //c, or //gs computer. Minor revisions may be necessary when using other versions of Logo.

First, let's teach the turtle to draw a square. Put the Logo language disk in the drive and turn on the Apple //e, //c, or //gs. When the screen displays WELCOME TO LOGO, type DRAW and press RETURN (make sure the CAPS LOCK key is down). You will see a small, triangular turtle in the center of the screen and a ? followed by a blinking cursor at the bottom (see Figure 5.13). Enter the following sequence of commands and watch the turtle draw a square. Press RETURN after each line.

```
FD 40
RT 90
FD 40
RT 90
FD 40
RT 90
FD 40
RT 90
```

FD 40 means FORWARD 40 units (or screen dots) and RT 90 means RIGHT TURN 90°. Type DRAW and press RETURN to make the turtle center itself on the screen and clear the monitor. Try to predict what the following sequence will draw. Note that the commands can be entered on the same line separated by spaces. Enter the sequence and test your prediction (to correct a mistake, press the ESC key to back up, then reenter the character).

```
FD 40 RT 120 FD 40 RT 120 FD 40
```

Notice that both sequences of commands repeat a pair of actions over and over. Logo lets the user carry out repetitions by employing the REPEAT command. Enter

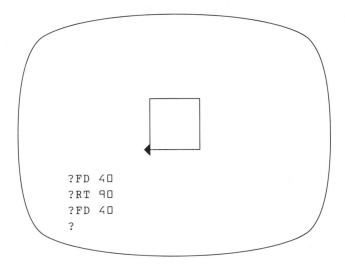

FIGURE 5.13 Logo Square

DRAW and press RETURN. Enter the following sequence of characters, including spaces, exactly as shown:

```
REPEAT 3 [FD 40 RT 120]
```

Notice that the turtle draws the same equilateral triangle as before.

Try to use the **REPEAT** command to draw a square just like the one drawn earlier.

Logo allows the user to save one or more procedures in memory and, if desired, on disk for later use. To save a procedure temporarily in memory, first pick a one-word name (e.g., SQUARE), enter TO SQUARE and press RETURN. This tells Logo that you want it to remember the procedure named SQUARE. You are now in the Logo **editor**, a mode in which you can enter and later change, or edit, procedures. Enter the following procedure for making a square:

```
REPEAT 4 [FD 40 RT 90]
```

Use the ESC key to erase incorrect entries. Press CTRL–C (hold down the CONTROL key and press C). If you have entered the commands correctly, Logo will ask you to PLEASE WAIT, then the screen will display SQUARE DEFINED. This means Logo has memorized your procedure named SQUARE so you can use it later. (Logo only remembers until you turn off the computer unless you also SAVE the procedure on disk. See the Logo manual supplied with the program for directions.)

You can now draw the square by simply entering the procedure name SQUARE and pressing RETURN. Type SQUARE and press RETURN. Repeat the process several times. What happens if you do not type DRAW and press RETURN after each run? Why?

To save the procedure named TRIANGLE, type TO TRIANGLE, press RETURN, and enter the following statement:

```
REPEAT 3 [FD 40 RT 120]
```

Press CTRL–C after the procedure is entered. When Logo displays TRIANGLE DEFINED, run TRIANGLE. Enter DRAW and press RETURN. Then run SQUARE. Notice Logo can remember more than one procedure at a time. Run TRIANGLE again without typing DRAW (see Figure 5.14).

Now let us explore the three values specified in the procedures named TRIANGLE and SQUARE.

```
TRIANGLE  →   REPEAT 3 [FD 40 RT 120]
SQUARE    →   REPEAT 4 [FD 40 RT 90]
```

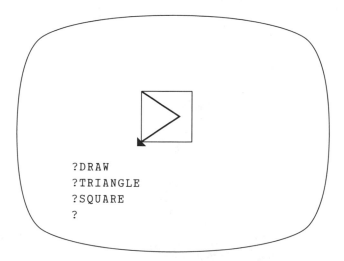

```
?DRAW
?TRIANGLE
?SQUARE
?
```

FIGURE 5.14 Logo Square and Triangle

The value preceding the left bracket indicates the number of times the commands inside the brackets should be repeated. The value following FD indicates the number of units the turtle will move forward. The value following the RT indicates the number of degrees the turtle should rotate to the right. The turtle can also move backwards (BK) and make left turns (LT). To make a pentagon (five-sided polygon) or other regular polygon with sides 40 units in length, we only need change the number of times the procedure is repeated and the amount the turtle turns at each vertex (see Table 5.2).

Notice in Table 5.2 that the turtle must turn a total of 360 degrees in its complete trip around a polygon (e.g., triangle—3 × 120 = 360; square—4 × 90 = 360). Therefore, the turtle must turn 72 degrees at each vertex to draw a pentagon (see Figure 5.15).

TABLE 5.2 Regular Polygons

Name	Sides	Length	Turtle Turn (Degrees)
Triangle	3	40	120
Square	4	40	90
Pentagon	5	40	?
Hexagon	6	40	?
Heptagon	7	40	?
Octagon	8	40	?
Nonagon	9	40	?
Decagon	10	40	?

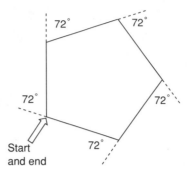

FIGURE 5.15 Pentagonal Turtle Trip

Complete Table 5.2. Write Logo procedures to draw each of the regular polygons listed. What makes the Logo sketch of the heptagon (seven-sided figure) unique? (Hint: Run HEPTAGON, then enter HT, for HIDETURTLE, and press RETURN.) Systematically increase the number of sides to 12, 15, 18, and 24. Notice that, for these polygons, it is necessary to reduce the side length to about 15 units to keep the sketch from wrapping around to the opposite side of the screen. Record your observations and discuss the results.

A wide range of classroom tested activities is available for Logo. See the reference and software sections at the end of this chapter for additional suggestions.

SUMMARY

Spawned by the demands of World War II and perfected as a by-product of the space effort, a microcomputer that currently costs as much as an electric typewriter is as powerful as large mainframe computers of only two decades ago. The availability of inexpensive electronic computing devices will play an increasing role in mathematics instruction into the next century. Hand-held calculators and microcomputers are being employed in basic skill instruction, concept development, and as problem-solving tools. Traditional computer-aided instruction (CAI) and BASIC and Logo programming applications are being complemented with computer-supported tutorials, simulations, software tools, and authoring systems. The new instructional applications take better advantage than CAI of the interactive capabilities of the computer yet require much less hands-on computer time to be effective than is required for programming instruction. Good software tutorials, simulations, and tools are as difficult to produce as other quality instructional materials and should therefore be carefully evaluated prior to purchase. Educational computing magazines, commercial directories, and local computer-using educator groups offer advice on the quality of commercial and public domain software products.

COURSE ACTIVITIES

1. Interview a teacher, a principal, and a parent about their views on the use of calculators and computers in elementary schools. Compile a list of positive and negative aspects of each. Should the school provide calculators for everyone? Should computer use be considered a basic skill? Should calculators be available during quizzes and standardized tests? What are the most effective uses of technology in the classroom? Write a short essay summarizing your conclusions about the use of technology in mathematics instruction.

2. Visit a school or other facility that uses microcomputers for instructional purposes. Describe the type of interactions observed. Are all the user inputs of the multiple choice variety? Does the student need to use paper and pencil to work problems? Are graphics used to assist with concept development or are they just decorative? Does initiative pass between the software and the user? Could the same task be as easily accomplished with paper and pencil? Discuss the results of your observations with your peers.

3. Review a piece of educational software using an appropriate evaluation form (see Figure 5.1). Share the results with other members of your class.

4. Check to see if a list of recommended mathematics software is available for use in local schools. How was this list compiled? Were local teachers involved in the selection? Were standard evaluation criteria employed? What is the distribution of the software products according to the user-initiative scale discussed in this chapter? Discuss your findings.

5. Review the calculator curriculum resource materials listed in the references. Using file cards describe, along with its learning objective, one appropriate calculator activity for each grade level K–6. Try out several of these activities with a peer or, if possible, with an appropriate group of children. Discuss the effectiveness of each activity. Exchange these activity cards with your peers and begin a curriculum resource file for mathematics teaching ideas.

6. Select an instructional objective that can be addressed using a calculator (e.g., greater than or less-than relationships). Write a lesson plan using the format presented in Chapter 2 (or other appropriate form). If possible, try out the lesson with a small group of children. Discuss the changes you would make if you taught the lesson again.

7. What effect do you think the availability of inexpensive calculating devices will have over the next decade on the content included in elementary mathematics curriculum? Will we stop teaching arithmetic? What will be the role of fractions and decimals? Will computers eliminate the need for textbooks? Teachers? Write an essay supporting your conclusions.

8. Using several centimeter-squared paper rectangles (20 cm × 30 cm), cut a series of identical squares from each corner and fold the resulting forms into boxes. Using a hand-held calculator or computer, construct a table showing the length of the edge of each square and volume of the resulting box. Graph the results. What is the size, to the nearest centimeter, of the removed square that creates the box with the greatest volume? Try to invent a general rule that relates the length and width of any rectangle to the size of the removed squares that maximizes the volume of the resulting boxes. Work in groups to solve this class-size problem.

9. If you have access to a computer, enter a BASIC program that generates the volume of boxes resulting from cutting equal squares from the corners of a one by two meter cardboard rectangle (adapt the example given in this chapter). Change the values in the program to generate tables showing the resulting volumes to the nearest 10 cm, 1 cm, and 1 mm.

10. Write a Logo procedure that draws a square with a triangular roof. Describe two other Logo tasks appropriate for the primary level and two for grades 3–6 (see references for Logo curriculum resources).

11. Read one of the *Arithmetic Teacher* articles listed in the reference section. Write a brief report summarizing the main ideas of the article and describe how the recommendations for instruction might apply to your own mathematics teaching.

MICROCOMPUTER SOFTWARE

Addition Logician Four addition drills and one review program for the Grade 3 level (MECC).

AppleWorks An easy to use, integrated data base, spread sheet, and word processing system that allows the transfer of files among the three tools (Apple Computer, Inc.).

Arith-Magic Three problems, *Diffy*, *Tripuz*, and *Magic Squares*, presented in a game format that gives students practice with computation and recognizing patterns (Quality Educational Designs).

The Bank Street Laboratory Tools for measuring and graphing variations in sound, light, and temperature (Holt, Rinehart & Winston).

Base Ten on Basic Arithmetic Multiplication and place value practice (MECC).

Botanical Gardens Allows the users to test variables affecting the growth of four fictitious plants and create four plants of their own to explore (Sunburst Communications).

Calliope An idea processor for children and adults that uses light bulbs to store and reorganize thoughts and graphics (Innovision).

Challenge Math Provides practice using basic whole number and decimal operations in an environment motivated by space intruders, a dinosaur-like creature, and a mysterious mansion (Sunburst Communications).

Creative Play: Problem Solving Activities with the Computer Explores problem solving in several interesting contexts (Math & Computer Education Project).

Delta Drawing An easy-to-use turtle graphics environment for grade K–2 children (Spinnaker).

Ecosystems An ecological simulation (Holt, Rinehart & Winston).

Fay: That Math Woman A set of lessons that provides computation practice for grades 1–3 using a number line model (Didatech Software).

Fraction Bars Computer Program Set of seven disks that provides practice solving problems involving fractions (Scott Resources).

FrEdPrompts A set of public domain writing environments that work with *FrEdWriter* (CUE Softswap/Steele Publishing).

FrEdWriter A public domain word processor for classroom with on-line help and a simple on-screen prompting feature (CUE Softswap/Steele Publishing).

Geometric preSupposer: Points and Lines Versatile tool that helps students with geometric constructions and measurements of figures and supports the testing of conjectures (Sunburst Communications).

Geometry Problems for Logo Discoveries Problem-solving activities employing turtle geometry (Creative Publications).

Gertrude's Puzzles Practice with classification in a flexible problem-solving environment (Learning Company).

Gertrude's Secrets Practice classifying objects in an easy-to-use microworld environment (Learning Company).

Graphing Equations Exploration of linear, quadratic, and trigonometric function graphs, includes *Green Globs* game (Sunburst Communication).

Interactive Writing Tools Prompted writing environment providing dynamic support for expository, letter, narrative, newspaper, and poetry writing (Encyclopaedia Britannica Educational Corporation).

Interpreting Graphs Provides experience relating graphs to real-world events (Sunburst Communications).

King's Rule The user tries to discover the rule relating sets of numbers at varying levels of difficulty (Sunburst Communications).

Logo A graphics-oriented programming language for children (Terrapin/Krell and Apple Corporation).

Logo Discoveries Problem-solving activities employing Logo programming and Logo graphics (Creative Publications).

The Marketplace Several simulations, including *Lemonade*, that explore the economic features associated with commerce (MECC).

Math Ideas with Base-Ten Blocks Computation practice using graphics of base-ten blocks (Cuisenaire Company of America).

MECC Graph A graph generating tool that automatically draws a bar graph with scales and titles for a set of values entered by the user (MECC).

Pinball Construction Kit Exploration that allows the construction of custom-built pinball machines using a flexible, graphics-oriented environment (Budge Co.).

PILOT Authoring system for the construction of multiple choice questions for lessons and tests (various publishers).

Place Value Place Two programs employing a multicolumn graphic calculator with which users can explore place value concepts involving various bases, addition, and subtraction in both tool and game formats (InterLearn Inc.).

Planetarium Simulated astronomic map (Lightspeed Software).

Problem-Solving Strategies Set of four problem-solving situations employing the solution strategies of making tables, looking for patterns, and guess-and-test (MECC).

Right Turn: Strategies for Problem Solving A problem-solving experience involving making transformations on a 3 × 3 colored grid and predicting the results of flips and turns (Sunburst Communications).

Robot Odyssey Problem solving in a simulated world using robots designed by the user (The Learning Company).

Rocky's Boots Exploration of electric circuits in a flexible, graphics-oriented environment (The Learning Company).

SemCalc Software tool that helps the user solve problems using unit analysis (Sunburst Communications).

Sensible Speller Matches word processor created text to large dictionary to help correct spelling errors (Sensible Software).

Shark Estimation Games Four estimation activities simulating a hunt for a hidden shark involving integer and decimal values on the number line and coordinate plane (InterLearn Inc.).

Snooper Troops A series of engaging detective cases that require use of reasoning skills (Spinnaker Software).

Stickybear Math A series of addition and subtraction drills with excellent graphics for Grades K–3 (Weekly Reader Family Software).

Strategies in Problem Solving: Dinosaurs and Squids Presents four types of nonroutine problems appropriate for intermediate and middle school students and a tutorial on how to employ solution strategies (Scott, Foresman & Company).

Survival Math A set of simulated activities requiring the use of arithmetic to solve real-world problems (Sunburst Communications).

The Writer's Assistant A word processing system for children and adults with on-line help and three levels of experience (Encyclopaedia Britannica Educational Corp. & InterLearn Inc.).

Writing Toolmaker Authoring tool to aid in the construction of software lessons involving on-screen prompts, menus, and graphics (InterLearn Inc.).

REFERENCES AND READINGS

Battista, M. (1987). MATHSTUFF Logo procedures: Bridging the gap between Logo and school geometry. *Arithmetic Teacher, 35*(1), 7–10.

Becker, H. (1985). *The second national U.S. survey of instructional uses of school computers*. Baltimore, MD: Johns Hopkins University, Center for the Organization of Schools.

Bell, E., & Bell, R. (1985). Writing and mathematical problem solving: arguments in favor of synthesis. *School Science and Mathematics, 85*(3), 210–221.

Billstein, R., Libeskind, L., & Lott, J. (1985). *MIT Logo for the Apple (Terrapin/Krell)*. Menlo Park, CA: Benjamin/Cummings.

Billstein, R., & Lott, J. (1986). The turtle deserves a star. *Arithmetic Teacher, 33*(7), 14–16.

Bitter, G. (1982). The road to computer literacy, Part III: Objectives and activities for Grades 4–6. *Electronic Learning, 2*(3), 44–48, 90–91.

California State Department of Education. (1985). *Mathematics framework for California public schools*. Sacramento, CA.

Campbell, M., & Fenwick, J. (1985). *Exploring with Logo*. Newton, MA: Allyn & Bacon.

Carter, B., & Leinwand, S. (1987). Calculators and Connecticut's eighth-grade mastery test. *Arithmetic Teacher, 34*(6), 55–56.

Comstock, M., & Demana, F. (1987). The calculator is a problem-solving concept developer. *Arithmetic Teacher, 34*(6), 48–51.

Corbitt, M. (1985). The impact of computing technology on school mathematics: Report on an NCTM conference. *Arithmetic Teacher, 32*(8), 14–18.

Driscoll, M. (1981). *Research within reach: Elementary school mathematics*. Reston, VA: National Council of Teachers of Mathematics.

Educational Software Evaluation Consortium. (1986). Educational software preview guide. San Diego, CA: San Diego County Office of Education, Region 15, Teacher Education and Computer Center.

Fennell, F., & Ammon, R. (1985). Writing techniques for problem solvers. *Arithmetic Teacher, 33*(1), 24–25.

Focus issue—Calculators. (1987). *Arithmetic Teacher, 34*(6).

Focus issue—Teaching with microcomputers. (1983). *Arithmetic Teacher, 30*(6).

Graves, D. (1978). *Balance the basics: Let them write*. New York: Ford Foundation.

Hansen, V. (Ed.) (1984). *Computers in mathematics education: 1984 Yearbook*. Reston, VA: National Council of Teachers of Mathematics.

Hembree, R., & Dessart, D. (1986). Effects of hand-held calculators in precollege mathematics education: A meta-analysis. *Journal for Research in Mathematics Education, 17*, 83–99.

Immerzeel, I., & Ockenga, E. (1977). *Calculator activities for the classroom*. Palo Alto, CA: Creative Publications.

Kantowski, M. (1983). The microcomputer and problem solving. *Arithmetic Teacher, 30*(6), 20–21, 58–59.

Lathrop, A. (1982). The terrible ten in educational computing. *Educational Computer, 2*(5), 34.

Leuhrmann, A. (1982). Part V. Computer literacy, what it is; why it's important. *Electronic Learning, 1*(5), 20, 22.

Levin, J., Riel, M., Boruta, M., & Rowe, R. (1985). Muktuk meets Jacuzzi: Computer networks and elementary schools. In S. Freedman (Ed.), *The acquisition of written language: Revision and response* (pp. 160–171). New York: Ablex.

Levin, J., & Souviney, R. (1983). Computers: A time for tools. *Newsletter of the Laboratory of Comparative Human Cognition, 5*(3), 45–46.

Mehan, H., Miller-Souviney, B., Riel, M., Souviney, R., Whooley, K., & Liner, B. (1986). *The write help*. Glenview, IL: Scott, Foresman.

Mercer, C., & Mercer, A. (1985). *Teaching students with learning problems*. Columbus, OH: Merrill.

Miller, S., & Thorkildsen, R. (1983). *Getting started with Logo*. Allen, TX: Developmental Learning Materials.

Miller-Souviney, B. (1985). *Computer supported tools for expository writing: One computer—twenty-eight kids*. Unpublished masters's thesis, University of California, San Diego.

Moore, M. (1984). *Geometry problems for Logo Discoveries*. Palo Alto, CA: Creative Publications.

Moore, M. (1984). *Logo Discoveries*. Palo Alto, CA: Creative Publications.

National Council of Teachers of Mathematics. (1980). *An agenda for action*. Reston, VA: Author.

National Council of Teachers of Mathematics. (1987). A position statement on calculators in the mathematics classroom. *Arithmetic Teacher, 34*(6), 61.

Papert, S. (1980). *Mindstorms: Children, computers, and powerful ideas*. New York: Basic Books.

Reys, B., & Reys, R. (1987). Calculators in the classroom: How can we make it happen? *Arithmetic Teacher, 34*(6), 12–14.

Riel, M. (1982). *Computer problem solving strategies and social skills of linguistically impaired and normal children*. Dissertation Abstracts International, #8216102. (University Microfilms No. 82–1b, 102)

Suppes, R., & Morningstar, M. (1972). *Computer-assisted instruction at Stanford, 1966–1968: Data, models, and evaluation of arithmetic programs*. New York: Academic Press.

Suydam, M. (1979). *The use of calculators in pre-college education: The state of the art*. Columbus, OH: Calculator Information Center.

Thompson, C., & Van de Walle, J. (1985). Patterns and geometry with Logo. *Arithmetic Teacher, 32*(7), 6–13.

Wiebe, J. (1981). Using a calculator to develop mathematical understanding. *Arithmetic Teacher, 28*(4), 37–39.

Witte, C. (1984). *Simple computer programs: My first programs*. Torrance, CA: Schaffer.

SECTION TWO

MATHEMATICS STRANDS

The following nine chapters address specific strands in elementary school mathematics and associated teaching practices. Each chapter contains

1 An introduction to the underlying psychological principles of learning mathematics.

2 A presentation of specific mathematics concepts and skills with examples of appropriate teaching methods.

3 A set of supplementary classroom activities appropriate for use with elementary students.

4 A set of related problem-solving experiences appropriate for use in elementary classrooms.

5 Examples of how calculators and computers can be effectively employed in instruction.

6 Suggestions on adapting instruction for children with special learning needs.

7 A set of course activities to help students and beginning teachers practice the instructional techniques presented in each chapter.

8 A list of related computer software, resource books, and references.

6

Geometry—A Beginning

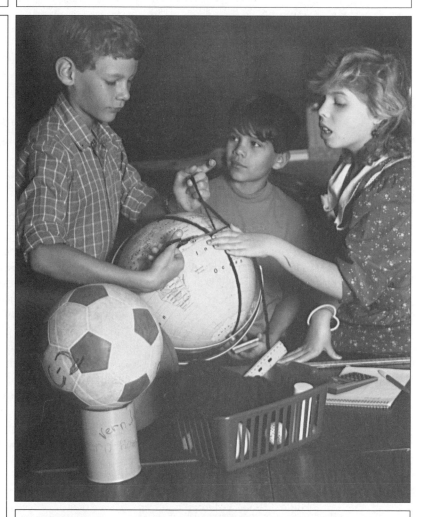

[The] universe... is written in the mathematical language [of] triangles, circles, and other geometric figures, without which it is humanly impossible to comprehend a single word.

—Galileo Galilei

Upon completing Chapter 6, the reader will be able to

1. Give examples of geometric knowledge characteristic of van Hiele's levels of thinking.

2. Describe four topological relationships Piaget suggests are used by young children to develop more complex geometric concepts and give examples of appropriate instructional activities.

3. List the characteristics of plane figures elementary children should recognize and give examples of appropriate instructional activities for each.

4. Describe activities to introduce to children the fundamental ideas of point, line, plane, and space.

5. Describe basic geometric figures and relationships associated with lines, line segments, and angles, and give examples of instructional materials that can be used to introduce each.

6. List the characteristics of common elementary space figures and give examples of activities to introduce each.

7. Describe the difference between congruence and similarity and give examples of appropriate classroom activities.

8. Describe the coordinate plane and plot a series of ordered pairs.

9. Describe the concept of line symmetry and give examples of appropriate classroom activities.

10. Describe reflection, translation, and rotation congruence and give an example of each.

Chapters 6 through 8 address the three fundamental strands of elementary mathematics education: **geometry**, **measurement**, and **number**. Learning about computation, patterns, functions, statistics, probability, and problem solving depends to a large extent on a thorough understanding of geometric, measurement, and number concepts and skills. Geometry is presented first because of its essential grounding in the child's environment. Measurement is introduced to provide the tools required to apply geometric principles in the real world. The subsequent development of number enables the powerful idea of symbolic representation to be employed in measurement and other problem-solving situations.

The term geometry originated in ancient Greece, meaning literally *earth measuring*. Children begin their study of geometry when they push a ball and display an "I did that!" smile. They soon discover that balls roll and blocks slide, that pointy things hurt and rounded objects are safer, that it is quicker to retrieve an object by crawling in a straight line than a curved path.

Knowledge of cognitive research on how children develop geometric concepts can be useful in designing classroom lessons. As discussed in chapter 2, Pierre van Hiele (1986), a Dutch mathematics educator, proposed several developmental levels of thinking through which students pass as they learn geometry and other mathematics concepts. The first level is characterized by the ability of children to recognize a shape in order to name a common object (e.g., That is a square table). A child demonstrating level two thinking is able to describe the characteristics of a known shape or object (e.g., The square table has four equal sides). Level three thinking involves classifying shapes or objects into groups according to their characteristics (e.g., All tables with four equal sides and 90° corners are square). Normally, children are expected to pass through the first three levels of thinking by the end of Grade 6. The fourth level is attained when students are able to employ deductive techniques in the proof of theorems (e.g., The sum of the interior angles of any triangle equals 180°), a capability that is generally expected of secondary students and adults (Hoffer, 1981).

Though van Hiele did not dispute the possibility of higher levels of thought, his work stresses to educators the importance of fully exploiting the potential of the first four levels. Elementary teachers can use the characteristics of the van Hiele levels 1–3, *recognizing*, *describing*, and *classifying*, as a general guide for sequencing geometry activities for children. Whenever possible, geometry activities should encourage more than one level of thought in order to provide learning opportunities for the range of abilities represented in a normal classroom. While Piaget described fundamental mental operations that characterize each level of development, van Hiele focused on techniques to stimulate children to move from one level of thought to the next.

FUNDAMENTAL SPATIAL RELATIONSHIPS

In their work on young children's conception of space, Piaget and Inhelder (1967) identified four fundamental spatial relationships that are used as tools for the later development of more complex geometric notions.

☐ *Proximity*—relative nearness of one object or event to another.
☐ *Separation*—an object or event coming between or separating two or more objects or events.
☐ *Order*—objects or events sequenced according to size, color, time, or other attribute.
☐ *Enclosure*—an object or event surrounded by two or more objects or events (enclosure can refer to one-, two-, or three-dimensional situations—see Figure 6.1).

These four relationships constitute the foundation of a branch of geometry called **topology**. The study of topology focuses on relationships among geometric figures that remain constant regardless of how the figures are stretched. For example, in traditional Euclidean geometry, named for the Greek mathematician Euclid, a square and a triangle are regarded as characteristically different figures. In topology, however, these figures would be regarded as equivalent because one can be stretched, or transformed, into the other without cutting or twisting either figure (see Figure 6.2). While Euclidean geometry focuses on the size of angles, number of sides, and relationships among components such as parallel or perpendicular lines, topology is concerned with issues of boundaries, enclosures, and transformations.

A **simple closed curve** is any figure drawn on a flat surface (plane) that encloses exactly *one* region. For example, a square and a triangle are simple closed curves

FIGURE 6.1

and would be regarded as identical in topology because each encloses one region. The Chinese Yin Yang symbol is a familiar figure that is not a simple closed curve and therefore is not topologically equivalent to a square, triangle, circle, or pentagon (see Figure 6.3).

Though adults easily identify the Euclidean properties that differentiate the square from the triangle, Piaget observed that young children seem to perceive the world topologically prior to developing a conventional, or Euclidean, understanding of spatial relations. This is one reason why young children initially have difficulty distinguishing between topologically equivalent figures such as a square and a triangle.

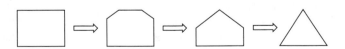

FIGURE 6.2

FIGURE 6.3
Chinese Yin
Yang Symbol

It is interesting to note that even Euclid's view may be an accurate representation of reality for only relatively modest distances. Einstein observed that his relativity theories, if true, would raise havoc with Euclidean concepts of spatial relations. For example, it would be possible to construct two parallel lines that, if extended far enough, would cross! This idea runs counter to our conventional understanding of geometric relationships. It is generally accepted today that space is curved. As astronomers view events far across the galaxy, light, which travels in the straightest possible line, seems to bend around objects of great mass like the sun and therefore might eventually cross.

Euclidean concepts match real-world measurements only if the size of the space involved is relatively small. We, like preschoolers, must adjust our geometric tools depending on the task at hand. Just as adults might be disoriented by the astronomer's view of interstellar space without a prior understanding of Euclidean geometry, children need careful guidance in consolidating their topological knowledge prior to the formal introduction of Euclidean concepts such as points, lines, planes, parallel constructions, congruence, and symmetry.

An additional topological concept, *continuity*, also seems to be related to the concepts of proximity, separation, order, and enclosure. Continuity is the notion that objects can be infinitely subdivided. In geometry we refer to a line as an infinite collection of points. A point is thought of as a location, with no size or shape associated with it. How can a line, which has measurable length, be constructed of points that have no size? Such is the paradox of infinity. Piaget observed that children do not fully grasp the notion of continuity, or infinite subdivision, until they reach the level of formal operations.

INFORMAL GEOMETRY

In Grades K–3, geometry activities focus on the consolidation of basic topological concepts and the informal introduction to geometric curves, shapes, and solids. Children at this age frequently display an incomplete notion of the properties of common geometric figures. For example, if a square is rotated 45°, it appears to be diamond-shaped. Such thinking focuses on irrelevant features of the figure (the orientation of the figure on the page). The observer may then conclude that the shape is no longer a square (see Figure 6.4).

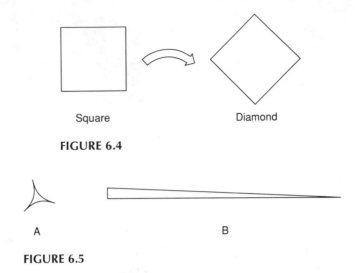

Square   Diamond

FIGURE 6.4

A   B

FIGURE 6.5

In a study of spatial relations by Burger, Hoffer, Mitchell, and Shaughnessy (1981), elementary students were asked to identify the triangles in a display of line sketches. Many readily accepted as triangles those figures constructed of three curved sides like sketch A in Figure 6.5, while rejecting figures like sketch B because they were too skinny. Even secondary students just beginning their study of high school geometry suffered from the same misconceptions.

Studies of this type point out the need for informal geometry experiences in the elementary and middle school grades. Students need opportunities to test their understanding of which features of geometric figures are important and which are not when classifying geometric figures (e.g., number of sides versus orientation on the page, or the measure of interior angles versus the lengths of sides).

Many textbooks portray geometric figures using stereotypic images (for example, generally showing triangles as isosceles or equilateral and parallelograms with a horizontal base). A way to help children focus attention on appropriate features is to display figures in nonstandard ways. Practice with such displays can help students apply systematic criteria when identifying a figure. Notice that all four choices accurately describe the shape shown in Figure 6.6.

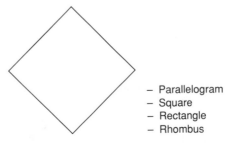

 – Parallelogram
 – Square
 – Rectangle
 – Rhombus

FIGURE 6.6 What Is the Name of This Shape?

FIGURE 6.7 The Child Is Under the Cat

Relative Position

Through topological experiences with objects in their environment (furniture, plants, string, rubber bands, pictures, blocks), children refine their understanding of relative position. Initially, children relate the position of objects to themselves. For example, they would say that the child in Figure 6.7 is under the cat in the tree.

Later, the relative positions of other objects in the environment can be investigated. Activities that actively engage children in experiences requiring verbalization of the proximity, separation, order, and enclosure relationships of objects to one another are particularly important. For example, the train in Figure 6.8 shows the following relative position concepts:

| Caboose | Tanker | Cattle car | Passenger car | Engine |

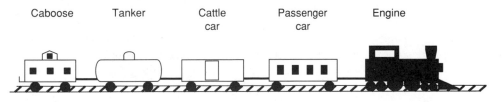

FIGURE 6.8

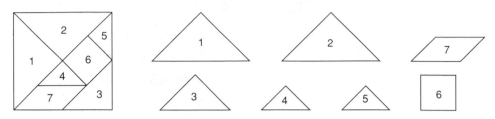

FIGURE 6.9

☐ *Proximity*—The passenger car is closer to the engine than the caboose.
☐ *Separation*—The cattle car separates the tanker from the passenger car.
☐ *Order*—The cars are in order from small to large.
☐ *Enclosure*—The engine and the caboose enclose the other cars on the track.

Contrasting terms describing opposites can also be introduced. Children should be encouraged to verbalize and record the results of their experiments. Terms should include:

over/under   right/left
first/last   inside/outside
near/far   on/off
together/apart   open/closed

Using Tangrams

An ancient Chinese puzzle called the **tangram** can be used as an instructional aid for exploring geometry. The seven-piece puzzle can be assembled to form a square, triangle, parallelogram, and trapezoid. Tangrams can be used to explore relative position and the spatial relations: proximity, separation, order, and enclosure. For example, the small triangle (4) in the square arrangement shown in Figure 6.9 is enclosed by the other tangram pieces.

Using the Geoboard

Another useful material for informal geometry activities is the **geoboard**. This versatile learning tool can be easily constructed by the teacher or older students as shown in Figure 6.10.

Children working alone or in pairs can construct designs on geoboards using colored rubber bands. These designs can then be used as examples to introduce the names of various geometric figures. Open and closed curves, triangles, rectangles, squares, and other polygons can be modeled using the geoboard. For example, children could be asked to make a design on their geoboard with exactly three sides. Some possible results are shown in Figure 6.11.

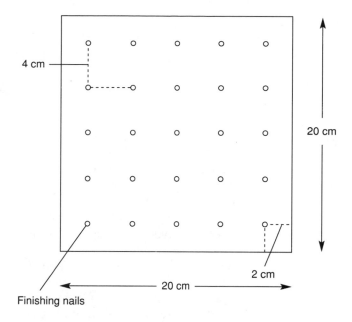

4 cm

20 cm

2 cm

20 cm

Finishing nails

FIGURE 6.10 Geoboard Construction Plan

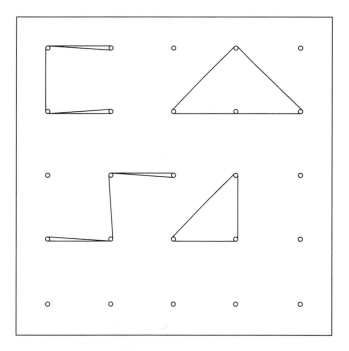

FIGURE 6.11 Three-sided Open and Closed Figures

Make a list of questions you could ask about the designs in Figure 6.11 to help children discover the characteristics of a triangle. What other geometric figures can be shown on a geoboard?

Working with Paths

The notions of point and line develop from work with position and motion. Objects are associated with a position, or point, in space relative to an observer or other objects. Paths connecting these points represent curved or straight line segments. Exploring relationships among networks of points and paths provides important topological experiences for later work with geometric figures. An activity children often find interesting involves testing a network of points and curves to see if it can be drawn without retracing a line or lifting the pencil. For example, is Figure 6.12 traceable?

If we begin at position B or D, part of the figure will remain untraced. If we start at point A or C, the figure is traceable. If a figure can be drawn by starting at some point without retracing a segment or lifting the pencil, we say the figure is traceable.

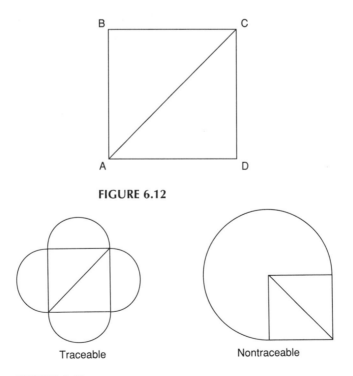

FIGURE 6.12

Traceable   Nontraceable

FIGURE 6.13

Test the two networks in Figure 6.13 to verify if each is traceable. Try to discover a rule based on the number of segments and vertices in a network that will enable you to predict if a figure is traceable.

Two observations aid the understanding of network-tracing problems. Notice that when starting at an *even vertex* (an even number of lines connect at a corner), the tracing must end at the same point. Also, if you do not start at an even vertex, it is impossible for the tracing to end there (see Figure 6.14).

Odd vertices have the opposite effect. If you begin at an odd vertex, the tracing will end elsewhere. If a tracing does not start at an odd vertex, it must end there (see Figure 6.15).

If a drawing begins at an odd vertex, it must end at another. Therefore, for a figure to be traceable it must have *exactly* two odd vertices or no odd vertices at all. Make up several networks and try to verify this conjecture. Can you predict at which vertex the tracing must start and end for each network?

Complete Table 6.1. Carefully observe the patterns in the table. Can you find the relationship among the number of vertices (*V*), regions (*R*), and/or edges (*E*) for all the figures tested? Test several new networks to see if your prediction is true for other cases. See if anyone else has invented a network that does not satisfy your prediction.

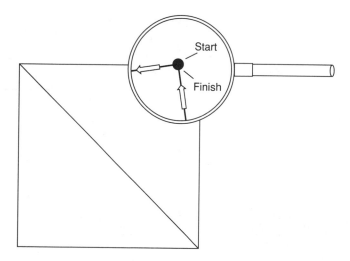

FIGURE 6.14 Even Vertex

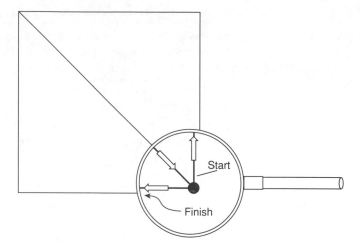

FIGURE 6.15 Odd Vertex

TABLE 6.1 Tracing Networks

Figure	Vertices (V)	Regions (R)	Edges (E)
Square	4	1	4
House with roof	5	2	6
Square with two diagonals	5	4	8
Your own network	?	?	?

Recognizing Shapes

Children begin early in life to classify objects in the environment according to their unique characteristics. In order to recognize objects and people, children must make increasingly finer distinctions among features. To further the development of classification skills, it is important for teachers to be precise when introducing new geometric concepts and terms. For example, a fundamental geometric figure is the **simple closed curve**. This figure lies in a plane (flat surface) and encloses exactly one region. However, sometimes children are confused since a geometric curve refers to a straight line as well as an arc. Several simple closed curves receive special attention in the study of geometry. These shapes, including the triangle, rectangle, and circle, display geometric properties and relationships that can be explored by elementary students. For example, Figure 6.16 shows that the number of vertices and edges can be used as a criterion for distinguishing among *polygons*, the simple closed figures composed of straight line segments.

Symmetry is an important concept that finds applications in several strands of mathematics. For example, we use symmetry to help us divide a pie into two equal pieces when introducing the idea of one half. Classification of geometric figures can also be refined by observing the symmetry of figures. A mirror can be used to

These are simple closed curves.

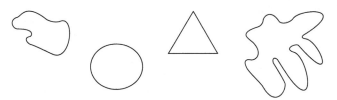

These have exactly three corners and are called triangles.

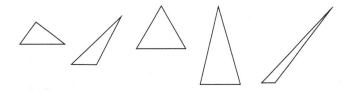

These have exactly four corners and are called quadrilaterals.

FIGURE 6.16

explore lines of symmetry in figures. Have children place a small plastic mirror on a picture of their face. To locate a line of symmetry, hold the mirror vertically and place its edge on a line running between the eyes until the image in the mirror and the part of the picture showing make a whole face (see Figure 6.17). Have children compare the results using the left side of their face and its reflection with that of the right side and its reflection. Try using other objects and pictures to find lines of

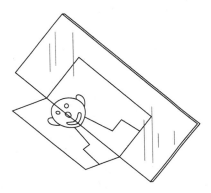

FIGURE 6.17 Mirror Image Showing Picture Symmetry

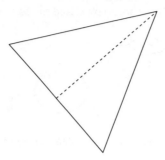

FIGURE 6.18 Isosceles Triangle

symmetry. A bulletin board display can be constructed to classify objects according to whether they contain zero, one, two, or more lines of symmetry.

Three-sided polygons are called *triangles*. Children can classify isosceles, equilateral, and scalene triangles by using symmetry. Give students a paper cutout of each of the three triangles and have them fold lines of symmetry. For example, if the triangle in Figure 6.18 was folded along the dotted line, the left side would exactly match the right. Such a line is called a *line of symmetry*.

A triangle with at least one line of symmetry is called *isosceles*. Some isosceles triangles have three lines of symmetry. These special isosceles triangles are called *equilateral* triangles (see Figure 6.19). Have the students list other differences between the two types of triangles. Triangles with no lines of symmetry are called *scalene* triangles (see Figure 6.20).

Can an isosceles triangle have exactly two lines of symmetry? Construct several paper triangles and fold lines of symmetry to find out. Explain your answer.

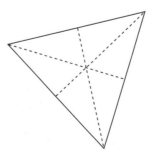

FIGURE 6.19 Equilateral Triangle

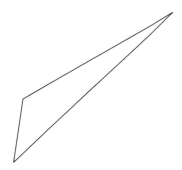

FIGURE 6.20 Scalene Triangle

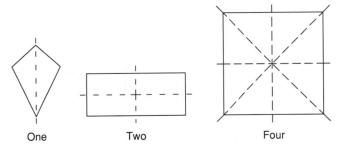

One   Two   Four

FIGURE 6.21 Quadrilateral Lines of Symmetry

Quadrilaterals can also be classified according to the number of lines of symmetry. Give students paper cutouts of quadrilaterals and have them fold lines of symmetry and write the number of lines on each figure (see Figure 6.21). What kinds of four-sided figures have exactly one line of symmetry? Two lines of symmetry? Do any have more than four lines of symmetry?

Using paper, try to construct a quadrilateral with exactly three lines of symmetry. What happens? Can you explain your result? Figures with equal sides and angles are called *regular polygons*. Complete Table 6.2 by determining the maximum number of lines of symmetry in each figure.

Notice in Table 6.2 that, as the number of sides of the regular polygons increases, the figure more closely approximates the shape of a circle. Since a circle is smooth (all its points equidistant from the center), we can think of it as a polygon with an unlimited (infinite) number of sides (see Figure 6.22). How many lines of symmetry does a circle have?

TABLE 6.2 Lines of Symmetry in Regular Polygons

Regular Polygon	Sides	Lines of Symmetry
Triangle (equilateral)	3	3
Quadrilateral (square)	4	4
Pentagon	5	5
Hexagon	6	?
Heptagon	7	?
Octagon	8	?
Nonagon	9	?
Decagon	10	?

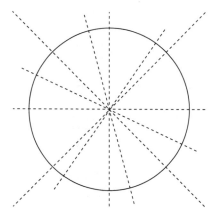

FIGURE 6.22 Circle Lines of Symmetry

Recognizing Space Figures

The properties of three-dimensional **space figures** are introduced at the elementary level. Children first experiment with geometric solids by building constructions of various shaped blocks to see which stack best and by identifying the shapes of the faces. Older children extend these early experiences by counting the number of planes of symmetry for various space figures and exploring the relationship among the number of vertices (corners), edges (line segments between vertices), and faces (flat surfaces).

Figure 6.23 shows six common, three-dimensional space figures—the *right rectangular prism* (cube, shoe box), *tetrahedron* (triangle based pyramid), *square pyramid* (square based pyramid), *sphere* (ball), *cone* (ice cream cone), and *cylinder* (can).

Grade 4–6 children can explore the symmetrical properties of space figures. A plane can be thought of as a flat surface that slices through a solid. If all pairs of corresponding points in the resulting two parts of the space figure are equidistant from the partitioning plane, this plane is called a *plane of symmetry*. For example, to be a plane of symmetry, any line passing through the space figure at points A_1 and A_2 must be equidistant from the corresponding point A on the plane (see Figure 6.24).

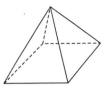

FIGURE 6.23 Space Figures

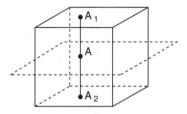

FIGURE 6.24 Plane of Symmetry

Complete Table 6.3 by determining the number of planes of symmetry in each space figure. The cube has nine planes of symmetry. Can you find all of them?

TABLE 6.3 Planes of Symmetry

Space Figure	Planes of Symmetry
Cube	9
Right rectangular prism	?
Tetrahedron	?
Square pyramid	?
Cylinder	?
Cone	?
Sphere	?

Developing Euler's Formula

An interesting problem-solving activity for Grade 5–6 children involves finding a general formula relating the number of vertices, faces, and edges of polyhedra, space figures composed entirely of flat surfaces. The Swiss mathematician Leonard Euler discovered this formula over two hundred years ago. Table 6.4 shows the number of vertices (V), faces (F), and edges (E) for the cube.

TABLE 6.4   Euler's Formula

Space Figure	Vertices (V)	Faces (F)	Edges (E)
Cube	8	6	$12 \rightarrow (8+6=12+2)$
Tetrahedron	?	?	?
Square pyramid	?	?	?
Your space figure	?	?	?

Complete Table 6.4 by counting the number of vertices (*V*), faces (*F*), and edges (*E*) for several common space figures. Try to rediscover Euler's formula relating *V*, *F*, and *E*. Does Euler's formula work for space figures with curved surfaces (e.g., cone, cylinder, sphere, etc.)?

Language Considerations

Terms used in mathematics have precise meanings. Often words used precisely in mathematics have more general or even ambiguous meanings in everyday conversation. The following sentences are examples of conversational and mathematical applications of common terms.

Normal Conversation	*Mathematical Usage*
That would be a nice *area* to live in.	The *area* of the farm is 20 hectares.
Add my name to the list.	*Add* the tax to the price.
Isabel will *sum* up the meeting for us.	The *sum* of the numbers is 12.
Chris always turns out a quality *product*.	The *product* of the values is 254.
It was just one *factor* in my decision.	A prime number has exactly two *factors*.
The repair was a complex *operation*.	Multiplication is a mathematical *operation*.

Think of ten more words that might cause confusion because of differences between conversational and mathematical usage. For each term, write sentences employing its conversational and mathematical meanings.

Children need opportunities to discuss the differences between conversational and mathematical uses of terms. After a term has been introduced in an activity, take care to reduce confusion by discussing the precise mathematical meaning of each term and identifying specific instances where the definition might conflict with common usage. For example, a line is often thought of as a relatively straight and narrow figure that varies in width according to its application, such as lined paper or the line down the middle of the road. Slightly more obscure are uses like railroad line and ocean liner.

The mathematical term *line* is, in fact, undefined in geometry, and its intuitive meaning may be difficult for most children to fully appreciate. An intuitive description of a geometric line is an infinite collection of points that has length but no width. Where a point can be thought of as pure location, a line is pure distance. Often the term line, which is infinitely long, is confused with line segment, a measurable section of a line.

The difference between the mathematical idea and the physical or popular representation of that idea may contribute to the confusion some children experience when learning mathematical concepts. The teacher needs to be aware of the potential conflict in meaning and be prepared to discuss appropriate examples to minimize confusion. It is likewise important that the teacher thoroughly understand the concept involved in order to anticipate potential points of confusion.

EUCLIDEAN GEOMETRY

Undefined Terms

Four fundamental undefined concepts in geometry are point, line, plane, and space. Piaget noted that many children below Grade 4 have difficulty understanding and applying these concepts. An intuitive understanding must be gradually fostered using children's existing perceptual notions as the basis for development.

A **point** can be thought of as a location. The period at the end of this sentence might represent a point. However, to describe an even more accurate position, conceive of this spot shrinking towards its center until it has no measure whatsoever—no dimension. Because the notion of a point in space can be only intuitively described, mathematicians say that the term point is *undefined*. In other words, the notion of a point must be intuitive.

Similarly, a geometric **line** is a one-dimensional figure consisting of an infinite collection of points that extends indefinitely in two directions. A line defined by points A and B is denoted as $\overleftrightarrow{AB}$, or it can be labeled with a single script letter when confusion is unlikely (see Figure 6.25).

A **plane** is an infinite collection of points that forms a flat surface extending indefinitely in all directions. A plane is a two-dimensional figure that encompasses an area but has no thickness. Figure 6.26 shows examples of three points determining a plane—three pencils supporting a book, and a camera tripod standing on the floor.

Space, the fourth undefined term, is the collection of *all* points. Space is three-dimensional and encompasses a volume (see Figure 6.27).

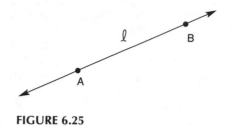

FIGURE 6.25

Think about a point inside your head directly between your ears. Imagine the path of a very, very thin laser beam exiting each ear. This path could represent a geometric line. With a partner, hold three sharpened pencils of different lengths pointing upward on a table (not in a straight line). Place a book so that its surface touches all three pencil points. This experiment can be used to show children that any three nonlinear points determine a unique plane. Why are tripods made with only three legs? Why do some tables wobble? Think of other metaphors that could be used to help children intuitively understand the concepts of point, line, plane, and space.

Defined Figures

Drawing on these intuitive notions, the elementary geometric figures can be defined. A **line segment** can be thought of as the shortest path between two points. A **ray** is part of a straight line that begins at some point and extends in only one direction like a ray of light from the sun. An **angle** is the figure created when two rays have a common endpoint, or **vertex**. To introduce these figures, have two children come to the front of the class to play the part of two points. Give each the end of a piece

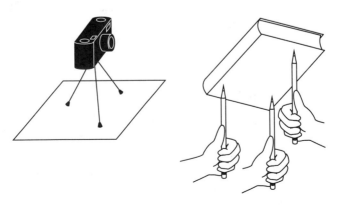

FIGURE 6.26

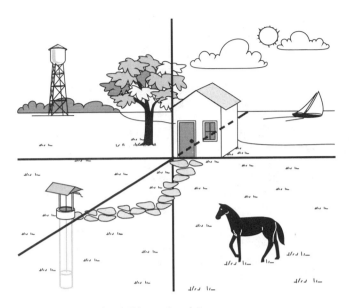

FIGURE 6.27 Three-Dimensional Space

of yarn about ten feet long and have them stretch it tight to model a line segment. The students can role play a ray by having one child hold the end of a ball of yarn (the endpoint), while a second child stretches the string from a second point about five feet from the first (in the direction of the door), and the teacher unrolls the ball in a straight line right out the door explaining that, if this were a real ray, the yarn would stretch out forever. Other geometric figures can be role played similarly (see Figure 6.28). As a follow-up activity, have children cut out pictures from magazines and write a description of how each object or action displays a line segment, ray, or angle. Make a bulletin board display using the pictures representing geometric figures and the students' written descriptions.

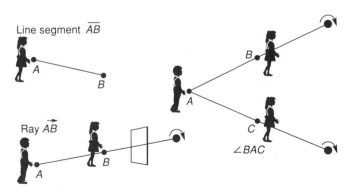

FIGURE 6.28

Comparing Angles

Joggers and backpackers carefully plan their route based on the inclines they expect to encounter. Intuitively, children also know that angles come in different sizes. Riding their bicycles up some hills is easier than others.

Children should have opportunities to physically compare the size of angles using cardboard figures, blocks, and other objects in the classroom. Students frequently confuse the length of the line segments representing the sides of an angle for its **angle measure**. Part of the reason for confusion is that physical angles are constructed of line segments that are only representations of rays of infinite length. In Figure 6.29, for example, some children might confuse what is being measured and claim that $\angle A > \angle B$ since the sides of angle A appear longer than those of angle B. Children need experience physically comparing angles to help them understand that the size of an angle is the measure of the spread between its sides, not the apparent length of its sides.

Directly comparing angles found in the environment helps students build an understanding of what is measured when we measure an angle. An **angle copier** like those used by carpenters can be constructed from two strips of heavy cardboard connected at one end by a brass fastener. Have students practice copying angles found in the classroom, trace each on a separate piece of paper, and label them with the name of the object measured. These angles can then be directly compared by laying one on top of the other and then ordered from small to large. Such informal experiences help students learn to distinguish between the length of the segments representing the sides of angles and the angle measure of their spread.

Construct an angle copier and measure ten angles found in the environment. Draw a sketch of each and put them in order from small to large.

In the upper grades, children learn to quantify angles by assigning a numerical angle measure (see "Measuring Angles" in chapter 7 for details). Ancient astronomers divided the circle into 360 equal pie-wedge shaped regions, perhaps because it was once thought that the year had 360 days. Each of these wedges was equal to the unit of angle measure we now call one **degree** (see Figure 6.30). The degree is denoted by a small superscript circle as in 90°. Note that units of temperature are denoted by the same term and symbol followed by a letter (F for Fahrenheit or C for Centigrade). The two uses are not related.

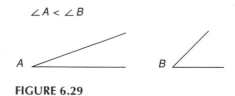

$\angle A < \angle B$

FIGURE 6.29

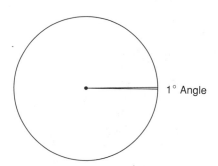

FIGURE 6.30 360 Degrees in a Circle

The **right** angle (90°) is familiar to most children. The lines in rectangular floor and ceiling tiles meet at right angles. Pictures of farms in the Midwest show a grid of roads that also meet at 90°. Have students identify objects in the classroom that contain line segments that meet at right angles. Other angles can be classified by their relationship to right angles. Angles with measures less than 90° are called **acute** angles. Those with measures greater than 90° are called **obtuse** angles. Have three students role play the construction of these angles for the class using lengths of yarn to represent the sides. Angle copiers can then be used to locate and label acute, obtuse, and right angles found in the classroom. **Straight, complementary, supplementary,** and **vertical angles** can be introduced in a similar fashion (see Figure 6.31).

Polygons and Circles

Grades 4–6 are introduced to advanced work with a special group of simple closed curves called **polygons**. A polygon is a simple closed curve made entirely of line

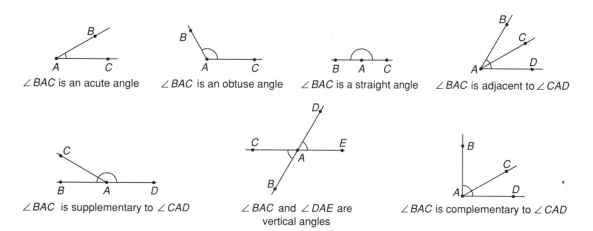

FIGURE 6.31

segments (no curved arcs). Regular polygons such as equilateral triangles and squares have equal sides and equal interior angles.

As the number of sides increases, regular polygons appear to approximate the shape of a circle. The pie-wedge shaped central angle for the 360-side polygon equals 1°. As the number of sides increases, the size of this angle diminishes. Intuitively, when the number of sides becomes very large (approaches infinity), every point on the figure will be equidistant from the center. As with polygons, **circles** consist only of the points along the edge of the figure, not the interior area.

This procedure for defining a circle can be modeled by giving each child in a class a piece of yarn five feet long. Clear an area of at least ten feet across in the front of the classroom and have one student sit in a chair in the center. Have three students come forward and the child in the center hold one end of the three pieces of yarn while the three radiate evenly around the center holding the other end of their string of yarn. Point out that the segments connecting the three children form a triangle. Have a fourth child come forward, hand one end of the yarn to the center child, and see if the four students around the edge can form a square. Continue until the entire class encircles the child in the center. Ask what shape the class has constructed. How can they tell they are all the same distance from the center? As a follow-up activity, have students draw pictures of objects and events that use circles or parts of circles (e.g., *radius, diameter, chord, arc, tangent*).

> Figure 6.32 shows a circle with its circumference, diameter, radius, a chord, an arc, and a tangent line labeled. Make a list of real-world situations that model each part of a circle (e.g., a spoke of a bicycle wheel approximates the radius).

Several relationships exist among the parts of a circle. For example, using a bicycle wheel you can show that the length of the diameter (distance across the hub) is equal to twice the length of the radius (length of a spoke). Give the class a set of round objects or pictures of circles and have them verify this relationship for various

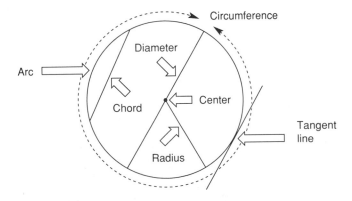

FIGURE 6.32

sized circles. Less obvious is the relationship between the circumference and the diameter of a circle.

> Demonstrate that the circumference of any circle is about three times the length of its diameter. Cut a piece of masking tape equal to the circumference of a jar lid. Compare this length to the diameter of the lid. You can show the relationship by tracing the lid three times and sticking the tape over the sketch as shown in Figure 6.33. Repeat the process using several different sized lids. Is the result always the same? How tall is a tennis ball can that holds three balls?

Dividing the circumference of any circle by its diameter gives a constant value called **pi** (π), which is slightly larger than 3. The value of π has been investigated for centuries. As early as 2000 B.C., the Babylonians and Egyptians were aware that the ratio of the circumference of the circle to its diameter was constant. Using relatively crude sketches, the Babylonians calculated π to equal 3.125. Later, the Egyptians calculated $\pi = (16/9)^2$, or approximately 3.16. Archimedes (240 B.C.) used regular polygons with an increasing number of sides to determine that π was between 223/71 and 22/7 (approximately 3.14163). Around 1665, Newton computed the value of π correctly to sixteen decimal places (3.1415926535897932) using the calculus techniques he invented to prove the Laws of Motion.

There were also pursuits of π that led to folly. In the eighteenth century, an aspiring mathematician, William Shanks, spent fifteen years computing the value of π to 707 decimal places. It was later discovered that he made an arithmetic error in one of the early computations, rendering all the subsequent decimal values incorrect. In 1897, Dr. Edwin Goodwin authored a bill introducing a new mathematical truth that was submitted to the State Legislature of Indiana. In his well-meaning effort to provide royalties to the Indiana State Department of Education through sale of his discovery to textbook publishers, he nearly caused legislation that set π equal to 9.2376, one of the worst estimations of π in four thousand years!

Parallel Lines

Parallel lines (those that lie in the same plane but never cross) appear frequently in the man-made environment but are found less often in nature. As shown in Figure 6.34, rectangular buildings contain many parallel lines.

Circumference of one jar lid

FIGURE 6.33

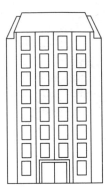

FIGURE 6.34

The forms of nature are often expressed in branching, circular, or elliptical patterns. Crystals are one of the few examples of parallel structures that occur naturally. To help draw attention to various geometric forms found in the environment, have children cut out pictures of objects and construct a collage showing parallel, branching, circular, and elliptical structures (see Figure 6.35).

The study of parallel lines enables Grade 4–6 children to explore the features of **parallelograms**, the set of quadrilaterals with parallel opposite sides. Parallelograms are classified into overlapping categories that differ by one or more features. For example, a *rectangle* is a parallelogram with 90° interior angles; a *rhombus* is a parallelogram with four equal sides; a *square* is a parallelogram with equal sides *and* 90° interior angles. To help children learn the characteristics that differentiate one type of parallelogram from another, prepare a deck of cards showing each figure labeled with the most restrictive name. For example, for the figure with four equal sides and 90° interior angles, use the label "square" even though it would also be correct to use the less restrictive labels—rectangle, rhombus, or parallelogram. Have one child at a time come to the front of the room and select a card from the deck. The class can then ask the student yes or no questions until someone identifies the figure. Keep track of the number of questions needed to arrive at the right answer. Repeat the game several times, or have the students work in small groups with individual

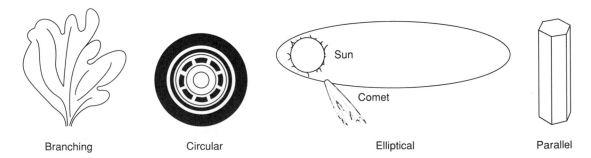

Branching   Circular   Elliptical   Parallel

FIGURE 6.35

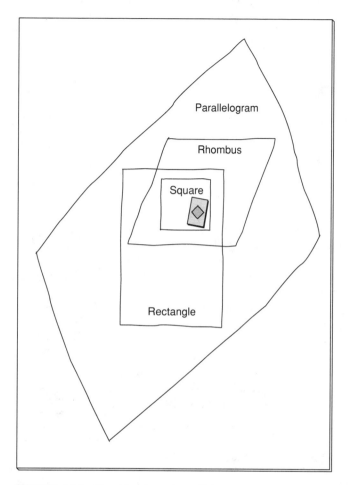

FIGURE 6.36 Classification of Parallelograms

decks of cards. Once the class guesses the name of the figure, have the child place the card in the correct position on the classification diagram shown in Figure 6.36.

Similarity and Congruence

Two polygons are said to be **similar** if their corresponding angles are equal. The lengths of corresponding sides of similar polygons are *proportional* (the ratios of the lengths of corresponding sides are equal). Children can explore this relationship by reproducing a triangle on the geoboard such as $\triangle ABC$ in Figure 6.37 and constructing a new $\triangle DEF$ with each corresponding side twice as long. Students can make paper copies of each angle to verify that the corresponding angles are equal.

When the lengths of the corresponding sides of two similar figures are equal, we say the figures are **congruent**. As with angles and line segments, two figures are congruent if they match exactly when superimposed. Examples of congruent figures

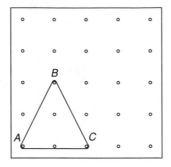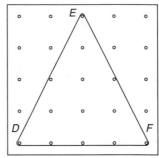

FIGURE 6.37 Similar Triangles

are found in many places. The tops of identical student desks are congruent, as are the covers of a book. In fact, modern mass production was made possible by the development of our ability in the early 1900s to consistently manufacture congruent parts for consumer goods and machines.

To give K–3 children practice identifying congruent figures, have the class collect thirty to forty cereal boxes (make sure some boxes have congruent faces). Children can test pairs of boxes to see if any of the faces match exactly. Ask children whether, if the fronts of two boxes match exactly, the backs will also be congruent. How about the corresponding sides, tops, and bottoms? As a follow-up activity, have children bring in pairs of congruent objects (socks, leaves, newspaper pages, etc.) and construct a bulletin board display of the collection.

COORDINATE GEOMETRY

As we have seen, Euclidean geometry does not allow the use of measurement as a tool in studying relationships among points, lines, and planes. Coordinate geometry, on the other hand, uses measurement to explore many of the same geometric concepts. This geometry depends on two fundamental ideas developed over three hundred years ago by the French mathematicians Descartes and Fermat. The first idea combined the notion of the geometric plane with that of the number line into what is called the **Cartesian coordinate plane**. The second development was the invention of a symbolic notation called **coordinate pairs**, which made it convenient to identify individual points in the coordinate plane. Figure 6.38 shows the coordinate plane consisting of two perpendicular number lines drawn on a flat surface intersecting at zero. The number lines are divided into equal units labeled with integer values. The horizontal number line is called the *x-axis* and the vertical number line is the *y-axis*. Fractional values between the integer points can also be used. Often, it is convenient to work with only the positive values in the upper-right quarter of the plane, the *first quadrant*.

As shown in Figure 6.39, a geoboard can be used as a model of a coordinate plane. Specific points on the coordinate plane can be named by first counting the

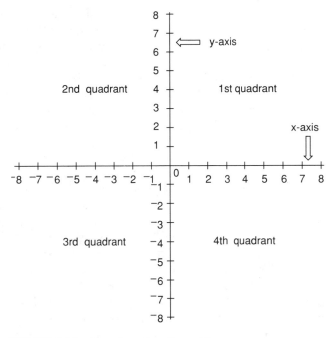

FIGURE 6.38 Cartesian Coordinate Plane

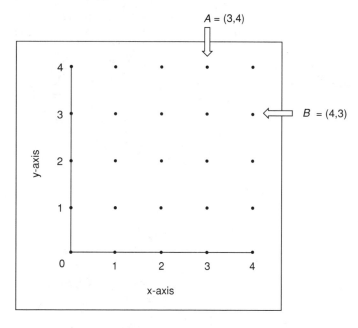

FIGURE 6.39 Graphing Ordered Pairs

number of spaces along the x-axis, followed by the number of spaces in the direction of the y-axis. For example, the position of point *A* in Figure 6.39 would be specified by counting three horizontal units along the x-axis, followed by four vertical units in the direction parallel to the y-axis. The symbol for point *A* is the coordinate pair (3,4). Notice that the order of these two values is important because each position identifies both a direction (axis) and a distance. Reversing the coordinates to (4,3) names a different point on the coordinate plane. Graphing points in the other three quadrants involves the use of negative numbers (integers), which will be discussed in chapter 13.

We can plot, or *graph*, a set of coordinate pairs by reversing the process. The first coordinate value (x) indicates the distance along the x-axis, and the second value (y) indicates the distance in the direction parallel to the y-axis.

Young children can be introduced to coordinate graphing by demonstrating its use on a transparent geoboard or a large grid transparency on an overhead projector. Give each student a geoboard and a quantity of very small rubber bands (those that just fit over a finger). It may be helpful for some children to use chalk to draw an x-axis and y-axis along the bottom row and left column of pegs on each geoboard. Select an ordered pair in the first quadrant (e.g., (2,3)) and ask a child to come forward to help you find its place on the geoboard. Have the child count the appropriate number of spaces along the x-axis and place a finger on the correct peg. Then have the child count the correct number of spaces in the direction of the y-axis and put a small rubber band over the peg representing the ordered pair. Have all of the children follow along on their own geoboards.

Once children know how to graph ordered pairs, they can get further practice by creating pictures from previously prepared lists of coordinates. Figure 6.40 shows the picture created by graphing a list of coordinate pairs. This list was compiled by first drawing a simple picture on the coordinate plane and then identifying coordinate

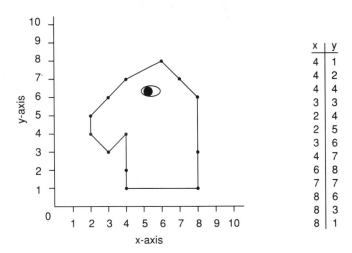

x	y
4	1
4	2
4	4
3	3
2	4
2	5
3	6
4	7
6	8
7	7
8	6
8	3
8	1

FIGURE 6.40

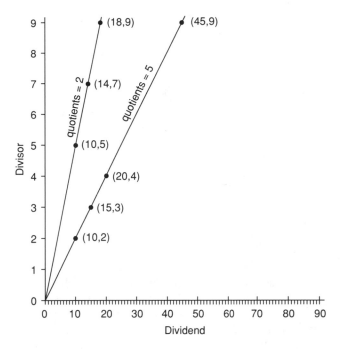

FIGURE 6.41 Graph of Division Table

pairs for points that roughly corresponded to the outline of the sketch. Children can also create lists of coordinate pairs in this way and give them to their classmates to complete.

Older children can use the coordinate plane to graph many types of relations. For example, children can graph the division (and multiplication) table. Graph the dividend along the x-axis and the divisor along the y-axis. As shown in Figure 6.41, each resulting line on the graph represents one set of division facts (in this case, quotients of 2 and 5).

Finish the graph of the division table shown in Figure 6.41. Construct a line for division facts with quotients 1–9. How could this graph be used as a multiplication learning aid?

CLASSROOM ACTIVITIES

A wide range of materials is available to assist with the development of geometric concepts. Most are inexpensive or can be easily constructed.

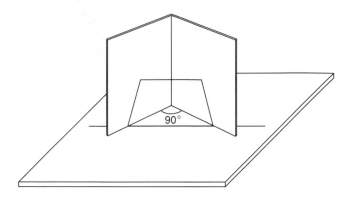

FIGURE 6.42 Hinged Mirror Image

Mirror Symmetry

Plastic rectangular mirrors (or metal photographic plates) about the size of a file card are a useful tool for exploring geometric shapes and symmetry. Tape the edges of two mirrors to form a flexible hinge with the mirrored surfaces facing inward. With a black marker, make a heavy straight line on a piece of white paper. Spread the mirrors at about a 90° angle and place them over the line so that you can see both the line and its reflection in the mirrors (see Figure 6.42). What shape do you see?

Vary the angle of the mirror and observe the polygons that are created. A square is created by a mirror angle of 90°. Mirrors can also be used to introduce symmetry, parallel lines, angle measure, and other geometric concepts.

> Use a protractor (see chapter 7) to measure the mirror angles for the regular polygons. Try to find a pattern that relates the mirror angle and the number of sides in the resulting polygons.

Perspective Drawing

In Grades 4–6, children often enjoy experiences with perspective drawing. Techniques have been developed over the centuries to allow the artist to provide an illusion of three dimensions using a two-dimensional drawing (see Figure 6.43). Three common techniques are used to give the illusion of depth.

- ☐ *Occlusion*: Objects in front block those behind.
- ☐ *Intersecting Parallel Lines*: Objects farther from the viewer are drawn smaller and positioned higher than those in front. Parallel lines seem to intersect at the horizon.
- ☐ *Foreshortening*: Circles become ovals and squares become rectangles.

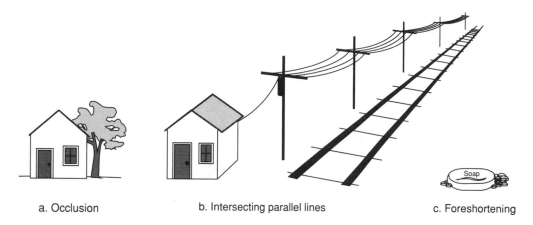

a. Occlusion　　　　b. Intersecting parallel lines　　　　c. Foreshortening

FIGURE 6.43

Give students opportunities to trace or sketch examples of each of these techniques. Prepare examples of sketches on two-centimeter (or one-inch) squared paper and have the students transfer the drawings to smaller or larger grids. By selecting key points on the original sketch and assigning an ordered pair on the coordinate plane, corresponding points can be graphed on a larger or smaller grid, creating a scaled copy of the original sketch. Perspective drawing can provide children with much needed experience in visualizing geometric relationships (see Figure 6.44).

Line Design

The illusion of making curves out of straight lines is an interesting coordinate geometry activity. One procedure is to use the first quadrant of the coordinate plane, connecting point (1,0) with point (0,10), (2,0) with (0,9), and so on. The final design, called an *envelope*, gives the illusion of a smooth parabolic curve (see Figure 6.45).

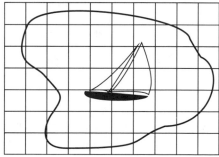
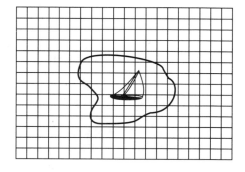

FIGURE 6.44　Graphic Transfer Using a Grid

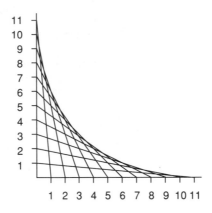

FIGURE 6.45 Line Design

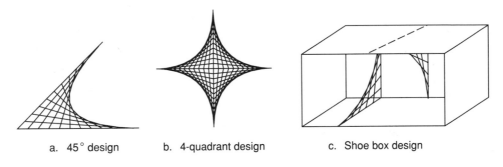

a. 45° design   b. 4-quadrant design   c. Shoe box design

FIGURE 6.46

Children can also punch holes in a file card and use colored yarn to make line designs.

A number of variations are shown in Figure 6.46 that can serve as art activities within a mathematics lesson. Vary the angle between the two axes to generate other curved shapes. Symmetric designs can be created using all four quadrants of the coordinate plane. Three-dimensional designs can also be constructed inside a shoe box using yarn or heavy thread. Line design activities provide opportunities for informal exploration of geometric concepts and relationships.

Paper Folding

Folding sheets of ordinary paper enables students to explore relationships among components of geometric figures. Many geometric shapes can be constructed in this way. More experienced students can invent their own procedures for constructing a figure and try to explain why they work. Others can try following sets of written instructions to create geometric figures.

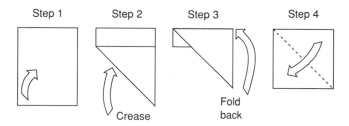

FIGURE 6.47 Folding a Square

To make a square, position the paper vertically and fold the lower left corner to right side of the paper until the bottom edge exactly matches the right edge. Crease the paper diagonally and fold the top rectangle back. The two triangles, when unfolded, form a square (see Figure 6.47).

A parabolic curve can be folded by placing a dot about two centimeters (one inch) from the bottom edge in the center of a piece of paper. Make repeated straight line folds by lifting the left or right bottom corner of the paper, placing bottom edge of the paper on the dot, pressing the fold flat, and creasing the paper. Adjust the folds so that various points along the bottom of the paper touch the center point. Repeat the process twenty to thirty times, and smooth the paper onto the table. Lightly trace the parabolic curve, which looks like the line designs in the previous activity (see Figure 6.48).

An ellipse (oval) can be folded using a similar technique beginning with a paper circle. Use a jar top or compass to construct a fifteen-centimeter (six-inch) paper circle. Place a dot about two centimeters from the edge. Make repeated folds by placing several points around the circumference of the circle on the dot and creasing the paper. The creases form an envelope pattern of an ellipse (see Figure 6.49).

A regular pentagon can be folded from a five-centimeter (approximately two-inch) strip of paper. Make a simple overhand knot and carefully adjust the strip to make creases to fold the knot flat. Cut off the ends of the strip to clearly show the pentagonal shape (see Figure 6.50).

Many figures can be constructed by folding paper. Students should try to explain why each folding procedure results in the intended figure. For example, how can you be sure that the first procedure actually creates a square with four equal sides

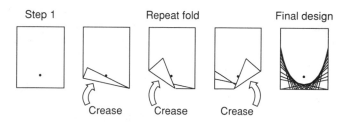

FIGURE 6.48 Folding a Parabola

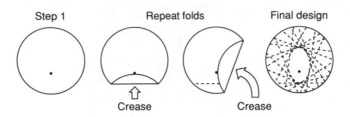

FIGURE 6.49 Folding an Ellipse

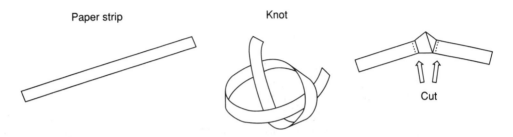

FIGURE 6.50 Folding a Pentagon

and four 90° angles? Other paper folding activities are described in the references listed at the end of this chapter (e.g., Goldberg, 1977).

Exploring the Möbius Strip

An interesting topological experiment for Grades 4–6 explores the curious properties of the **Möbius strip**. This activity introduces the type of issues that are studied in topology. Start with a 3 cm × 25 cm (1 in × 11 in) strip of paper. On one side of the strip, in the center, draw a small square. On the other side draw a small circle. Because the strip has two sides, it is impossible to connect the circle and square with a line without going around the edge. Hold the ends of the strip, make a 180° twist, and fasten the ends with tape to form a loop. Start at the square and draw a line along the length of the strip. Surprisingly enough, the square and circle can be connected without lifting the pencil (see Figure 6.51).

This loop is named for the German mathematician who explored its properties over one hundred years ago. As we discovered above, the Möbius strip is a structure that has only *one* surface and *one* edge. Try cutting along the line down the center of the strip until you return to the starting point. What structure results?

Letter Topology

Which letters are topologically equivalent? Using block capital letters, imagine stretching the letter *A* into the letter *D* without cutting or reconnecting. Because each

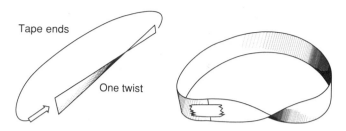

FIGURE 6.51 Möbius Strip

FIGURE 6.52 Topologically Equivalent Letters

encloses one region, *A* and *D* are considered topologically equivalent (see Figure 6.52).

Because the letter *B* encloses two regions, it can not be stretched into a *D* without cutting a line. Letters can be grouped into three topological classifications: zero regions like *K*, one region like *O*, and two regions like *B*. Have children group the twenty-six English letters into the three topological classes and make a graph of the results. Which class contains the most letters? The fewest? Would the small letters create a different graph? Why?

Straw Space Figures

Plastic straws (ten to twenty centimeters long) and string or heavy thread are all you need to make three-dimensional space figures. A **tetrahedron** can be constructed from six straws. First, pass a piece of string through three straws. Form a triangle and tie a snug knot. Tie a piece of string to each vertex, pass each through a straw, and tie the three ends together. How many faces comprise a tetrahedron? What is their shape? Are they congruent? What is the angle between any two faces? A model of a **cube** can be made in a similar manner. The resulting structure is not rigid like the tetrahedron. Such experiences help children learn that all faces of a construction must be triangular for it to be rigid (see Figure 6.53).

Try to construct other rigid space figures using straws of the same length. Are all the faces congruent? If they are, what is the angle between any two faces?

Compass and Straight Edge Constructions

You can introduce Grade 4–6 children to the construction of geometric figures with the **compass** and **ruler** or straight edge (a ruler without scale markings). Children

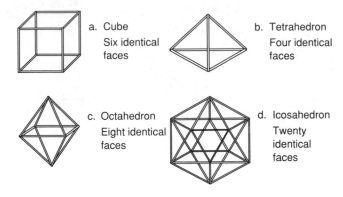

a. Cube
Six identical faces

b. Tetrahedron
Four identical faces

c. Octahedron
Eight identical faces

d. Icosahedron
Twenty identical faces

FIGURE 6.53

require practice to use a compass well. Informal experiences creating designs and copying figures using a compass should precede work with constructions. The compass can be used to mark off equal lengths, and the straight edge makes it easy to draw (but not measure) line segments. With these two instruments, it is possible to construct a 90° angle, *bisect* (divide into two equal parts) an angle or a segment, draw a square, an equilateral triangle, a circle, and other geometric figures. In Figure 6.54, the small arcs indicate where the compass has been used to mark off distances.

PROBLEM-SOLVING EXPERIENCES INVOLVING GEOMETRY

Geometry offers a range of problem-solving opportunities for students of all ages. In secondary school geometry, students are expected to systematically describe spatial relationships and prove logical conjectures involving fundamental assumptions called *axioms*. At the elementary level, students informally explore geometric relationships to provide the basis for geometric concepts and the definitions of mathematical terms. Though formal proofs are generally inappropriate at this level, many children can develop intuitive generalizations based on the results of systematic experiments.

Developing Pick's Formula for Area on a Geoboard

Rubber bands can be stretched between nails on the geoboard grid to form polygons. The distance between two adjacent horizontal or vertical nails is defined as one unit of length. A square with an edge length of one has an area of one square unit (see Figure 6.55).

It is possible to determine the area of any polygon on a geoboard by counting the number of nails touching the perimeter and the number inside the figure. **Pick's formula** for determining the area A of a polygon based on the number of perimeter nails P and interior nails I is

$$A = 1/2P + I - 1$$

a. 90° angle

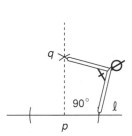

b. Bisect ∠P

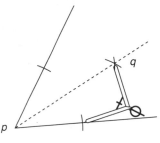

1. Mark off equal lengths from point *p* on line ℓ.
2. Mark off equal lengths from these arcs giving point *q*.
3. Line connecting *p* and *q* forms 90° angle with line ℓ.

1. Mark off equal lengths along sides from vertex *p*.
2. Mark off equal lengths from these arcs giving point *q*.
3. Line connecting points *p* and *q* bisects ∠P.

c. Square

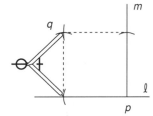

d. Equilateral triangle

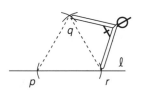

1. Construct line *m* perpendicular (90°) to line ℓ as in example *a*.
2. Mark off equal lengths along lines ℓ and *m* from vertex *p*.
3. Mark off the same length from these arcs giving point *q*.
4. Connecting the arcs on lines ℓ and *m* with point *q* gives a square.

1. Mark off any length on line ℓ.
2. Mark off same length from these arcs giving point *q*.
3. Connecting the arcs on line ℓ with point *q* gives an equilateral △ *pqr*.

FIGURE 6.54

There are two construction methods for computing area on the geoboard— *subdividing* and *surrounding*. Figure 6.56 shows how a figure can be subdivided into smaller parts with areas that are easy to compute. The area of the original figure is found by calculating the sum of its parts.

The polygon shown in Figure 6.57 cannot be easily subdivided into parts. However, it can be surrounded by a larger rectangle, the area of which is easy to determine. The area of the original figure is found by subtracting the area of the shaded areas outside the figure from that of the surrounding rectangle.

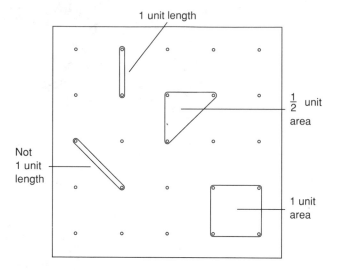

FIGURE 6.55

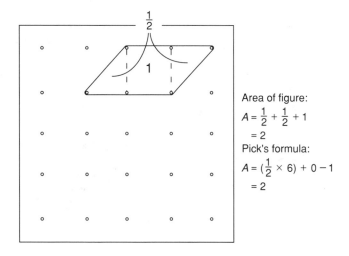

Area of figure:

$$A = \frac{1}{2} + \frac{1}{2} + 1$$
$$= 2$$

Pick's formula:

$$A = (\frac{1}{2} \times 6) + 0 - 1$$
$$= 2$$

FIGURE 6.56

> Verify that Pick's formula gives the correct area of polygons found by computing the area of several figures on the geoboard using the formula and one of the two construction methods.

Verifying the Pythagorean Theorem on the Geoboard

An understanding of Pick's formula enables older students to experiment with another famous geometric relationship, the **Pythagorean theorem**. As shown in Figure 6.58,

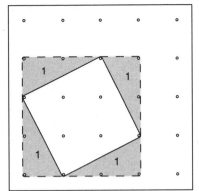

Area of figure:
$A = 9 - (1 + 1 + 1 + 1) = 5$
Pick's formula:

$A = (\frac{1}{2} \times 4) + 4 - 1$

$\quad = 5$

FIGURE 6.57

his theorem specifies a relationship among the three sides of right triangles: ($A^2 + B^2 = C^2$).

On a geoboard, it is possible to show a right triangle with squares constructed on each side. This physical representation of the Pythagorean theorem demonstrates for one example that the combined areas of squares constructed on sides A and B are equal to the area of the square constructed on side C (see Figure 6.59).

> Verify the Pythagorean theorem by constructing a right triangle on a geoboard and constructing squares on each side. Using Pick's formula, show that the sum of the areas of the squares constructed on sides A and B is equal to the area of the square constructed on side C. Repeat the problem several times using different right triangles. For some examples, it may be necessary to place two or more geoboards together in order to enlarge the array.

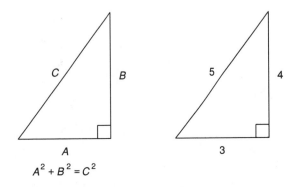

$A^2 + B^2 = C^2$

FIGURE 6.58 Pythagorean Theorem

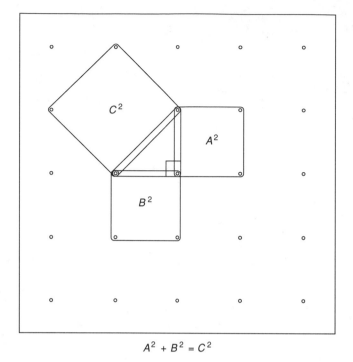

$$A^2 + B^2 = C^2$$

FIGURE 6.59 Pythagorean Theorem on Geoboard

COMPUTER APPLICATIONS

Geometry Software Tools

A number of excellent software products are available to aid the study of geometry. Two particularly useful software tools for elementary students are *Logo* (discussed in chapter 5) and the *Geometric Supposer*. Other products, such as *Delta Drawing* for younger children and *Graphing Equations* for upper elementary and secondary students, can also be useful additions to the mathematics classroom.

The Geometric Supposer: Triangles. As a computer software tool, the *Geometric Supposer: Triangles* (also available for points and lines, quadrilaterals, and circles) allows students to explore relationships among parts of triangles by developing and testing conjectures. The software presents a random example of a selected type of triangle (right, acute, obtuse, isosceles, or equilateral) upon which the user can specify various constructions (e.g., the midpoint of side AB or the bisector of $\angle BAC$) and measurements (e.g., the length of segment $\overline{AB}$ or the area of $\triangle ABC$). The graphic constructions and measurements are computed and displayed by the program.

For example, Grade 4–6 students can use the *Geometric Supposer* to discover that, for any triangle, the sum of the interior angles equals 180°. The user selects

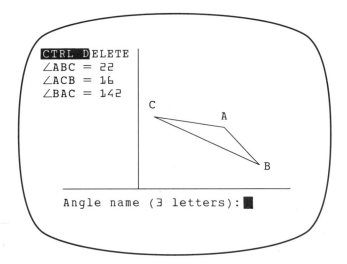

FIGURE 6.60 Geometric Supposer: Triangles. *Note:* From "Geometric Supposer: Triangles" [Computer Program] by Sunburst Communications, 1985. Copyright 1985 by Sunburst Communications. Reprinted by permission.

any type of triangle (e.g., obtuse) and asks the *Supposer* to measure each of the three angles and display the results on the screen (see Figure 6.60). The three measures can be added by the computer, on paper, or using a calculator.

Repeating this procedure for many new triangles randomly selected by the program builds confidence that the conjectured relationship holds for all triangles.

The *Geometric Supposer* software tools allow students to explore geometry concepts by developing conjectures that can be systematically tested on a series of examples. The process gives students experience doing mathematics that will prepare them for formal proofs. Additional applications of these versatile tools are presented in the next chapter on measurement.

ADAPTING INSTRUCTION FOR CHILDREN WITH SPECIAL NEEDS

The special-needs learner often profits from additional experiences with topological activities like those described in this chapter. When these concepts are not fully developed in the lower grades, children may have problems developing the concepts of symbol recognition, relative position, and continuity. Ensuring adequate experiences for K–2 students and reintroducing topological activities for older special-needs students will help minimize these and other perception problems that affect mathematics learning.

Children who display visual perception problems benefit from use of extra-large geometric models and teaching aids. Large graph paper, geoboards, and shape blocks

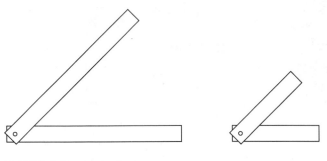

FIGURE 6.61 Angle Copier

help such students participate more effectively in regular geometry lessons. For example, the difficulty experienced by students with normal sight in measuring and comparing angles can be further complicated for the visually impaired. As observed earlier, students often confuse the measure of an angle's spread with the apparent length of its sides.

To help visually impaired students understand the concept of angle measure, the teacher can employ physical models that tap different sensory modes of learning. For example, use an angle copier as described earlier and make a second one with two strips about half the length of the first (see Figure 6.61).

These models can then be used to *copy* angles found in the classroom. Using both to copy the same angle demonstrates that the copiers must display the same angle measure even though one has longer legs. To compare angles, use each copier to record the different angles a door makes with a sill (closed = 0° and perpendicular to the wall = 90°). Demonstrate that the amount of spread between the sides, not the length of the legs, determines the measure of an angle. After recording two different size angles, place the two copiers under a sheet of tagboard on a table. Select a student to come forward, reach under the cover, and judge the larger of the two angles by touch alone. Adjust the two angles for each turn. Remove the cover to reveal the solution after each prediction.

Other geometric figures and relationships can be demonstrated using large models and tactile experiences. A circle can be modeled by having a student hold one end of a jump rope while another student walks around the first keeping the rope tight. Similar activities can be used to model a triangle, square, parallel lines, and other geometric figures and relationships.

Transformational Geometry

An alternate set of transformational tools is available to explore geometric concepts. Gifted and talented students in particular may find this approach a challenging way to extend their understanding of geometry. Transformational geometry is concerned with the relative position, congruence, and symmetry relations among figures in space. Three different types of motions, or **congruence relations,** are defined.

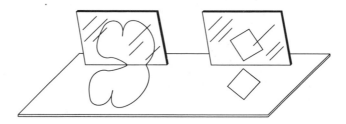

FIGURE 6.62 Mirror Reflections Showing Congruent Figures

- □ *Reflection* congruence—flip
- □ *Translation* congruence—slide
- □ *Rotation* congruence—turn

Reflection congruence was previously introduced during the discussion of lines of symmetry. Congruence is defined as the matching of all points under a reflection along a line of symmetry. For example, notice the placement of a mirror on the drawing in Figure 6.62. The original drawing is said to be reflection congruent to its image in the mirror. Every point in the drawing maps onto its image in the mirror. This process is also called a *flip* because the drawing is flipped across a line of symmetry onto its image in the mirror.

> Using a small, rectangular mirror, explore the symmetric properties of various geometric figures and other graphics and pictures. Demonstrate the number of lines of symmetry of various triangles, quadrilaterals, other polygons, and a circle.

Two figures are said to be *translation congruent* if, without twisting or rotating, one figure can be slid over the other and all points match exactly. This rigid motion is called a translation, or a *slide* (see Figure 6.63).

Finally, two objects are said to be *rotation congruent* if, by turning one figure about a point, it can be made to match exactly all the points in the second figure without using flipping or sliding motions. A rotation is also called a *turn* (see Figure 6.64).

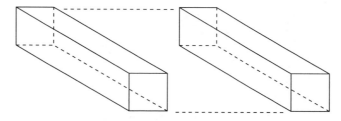

FIGURE 6.63 Translation of Congruent Figures

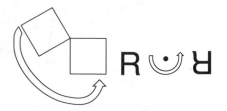

FIGURE 6.64 Rotation of Congruent Figures

Some Euclidean concepts can be easily demonstrated using transformational techniques. For example, two lines are parallel if they are slide congruent; two angles are equal if their corresponding sides are slide or rotation congruent; two triangles are similar if the lines that extend each pair of corresponding sides are slide congruent; and a figure is symmetric if it is flip congruent to itself (see Figure 6.65).

The letters of the alphabet can be used to practice the transformational geometry concepts of flips, slides, and turns. For example, the letters *b* and *d* have reflection symmetry, and *d* and *p* have rotation symmetry (see Figure 6.66). Any letter has translation symmetry to itself (e.g., *m* → *m* in the word *mathematics*). Make a list of words that contain examples of letter pairs that display these three transformations (e.g., *u* rotates to *n* in the word *under*).

Figure 6.67 shows how to use a mirror to find which block capital letters have one or more lines of reflection congruence. It also shows that the letters *M* and *W* have rotational congruence. Find all the block capital letters that have at least one line of reflection congruence. Find pairs of letters that have rotational congruence.

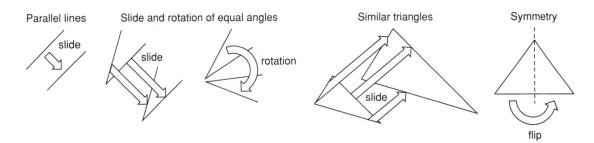

Parallel lines · Slide and rotation of equal angles · Similar triangles · Symmetry

FIGURE 6.65 Transformations

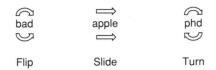

bad	apple	phd
Flip	Slide	Turn

FIGURE 6.66 Letter Transformations

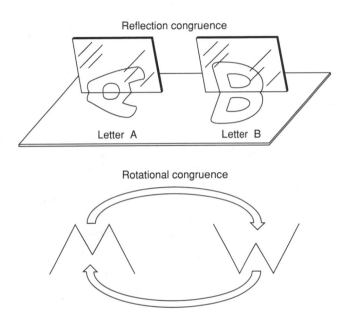

FIGURE 6.67

SUMMARY

The study of geometry is an integral part of mathematics education at all grade levels. According to van Hiele, children pass through four levels of thinking when learning geometric concepts: recognition of figures, description of figure characteristics, classification of figures, and deductive proof of theorems. Elementary students are expected to accomplish the first three of these levels of thinking by the end of Grade 6. Piaget observed that young children begin learning about geometric concepts by employing the topological relationships of proximity, separation, order and enclosure. Understanding of continuity, or infinite subdivision, also seems to play a role in the development of spatial relations. The study of geometry begins at the K–3 level with informal investigations of topological relations leading to an understanding in Grades 4–6 of the Euclidean concepts of point, line, line segment, angle, and the classification of geometric figures such as polygons and circles. The study of topology is primarily concerned with the boundaries, enclosures, and transformations of geometric figures and surfaces. This view of geometry explores those elements that remain constant when figures are stretched or transformed into other shapes without twisting or cutting. Euclidean geometry includes the study of points, lines, angles, and the relationships among components of geometric figures such as parallel lines and similar triangles. Other important elementary concepts include symmetry, similarity, congruence, and the classification of space figures. Work with topological concepts can help children, particularly special-needs learners, develop basic understandings that will improve their work in other areas of mathematics and language. Exploring figures using transformational geometry techniques may provide additional access for students, especially gifted and talented children, to the important concepts of symmetry, similarity, and congruency.

COURSE ACTIVITIES

1. Invent a paper-folding procedure to construct an equilateral triangle. Prove that the resulting triangle is really equilateral.

2. Exactly five space figures (solids) can be constructed from identical regular polygons. These objects are called the *Platonic solids*. The faces of these space figures can be equilateral triangles, squares, or pentagons. Cut several equilateral triangles out of tag board and tape them together to make three of the space figures in Figure 6.68. Then cut out six squares and twelve pentagons to construct the final two space figures. When you are taping together the first few pieces of each structure, try to imagine why there are only five Platonic solids. Discuss these conjectures with your peers.

3. Cut out several strips of paper 5 cm × 40 cm (2 in × 16 in) and draw a line down the middle of one strip on both sides. Twist the strip 180°, fasten the ends into a Möbius strip, and cut along the line. Notice you get a new loop twice as long as the original. Using another strip of paper, divide it into thirds by drawing two lines along its length on the front and back. Again fashion it into a Möbius strip and begin cutting along either line. This time you get one long loop and one short loop. Continue dividing strips into fourths, fifths, and so on,

and keep track of the resulting number of large and small loops (you may need to use wider strips for the larger number of cuts). Complete Table 6.5 and look for a pattern in the results.

4. Construct an equilateral triangle with a compass and straight edge. Using each vertex as the center, draw an arc that exactly passes through the other two vertices. The

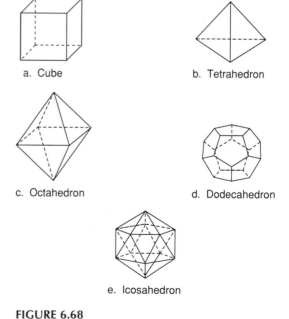

a. Cube

b. Tetrahedron

c. Octahedron

d. Dodecahedron

e. Icosahedron

FIGURE 6.68

TABLE 6.5 Moebius Strip Cuts

Cut Strip In	Short Loops	Long Loops
1	1	0
$\frac{1}{2}$	0	1
$\frac{1}{3}$	1	1
$\frac{1}{4}$	0	2
$\frac{1}{5}$	?	?
$\frac{1}{6}$	?	?

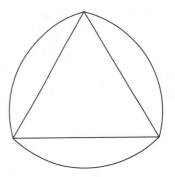

FIGURE 6.69 Constant Width Curve

final figure, shown in Figure 6.69, is called a *constant width curve*.

Make two copies of this shape from heavy corrugated cardboard. Push a pencil through the center of each to form an axle. Place a book on top and push it slowly. Observe how the book moves as you push it across the table. Are you surprised that the book does not bob up and down but moves as smoothly as if riding on a cylinder? Why?

5. Develop a lesson plan based on one of the geometry activities suggested in this chapter. Collect the necessary materials and implement the lesson with your peers as students or, if possible, with a group of children in a classroom setting. Have a friend observe the interaction and take notes on what goes well and what problems arise. Discuss the experience with peers and list suggestions for improvement.

6. Review a source of geometry classroom activities (see reference list). Start an idea file of lesson activities. Organize the entries by grade level and topic. Index cards stored in a small recipe box work well. Design a standard format for listing the necessary information by including a lesson objective, grade level, time allotment, materials, procedure, and evaluation. Continue adding to this file for each of the following chapters. Share and update the collection with your peers periodically.

7. Read one of the *Arithmetic Teacher* articles listed in the reference section. Write a brief report summarizing the main ideas of the article and describe how the recommendations for instruction might apply to your own mathematics teaching.

MICROCOMPUTER SOFTWARE

Bumble Plot Coordinate geometry simulation (The Learning Company).

Cactus Plot A function graphing tool (Cactus Software).

Creative Play: Problem Solving Activities with the Computer Explores problem solving in several interesting contexts (includes *Hurkle*, a coordinate geometry simulation) (Math and Computer Education Project, University of California, Berkeley).

Delta Drawing An easy-to-use turtle graphics environment for Grade K–2 children (Spinnaker).

Geometric Supposer A series of versatile tool environments that support the testing of conjectures about spatial relationships involving points, lines, triangles, quadrilaterals, and circles (Sunburst Communications).

Geometry Problems for Logo Discoveries Problem-solving activities employing turtle geometry (Creative Publications).

Graphing Equations Exploration of function graphs by connecting green globs (Conduit Software).

Juggler's Rainbow Practice identifying direction and patterns for Grade K–2 children (The Learning Company).

Logo A graphics-oriented programming language for children (Terrapin/Krell and Apple Corporation).

Logo Discoveries Problem-solving activities employing Logo programming and Logo graphics (Creative Publications).

REFERENCES AND READINGS

Abbott, E. (1952). *Flatland*. New York: Dover.

Bearden, D., Martin, K., & Muller, J. (1983). *The turtle's sourcebook*. Reston, VA: Reston.

Beckmann, P. (1982). *The history of pi*. Boulder: Golem Press.

Bidwell, J. (1987). Using reflections to find symmetric and asymmetric patterns. *Arithmetic Teacher, 34*(7), 10–15.

Biggs, E., & MacLean, J. (1969). *Freedom to learn: An active learning approach to mathematics*. Don Mills, Ontario: Addison-Wesley.

Boyle, P. (1970). *Graph gallery*. Palo Alto, CA: Creative Publications.

Brydegaard, M., & Inskeep, J. (Eds.). (1970). *Readings in geometry from The Arithmetic Teacher*. Reston, VA: National Council of Teachers of Mathematics.

Burger, W. (1982). Graph paper geometry. In L. Silvey & J. Smart (Eds.), *Mathematics for the middle grades: 1982 Yearbook* (pp. 102–117). Reston, VA: National Council of Teachers of Mathematics.

Burger, W. (1985). Geometry. *Arithmetic Teacher, 32*(6), 52–55.

Burger, W., Hoffer, A., Mitchell, B., & Shaughnessy, M. (1981). Goals for geometry: Can we meet them? *Oregon Mathematics Teacher*, November–December, 12–15.

Cangelosi, J. (1985). A "fair" way to discover circles. *Arithmetic Teacher, 33*(3), 11–13.

Goldberg, S. (1977). *Pholdit*. Hayward, CA: Activity Resources.

Henderson, K. (Ed.). (1973). *Geometry in the mathematics curriculum: 1973 Yearbook*. Reston, VA: National Council of Teachers of Mathematics.

Hiebert, J. (1984). Why do some children have trouble learning measurement concepts? *Arithmetic Teacher, 31*(7), 19–24.

Hoffer, A. (1981). Geometry is more than proof. *Mathematics Teacher, 74*, 11–18.

Jacobs, H. (1970). *Mathematics, a human endeavor*. San Francisco: W. H. Freeman.

Kim, S. (1981). *Inversions: A catalog of calligraphic cartwheels*. New York: BYTE Books, McGraw Hill.

Laycock, M. (1970). *Straw polyhedra*. Palo Alto, CA: Creative Publications.

Lindquist, M. (Ed.). (1987). *Learning and teaching geometry, K–12: 1987 Yearbook*. Reston, VA: National Council of Teachers of Mathematics.

Mathematics resource project. (1978). Palo Alto, CA: Creative Publications.

Payne, J. (Ed.). (1975). *Mathematics for early childhood: 1975 Yearbook*. Reston, VA: National Council of Teachers of Mathematics.

Pearcy, J., & Lewis, K. (1967). *Experiments in mathematics, stage 1, 2 and 3*. Boston: Houghton Mifflin.

Piaget, J., & Inhelder, B. (1967). *The child's conception of space*. New York: W. W. Norton.

Poggi, J. (1985). An invitation to topology. *Arithmetic Teacher, 33*(4), 8–11.

Ranucci, E. (1973). *Seeing shapes*. Palo Alto, CA: Creative Publications.

Ranucci, E., & Teeters, J. (1977). *Creative Escher-type drawings*. Palo Alto, CA: Creative Publications.

Seymour, D. (1976). *Tangramath*. Palo Alto, CA: Creative Publications.

Seymour, D., & Schadler, R. (1974). *Creative constructions*. Palo Alto, CA: Creative Publications.

Seymour, D., Silvey, L., & Snider, J. (1974). *Line designs*. Palo Alto, CA: Creative Publications.

Shaw, J. (1984). Mathematics scavenger hunt. *Arithmetic Teacher*, *31*(7), 9–12.

Special issue: Geometry. (1979). *Arithmetic Teacher*, *26*(6).

Trafton, P., & Leblank, J. (1973). Informal geometry in grades K–6. In K. Henderson (Ed.), *Geometry in the mathematics classroom* (pp. 11–51).

Reston, VA: National Council of Teachers of Mathematics.

Van Hiele, P. (1986). *Structure and insight: A theory of mathematics education*. New York: Academic Press.

Wilmot, B. (1985). Creative problem solving and red yarn. *Arithmetic Teacher*, *33*(4), 3–5.

Young, J. (1982). Improving spatial abilities with geometric activities. *Arithmetic Teacher*, *30*(1), 38–43.

7

Measurement—
Quantifying Space

Triangle is structure and structure is triangle. Period. It's what the whole Universe is based on.

—Buckminster Fuller

Upon completing Chapter 7, the reader will be able to

1. Describe the concept of measurement to children.

2. Give examples of length, weight, area, volume, time, and temperature measurements encountered in real-world situations.

3. Describe three basic cognitive abilities associated with the development of measurement concepts and give appropriate examples for length, weight, area, volume, and time.

4. List three basic instructional principles for teaching measurement.

5. Describe three levels of measurement that can be used to organize lessons and give examples of activities for each.

6. Define the concept of measurement error and give an example of an appropriate activity to introduce the concept.

7. Describe the structure of the metric system and its role in the mathematics curriculum and society.

After awarding the prize for the largest hog at the county fair, a patronizing out-of-state judge asked the winning farmer how he could weigh a pig that big. The farmer replied, "First you get the hog to sit on one side of a balance scale. That's the hard part. Then you count up the bricks you need to pile on the other side to make it perfectly level. When you're all done, you look around for some smart feller like you to guess how much a brick weighs!"

Our study of elementary mathematics continues with the topic of measurement. As indicated in the anecdote above, measurement is the process of assigning a value to a particular attribute of an object. Measurement generally involves work with continuous quantities. For example, an object can be weighed by counting the number of units (pounds, kilograms, marbles, bricks) needed to counterbalance it on a simple scale. However, it is impossible to divide the object into identical parts, so no measure will be exact. A box of eggs is an example of a discrete quantity because the number of eggs (the unit of measure in this case) can be exactly determined.

Measurement concepts are fundamental to the development of many mathematics skills. Comparison, estimation, and counting, which provide the basis for understanding whole and rational numbers, can be efficiently developed through concrete measurement experiences. For example, calculating the area of rectangular regions provides a concrete model for multiplication and division; the study of the metric system of measures and U.S. money can facilitate the use of decimals; and relative position, ratio, and proportion concepts arise naturally from scaling activities with maps.

Learning measurement can be a pleasant experience for children if a few basic teaching techniques are taken into account.

☐ Ask students to estimate measures prior to actual measuring experiences.

☐ Plan activities that allow students to record the results of their measurements by constructing individual and class graphs.

☐ Introduce problem-solving experiences to help students apply measurement skills to real-world situations.

These principles can be employed when exploring the measurement tasks encountered in everyday activities. For example, to buy a shirt for a friend, we either compare two objects directly (have the friend try on the shirt to see if it fits), compare an object to an arbitrary unit (if the friend isn't there, hold it up to someone that looks about the same size), or we select the correct size based on a standard unit. These three levels of measurement will be used to introduce each measurement attribute in the following sections.

Estimation should be emphasized during all measurement activities. Students should be encouraged to predict the outcome of comparisons before they actually carry out the experiments. When arbitrary and standard measurement units are introduced, students can record preliminary estimates to compare with actual counts of unit measures. Through this process of prediction (estimation) and verification (measurement), children develop a better understanding of which attribute is being measured and the relative size of the amounts involved. As children become proficient at estimating, these preliminary approximations also serve as a guide for assessing the reasonableness of solutions to measurement problems.

MEASURING LENGTH

The three different types, or levels, of measurement can serve as a guide for organizing sequences of instructional activities.

☐ Engage the students in *direct comparison* experiences.

☐ Introduce *arbitrary (nonstandard) units* of measurement.

☐ Introduce *standard (metric and customary) units* of measurement.

The idea of distance is encountered early in the development of most children. As soon as they begin to gain control over their arm movements, children develop their eye-hand coordination in order to reach for objects. During the child's first eighteen months, these frequently unsuccessful attempts to gain access to the environment are gradually refined as the ability to compare and estimate distances improves. This concept of **length** is referred to as the *measure of extent or distance.*

Comparison

It is difficult to develop a practical understanding of what measuring the length of something means without first making direct comparisons between objects. Children must refine general comparative terms such as *big* and *small* to those more specifically related to length, such as *long*, *short*, *far*, and *near*.

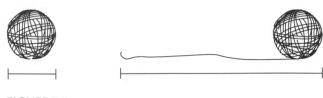

FIGURE 7.1

Fortunately, opportunities for comparing lengths abound in everyday life. Often, objects that are similar (e.g., brothers and sisters) are compared with each other. To develop a facility for measuring length, it is necessary to compare dissimilar objects (a feather and a pencil) and arrange more than two objects in sequence according to length. This process, called **seriation,** involves sequencing more than two objects according to some measurable attribute.

The task confronting you is to extend and refine the naturally occurring concept of length and develop the child's necessary language skills. For example, what does it mean to measure the length of a ball of string? We could interpret the length to be the diameter of the ball or the length of the string that makes up the ball (see Figure 7.1).

Distance is often discussed in terms of travel time. In Papua New Guinea, people describe the distance from one village to another downstream as "six days by canoe" (eight days upstream to return). In our culture, we might tell a friend that the distance from San Diego to Los Angeles is about "two hours." Technically, these time-related descriptions are not measures of length. Distance does not vary, while a leaky canoe or tire could alter travel time considerably.

In addition to seriation, Piaget observed that two other fundamental understandings are important to the measurement process: **transitivity** and **conservation.** We use transitivity when seriating a match, a pencil, and a screwdriver. Suppose the pencil is longer than the screwdriver, and the screwdriver is longer than the match (see Figure 7.2). Without having to place the match next to the pencil, the **principle of transitivity** (transference of a relation) tells us that the pencil is also longer than the match.

While this example may seem trivial, consider the more common situation where the three objects are supposed to be equal and are separated by considerable distances. When building a window for a house, it is necessary to measure the opening A with tape B and use this measure to construct frame C. When the frame is completed, one would expect it to be equal in size to the measured opening. This is an example of how fundamental the transitivity principle is to most measurement applications. If two objects are found to have the same measure using a third object, such as a tape-measure, balance, or clock, one would expect the two objects to be equal when compared directly. The transitivity principle also implies that if $A > B$ and $B > C$, then $A > C$. Other transitive relations include $=$, $<$, *is-over*, and *is-inside-of*.

In order for transitivity to be meaningful, children must realize that the measure of an attribute (in this case length) does not vary when it is moved about in space.

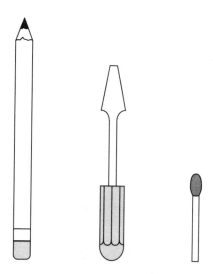

FIGURE 7.2 Principle of Transitivity

Piaget called this invariance principle **conservation.** Look at the two sketches in Figure 7.3. Is the length of segment $\overline{AB}$ equal to $\overline{CD}$, or is one longer?

Break a toothpick equal to the length of segment $\overline{AB}$. Move the stick to segment $\overline{CD}$ to verify that the two segments are equal. This optical illusion may surprise you. For many children in the early elementary grades, however, no contradiction would be apparent. After estimating that $\overline{AB}$ is longer than $\overline{CD}$, they might agree that the stick was as long as both $\overline{AB}$ and $\overline{CD}$, but insist that $\overline{AB}$ was still longer than $\overline{CD}$! Somehow, the overwhelming perception afforded by this optical illusion requires the stick to shrink in the minds of these children as it is moved from $\overline{AB}$ to $\overline{CD}$. In general, conservation refers to holding a measurement attribute (length, weight, time) invariant through some transformation.

Novice measurers at all grade levels should be given ample opportunity to confront such contradictions through hands-on experiments as they are introduced to each new attribute (length, weight, quantity, area, volume, time, temperature). Through such experiences, children develop confidence in systematic comparison and counting procedures over more subjective perceptual methods of quantifying the environment.

Through comparison and seriation activities, the child learns to attend to the attribute in question (in this case, length) and ignore other correlated measures (e.g., time and weight). Some children may require additional comparison experiences because they lack a complete understanding of seriation, conservation, and transitivity.

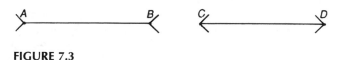

FIGURE 7.3

This lack of understanding may also interfere with their effective use of measurement instruments in measurement activities.

Arbitrary Units

The second step in learning to measure length emerges with the need to know *how much more*. Although direct comparison allows students to determine whether an object is longer or shorter than another, this procedure is useless when a more accurate description is required. For example, while it helps to know that your foot is longer than your brother's but shorter than your mother's, this alone is not sufficient information to order a new pair of running shoes from a catalog. To determine how much more one object is than another, one must define a **unit of measure.** A unit of measure is a quantity to be used as a standard for measuring objects and events that is agreed upon by a group of people.

For example, in Mexico men frequently measure a pair of pants for fit by inserting their forearm into the waistband. If the pants fit from elbow to closed fist, the pants are the correct size. Other traditional units include the arm-span, pace, hand, and piece (as in "down the road a piece"). While these units lack specificity and generalizability, they provide an excellent opportunity to focus on the need for unit measures, the need for various sized units to improve the precision of measures, and, ultimately, the need for standard units to facilitate unambiguous communication among groups of people. Any object can be used as a unit measure of length. Some handy items are paper clips, new pencils, chalk, crayons, handspans, paces, and footprints. Smaller units include grains of rice, the width of a little-finger print, and the thickness of a piece of paper. Children can measure objects using arbitrary, or traditional, units and record the results on individual or class graphs (see chapter 14 for a description of various types of elementary graphs and their applications). In Grades K–2, it is convenient to use larger nonstandard units such as handspans and footprints to measure length. Older children can effectively use smaller arbitrary units such as paper clips and beans.

Students will notice that, in most cases, the length of the object is not an exact number of units. Notice in Figure 7.4a that a bit of the pencil sticks out beyond the last paper clip used to measure it. This inaccuracy is called **measurement error.** If one more clip is added to the others, it sticks out a bit past the end of the pencil (see Figure 7.4b).

Is the pencil five or six paper clips long? Normally, one chooses the measure that looks the closest, in this case five paper clips. Picking the one that is closest (the one less than one half of a paper clip off) generates an error of, at most, one-half the measurement unit. When the length of an object is measured, its *actual* length is at most one-half unit above or below the stated measure. Measurement error is equal to one-half the unit of measure used.

The concept of measurement error, or greatest possible error, is introduced in the elementary grades. This concept is used later in the development of *relative error*, the ratio of measurement error to total length. For example, measuring the length of a pencil in paper clips results in a higher relative error than measuring the distance across a classroom in paper clips. In both cases, the measure could be off by at

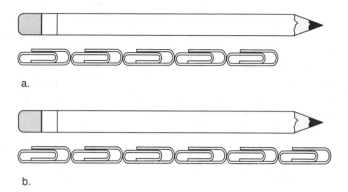

FIGURE 7.4

most one-half a paper clip. However, a one-half paper clip error in measuring a pencil represents less relative accuracy than the same error in measuring the longer distance across the room. Selecting appropriate units of measure keeps relative error at a reasonable level.

If a more precise measure is required, you must use a smaller unit. Suppose a child wanted to make a leather pencil case and needed a more precise measurement of length so as not to waste expensive leather. Two units of measure could be used— one large (paper clip) and one small (grain of rice). The pencil could be measured using the large unit first and the small unit for the amount left over (five clips and three grains). The error would be no more than one-half the smallest unit, assuming the grains of rice are all the same length (see Figure 7.5).

Additional error can arise from procedural violations (gaps between paper clips) or from nonuniform units (different sized rice grains). No measure is exact, but measures can be made more accurate by reducing the size of the measurement unit and by improving the consistency of the procedures and instruments.

Standard Units

Finally, students should be introduced to standard units of measure. In 1975, the Voluntary Metric Conversion Act was signed into law, legalizing the use of the **SI Metric System** *(Système International d'Unités)* for all commercial purposes in the United States. Following the initial spurt of excitement, interest in conversion to the metric system has leveled off. There has been progress, however, in expanding the use of metric units in some sectors of our society. Mathematics teachers can play

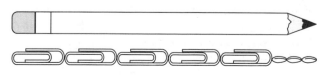

FIGURE 7.5

an important role in educating children and parents about the value of adopting a worldwide system of measures. Because of international trade requirements, the automobile, aerospace, and beverage industries now design their products using metric standards. The European Common Market countries have adopted a policy that all goods sold within their borders must be marked in metric units. This policy affects all U.S. exports to the region. In the United States, almost all foodstuffs are now labeled with both metric and customary units. Distances on road signs along the borders with Mexico and Canada and within our national parks are often posted in both kilometers and miles. The United States remains, however, virtually the only country in the world that has not established a specific time line for implementation of the metric system.

The schools currently lead society in its effort to adopt the metric system for general use. Clearly, this process will take substantially longer than the original ten-year estimate suggested at the time of the bill's passage. However, many states have adopted the recommendation of the National Council of Teachers of Mathematics that SI metrics be introduced at the elementary level as the primary system of measures, and that the customary units receive a more historical treatment. For the near term, however, it remains important for school children to have a working knowledge of both the metric and customary systems of measures.

Activities described in this book stress the use of metric units because they are likely to be less familiar to the reader. Customary units are also included as appropriate. Tables of customary measures are readily available in standard textbooks and resource guides. Teachers should be guided by local curriculum policy in deciding the proportion of metric and customary units they introduce in the mathematics curriculum. The instructional methodology remains essentially the same, however, regardless of the standard units involved.

A standard system of units does not necessarily provide more precise measures. Like a common language, however, its use tends to reduce misunderstandings among groups of people. The metric system, adopted by the French in 1790, is well on its way toward worldwide adoption. It enjoys acceptance much like the Hindu–Arabic numeration system that has become the standard in nearly every country in the world. Imagine the possible problems if countries around the world had not only different currencies but different numeration systems as well.

Though the metric system is no more precise than other systems of measures, it is logically structured and easy to learn. The units are related by powers of 10. The base unit of length is the **meter** (about the height of an average kindergartner). Shorter units are derived by dividing the meter into 10 parts *(decimeter)*, 100 parts *(centimeter)*, 1000 parts *(millimeter)*, and so on. Conversely, larger units are constructed by multiplying the meter by 10 *(dekameter)*, 100 *(hectometer)*, 1000 *(kilometer)*, and so on. Table 7.1 lists the metric prefixes and their relationship to the base unit. Larger (mega, giga, tera, peta, and exa) and smaller (micro, nano, pico, femto, and atto) unit prefixes are also defined but are not commonly introduced at the elementary level.

The symbol for each unit is written in small letters with no period and is separated from the number of units by a space (e.g., 12 km for 12 kilometers). For the commonly used metric units, only the symbols for the liter (L) and degree Celsius (°C) are capitalized.

TABLE 7.1 Metric Prefixes

Prefix	(Power of 10)	Meter (m)		Gram (g)		Liter (L)	
kilo	(1000)	kilometer	(km)	kilogram	(kg)	kiloliter	(kl)
hecto	(100)	hectometer	(hm)	hectogram	(hg)	hectoliter	(hl)
deka	(10)	dekameter	(dam)	dekagram	(dag)	dekaliter	(dal)
base unit	(1)	meter	(m)	gram	(g)	liter	(L)
deci	(1/10)	decimeter	(dm)	decigram	(dg)	deciliter	(dl)
centi	(1/100)	centimeter	(cm)	centigram	(cg)	centiliter	(cl)
milli	(1/1000)	millimeter	(mm)	milligram	(mg)	milliliter	(ml)

These prefixes are applied to each basic unit of metric measure (length—meter; weight—gram; capacity—liter). Compare the orderly metric structure to that of the customary system. Notice in Table 7.2 that no pattern can be deduced in the customary system that would assist in memorizing the relative unit sizes and their names. To further compound the problem, liquid and dry volume measures are based on different sized units (pints), and an ounce of liquid capacity is unrelated to the ounce used for weighing.

Metric units are interrelated in other logical ways as well. If you build a box ten centimeters on an edge, its capacity will be almost identical to one **liter.** The water required to fill a one-liter box weighs one kilogram. Originally, the liter was supposed to equal one cubic decimeter (a box ten centimeters on an edge). It was later discovered that, due to slight weighing errors in the 1800s, one cubic decimeter of water weighed slightly more than a kilogram, so the liter was slightly reduced in size in order for one liter of water to weigh exactly one kilogram. The difference between a liter and a cubic decimeter is so small that it can be ignored for most purposes (see Figure 7.6).

Four metric units of length are commonly used at the elementary grade level:

☐ kilometer (km)
☐ meter (m)
☐ centimeter (cm)
☐ millimeter (mm)

TABLE 7.2 Customary Units of Measure

Length	Weight	Capacity
12 inches = 1 foot	437.5 grains = 1 ounce (dry)	3 teaspoons = 1 tablespoon
3 feet = 1 yard	16 ounces = 1 pound	16 tablespoons = 1 cup
5.5 yards = 1 rod	100 pounds = 1 hundredweight	2 cups = 1 pint
40 rods = 1 furlong	20 hundredweights = 1 ton	2 pints = 1 quart
8 furlongs = 1 mile		4 quarts = 1 gallon
		2 gallons = 1 peck
		4 pecks = 1 bushel

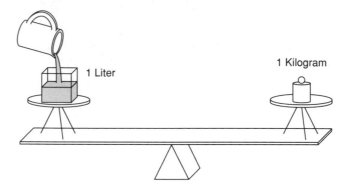

1 Liter

1 Kilogram

FIGURE 7.6

Each unit should be introduced using concrete experiences. One method is to have the students construct their own meter sticks. Meter sticks for Grade 2–3 students might have only decimeter markings. For Grades 4–5, meter rulers should show centimeter and decimeter markings, and for Grades 6 and above, millimeter, centimeter, and decimeter divisions. Rulers with millimeter markings should be purchased in order to ensure accuracy. Kindergarten and Grade 1 students should focus on direct comparison and arbitrary (nonstandard) unit measures.

You can help your students learn to measure any attribute by planning instructional activities consistent with the following three levels of measurement.

1. Through the use of *direct comparison* experiences, children learn to differentiate one attribute from all other correlated measures. Children also learn to put a series of objects in order according to the specified attribute and record the results using pictures and graphs (e.g., ordering leaves by area instead of length).
2. *Arbitrary (or nonstandard) units* are introduced when more information about a measurement is required. The introduction of units makes it possible to determine how much longer, heavier, or bigger one object is than another and allows the idea of measurement error to be introduced (e.g., using paper clips to measure and compare shoe lengths).
3. The introduction of *standard units* of measure (metric and customary) facilitates communication among people, and, in the case of metric measures, provides a logically structured system to ease learning and use (e.g., using meters and centimeters to measure and compare students' heights).

Each of the measurement concepts commonly taught at the elementary level can be introduced using this three-step approach. With young children, it is recommended that instruction concentrate on comparison and, as appropriate, the use of arbitrary unit measures. Depending on their level of experience and ability, older children may require only a brief introduction to a new measurement attribute using comparison activities, then move rapidly to using arbitrary and standard units of measure. However, when a new attribute is introduced, it is unwise to move too

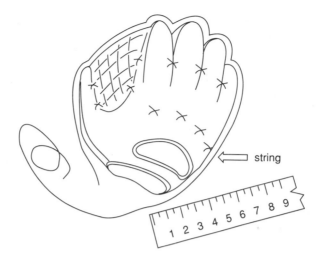

FIGURE 7.7 Measuring Perimeter

rapidly through these levels of measurement. Standard metric and customary units of measure can be introduced using similar activities, though greater attention must be paid to component units when working with feet, pounds, and quarts because no patterns are available to aid in memorization as with the metric system.

Measuring Perimeter

The distance around a plane figure is its **perimeter.** Children frequently confuse the perimeter length measure with the area of a figure. To help children remember that perimeter is a measure of length, not area, remind them that the prefix *peri* means *around*, as in periphery (the outside boundary) and periscope (a device that looks around something). Children can measure the perimeter of various objects in the classroom and on the playground. The perimeter of polygons such as the cover of a book can be calculated by measuring each segment and adding the results. The perimeter of such curved or irregular figures as a baseball glove can be found by using a piece of string to duplicate the distance around the object and then measuring the string with a nonstandard or standard unit of length. An interesting perimeter activity for older students is to have them draw a simple sketch using an estimated one-meter line and check the result using string and a meter stick (see Figure 7.7).

MEASURING WEIGHT

Early in their development, children learn to differentiate the **weights** of objects. Manipulating toys and other objects provides opportunities for comparison experiences. When children enter school, they can generally determine the heavier of two objects if the difference between the weights is great enough. However, some children may initially confuse the apparent size of an object (length or volume) with its weight, calling the larger one heavier.

Scientists prefer to use the concept of **mass** instead of weight. Mass refers to the amount of substance an object contains, while weight is the amount of force required to support it. At sea level on earth, the measure of an object's mass nearly equals its weight, hence the confusion between the two ideas. Mass remains constant everywhere in the universe, whereas weight changes according to environmental conditions. For example, a brick would weigh less on the moon than on earth, but its mass would remain constant.

At the elementary level, the two terms can generally be used interchangeably. Many teachers prefer the term weight, because it is commonly used outside of school. It becomes important to distinguish between mass and weight when students begin studying chemistry and physics in secondary school.

Comparison

To help children differentiate the weight of an object from its other attributes, it is necessary to provide an objective method for comparing the weights of items to each other and, later, to mass units. Initially, children may misunderstand the complex relationship among the *density, volume,* and *weight* measures of an object. Experience weighing objects differing in shape, size, and density helps children apply these concepts to real-world problems.

Students should be encouraged to *estimate* which of two objects is the heavier and then to test their prediction using a pan balance. Even with materials of the same density, younger children often demonstrate confusion between weight and other size attributes. For example, Piaget observed that many children under age seven contend that the weight of a piece of modeling clay changes when it is rolled into a sausage or a ball. Children are said to *conserve* weight if they understand that a change in the shape of an object does not by itself alter its weight. Experience comparing and weighing objects with a pan balance enables children to verify the accuracy of their initial perceptions when measuring and thereby enhance the development of the conservation concept.

An effective technique for helping children differentiate among measurement attributes uses *extremes*. Choose a large, light object like a styrofoam block and a small, heavy object like a stone. Due to the gross contrast between the volume and weight measures, the learner is forced to overcome the initial tendency to focus on volume when comparing the weight of the two objects. Using a balance makes it easier for children to accommodate the contradictory observations (see Figure 7.8).

The use of extremes in measuring establishes contradictions that can be exploited in the development of a wide range of mathematics concepts.

Arbitrary Units

Once the concept of weight has been adequately developed using comparison experiences, children should weigh objects using arbitrary units. Uniform materials such

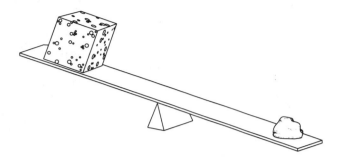

FIGURE 7.8

as marbles, blocks, or lima beans are excellent units of weight. Children can estimate the number of marbles required to balance a shoe and then use a pan balance to check their accuracy. The results of the estimates and weighings can be recorded on a class graph. Activities of this type lead to lessons about standard units of weight.

Standard Units

A metric unit of mass is the **gram.** As shown in Table 7.3, larger units are derived by combining 10, 100, or 1000 grams (or larger powers of 10). Likewise, smaller units are derived by dividing the gram by 10, 100, or 1000 (or larger powers of 10). To help children understand this relationship, the metric units can be conveniently related to the base-10 place value table.

Metric Units and Place Value

kg	hg	dag	g	·	dg	cg	mg
1000	100	10	1	·	0.1	0.01	0.001

The most common metric mass units for elementary use are the *gram* and *kilogram*. Upper grade students can also be introduced to milligrams and tonnes

TABLE 7.3 Metric Mass Measures

1000 milligrams (mg) = 1 gram (g)
100 centigrams (cg) = 1 gram (g)
10 decigrams (dg) = 1 gram (g)
1 gram (g) = 1 gram (g)
10 grams (g) = 1 dekagram (dag)
100 grams (g) = 1 hectogram (hg)
1000 grams (g) = 1 kilogram (kg)
1,000,000 grams (g) = 1 megagram (Mg) or tonne (t)

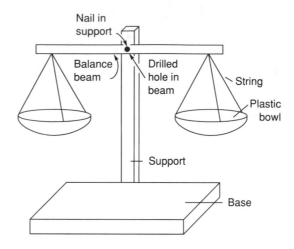

FIGURE 7.9 Pan Balance

through examples such as medicine bottles showing ingredients measured in milligrams and pictures of grain used as cattle feed measured in tonnes. It is possible to construct a simple pan balance like the one shown in Figure 7.9 that will measure to an accuracy of ten grams (one dekagram). A commercial balance is needed to measure objects to the nearest gram.

To practice measuring the weights of objects, children can be given a list of objects in the classroom that must be weighed using a balance and metric or customary mass pieces. Students can work in groups of two to four and record their results on a worksheet, or the activity can be set up as a learning center for individuals or pairs to complete over the period of several days. When all the students have completed the activity, compare the results by listing each group's measurements of the objects. Ask the students to explain why some measurements of the same object varied across groups. What was the highest and lowest measurement for each object? What is the correct answer? Discussion of questions like these helps children understand the approximate nature of measures. No one answer is correct. If a more precise measurement is desired, smaller mass units must be employed. No matter how small the unit, however, we can never know the exact weight of any object.

MEASURING AREA

The concept of area can be confusing to many children. Traditionally, area has been introduced using rectangular figures and defined as the product of the length and width. Later, children seem to associate area with the edge lengths of figures. This frequently leads to confusing the linear measure of perimeter with **area**, the measure of the surface. The situation can be further confused by the common practice of

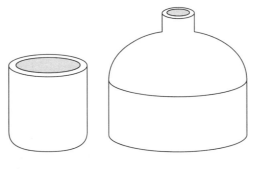

FIGURE 7.10

equating area to some operational measure, for example, a liter of paint covering ten square meters of wall. To help avoid such confusion, children can be introduced to the concept of area directly, without initial references to shortcut formulas.

Comparison

The areas, or painting surfaces, of various objects can be compared. For example, the area of the opening at the neck of a jar can be compared to other jars by seriating the lids. The height and capacity of the jar are not necessarily good indicators of the size of the opening (see Figure 7.10). Children should practice predicting, and then directly comparing, the areas of nonrectangular flat surfaces such as leaves and footprints.

Conservation of area is an important cognitive ability for children to possess if they are to successfully apply the concept of area measurement to everyday problems. Piaget observed that children *conserve area* if they understand that partitioning a surface and rearranging the pieces does not alter the overall area. For example, the two large triangles in a set of tangrams can be arranged as a square. Some children believe that rearranging these pieces into a large triangle, as shown in Figure 7.11, changes the total area.

In an interesting conservation activity for older children, students arrange all seven tangram pieces into a square, parallelogram, trapezoid, and triangle. Each of these shapes can be constructed using all seven tangram pieces and with no gaps inside the figure. Ask if the areas (painting surfaces) of all the finished shapes are equal (see Figure 7.12).

FIGURE 7.11   Conservation of Area

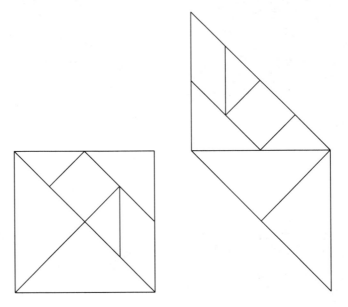

FIGURE 7.12 Making Shapes with Tangrams

Arbitrary Units

The measure of area can be quantified using arbitrary units. For example, thumbprints can be used to measure the area of two jar lids. Children first estimate the number of prints to cover the surfaces of each lid, and then carry out the measurement. Encourage students to write down their estimates *before* they measure. Of course, individuals will arrive at different measures of area for the lids depending on the size of their thumbs and how carefully they pack their prints together.

To help children remember the difference between the perimeter and area, give each child twenty-four square tiles. Ask them to arrange the tiles into a rectangle and record the area and perimeter (see Figure 7.13). The perimeter can be found

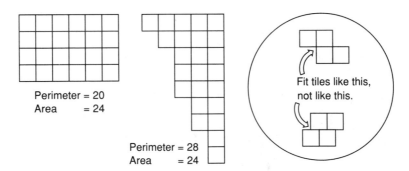

FIGURE 7.13 Perimeter and Area Using Tiles

by counting the distance around the figure (the edge of a tile is one unit of length). The area can be found by counting the total number of tiles (each tile is one unit of area). Which rectangle has the largest perimeter? The smallest perimeter? The largest area? The smallest area? Are there other shapes using all twenty-four tiles that have a larger perimeter? A smaller perimeter? Do the same activity using thirty-six tiles. Which figure has the smallest perimeter?

Standard Units

In order to improve the consistency of measures, identical units must be used. **Square centimeters** (or square inches) cover a flat surface with no gaps. These units are also related to linear measures that facilitate the later development of area formulas.

Students should work initially with nonrectangular or irregular figures and cover each surface with centimeter squares or grid overlays made from plastic film.

First estimate, then measure, in square centimeters, the area of your footprint and handprint as shown in Figure 7.14.

Correctly proportioned rectangular figures that are easily covered with unit squares can be used to introduce the shortcut formula—*area equals length × width*. Care should be taken to introduce the area formula only after children thoroughly understand the concept. In general, students should come to use the formula on their own after discovering its efficiency for counting interior unit squares (see Figure 7.15). Later, this method for finding area will be used to develop the array concept of multiplication and division.

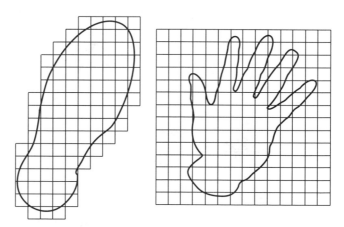

FIGURE 7.14

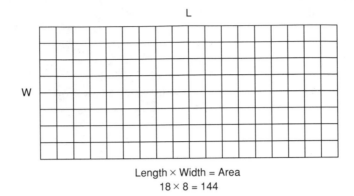

Length × Width = Area
18 × 8 = 144

FIGURE 7.15

> Using several rectangles, demonstrate that the formula (area = length × width) is simply a quick method of counting the unit squares.

The formula for the area of a triangle can be shown by cutting triangles into two pieces and rearranging them into a rectangle (see Figure 7.16).

Similarly, the formulas for the areas of parallelograms and trapezoids can be shown by cutting each figure into two pieces and rearranging them into a rectangle (see Figure 7.17).

> Demonstrate that these formulas hold for right and scalene triangles, rhombuses, and nonisosceles trapezoid shapes. Make paper models of several of these figures, cut them into two or more parts, and reposition the pieces to verify that the formulas give the correct areas.

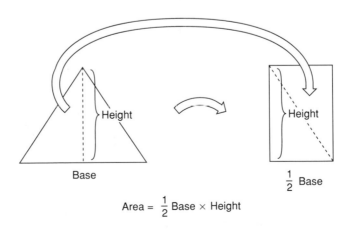

Area = $\frac{1}{2}$ Base × Height

FIGURE 7.16 Area of a Triangle

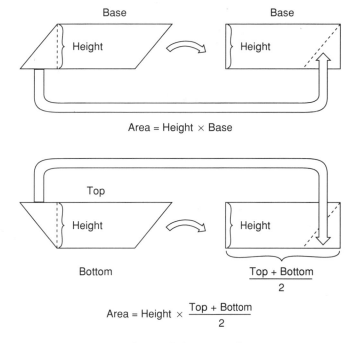

FIGURE 7.17   Area of a Parallelogram and a Trapezoid

The formula for the area of a circle can be motivated for Grade 5–6 children by dividing a circle into several equal wedge-shaped regions and repositioning them into the parallelogram-like figure shown in Figure 7.18. Notice that the scalloped base of the parallelogram is actually one-half of the circumference of the original circle, as the circumference makes up the top and bottom of the figure. Also note that the

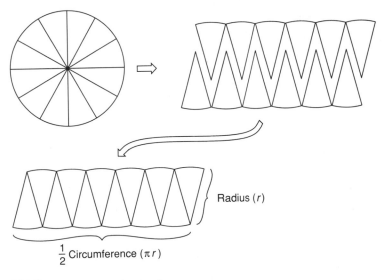

FIGURE 7.18   Area of a Circle

edge of the wedge-shaped region is the radius of the original circle. By cutting the circle into smaller and smaller regions, the base will look more and more like a straight line and the parallelogram more and more like a rectangle. The length of the parallelogram approaches one-half the circumference of the original circle, and its height approaches the circle's radius, giving the formula

$$\text{Area of Circle} = \tfrac{1}{2} \times \text{Circumference} \times \text{Radius}$$

$$= \tfrac{1}{2} \times (\pi \times 2 \times r) \times r$$

$$= \pi \times r^2$$

Measure the area of a circle using the parallelogram construction method, the unit-square estimation method, and the formula. Compare the results of the methods. How could you improve the accuracy of the parallelogram and unit-square measures?

MEASURING VOLUME

Children generally have many opportunities for comparing the capacity of containers before they enter school (building sandcastles, helping with cooking, or washing dishes). To supplement these experiences in the K–1 classroom, an informal learning center can be established using plastic containers of various sizes and shapes and a tub full of rice or clean sand. Children work in pairs or small groups filling and comparing the sizes of containers. Specific task cards can be designed showing pictures of containers with instructions to find out how many of a small container is required to fill a large one.

Technically, **capacity** measures how much a container can hold, and **volume** measures how much space is occupied by an object (e.g., a brick has a volume but no capacity). At the elementary level, however, little is lost by using these terms interchangeably.

Comparison

Comparing the capacity of differently shaped bottles is an excellent way to help children understand the concept of volume. Children (and many adults) can be confused by the visual illusion that taller bottles have a greater capacity than shorter ones. They do not recognize that an increase in height may be compensated for by a decrease in a bottle's cross section. For example, if a one-liter bottle of window cleaner were poured into a plastic bag and sealed with no air inside, the liquid would form a sphere about the size of a large grapefruit. A package of this size would certainly look less impressive displayed on a market shelf.

A lot of dishwashing liquid is sold due to the same tall-is-more illusion. Children are better able to understand and apply volume measures if they can conserve volume. Piaget observed that children *conserve volume* if they understand that the amount of a substance does not vary by changing the shape of the container in which it is placed (e.g., by pouring a serving of lemonade from a short, fat glass into a tall, thin container).

Arbitrary Units

The measure of the capacity of a container can be quantified using arbitrary units. Any uniform medium, such as marbles, dry beans, or rice, can serve as arbitrary volume units. Encourage students to estimate the number of volume units before they measure a container with arbitrary units. Counting the number of items that can be packed into a container provides a reasonable estimate of capacity. This also offers the opportunity to discuss volume measurement error. In general, smaller units conform more closely to the shape of the container and reduce the unaccounted-for capacity by reducing the air space between the units.

Standard Units

Metric and customary units can be introduced using centimeter cubes (ones-units from base-10 blocks or Centicubes) or cubic-inch blocks. Counting the number of cubes that can be carefully packed into a container gives a reasonable measure of capacity. More accurate measures of volume can be made by filling containers with water using a graduated cylinder calibrated in metric or customary cubic units. A plastic bag can be used to line nonwaterproof containers.

> Estimate the volume of a coffee cup in cubic centimeters. Using a supply of centimeter cubes, measure the approximate volume of the cup. Measure the volume using water and a graduated cylinder scaled in cubic centimeters. Compare the results of both techniques. Which method do you think gives the most accurate measure? What are the potential instructional advantages of each?

The volume of solid objects can be measured using **Archimedes' displacement principle.** Supposedly, in the third century B.C., Archimedes was asked to figure out a way to determine the exact volume of the king's crown as a way to show that it was 100 percent gold (the king was perhaps a bit paranoid). One day, while taking a bath, he noticed that the water level seemed to rise in proportion to the amount of his body that was submerged. By measuring the difference in water level, it was easy to compute the amount of water pushed out of the way (displaced) by his body or, more importantly, the crown.

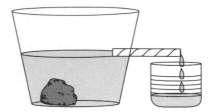

FIGURE 7.19 Displacement Pail

A simple displacement apparatus can be constructed, as shown in Figure 7.19, using a plastic pail, a straw, and a graduated cylinder. By filling the pail to the overflow tube, the volume of any nonporous object can be measured by carefully placing it in water and measuring the amount of water pushed out of the pail (displaced).

> Estimate the volume of five large stones and place them in order from small to large. Using the displacement technique, measure the volume of each and compare the results with your estimates.

For everyday use, volume is measured in metric units called **liters** (or cups, quarts, and gallons in customary units). Larger and smaller units are derived by multiplying and dividing the liter by powers of 10. The most common units are the liter and milliliter (1/1000 liter).

> Build a liter container by taping together five (10 cm × 10 cm) cardboard squares. Also, construct a milliliter box using five (1 cm × 1 cm) squares. Using base-10 unit blocks or plastic Centicubes, verify that one thousand milliliter boxes can be packed into a liter cube.

Notice that one liter (L) is another name for one cubic decimeter (dm^3) and that one milliliter (ml) is equal to one cubic centimeter (cm^3). The liter system was introduced to provide units of volume more convenient for everyday use than those offered by the cubic units (see Table 7.4).

Volume Formula for Rectangular Containers. A formula that gives the volume of rectangular containers can be derived from packing activities. Using centimeter squared paper, have children construct a box 5 cm × 6 cm × 4 cm. Count the number of cubes needed to cover the bottom layer leaving no gaps. Notice that the number of cubes is equal to the product of the length and width of the box—the area of the bottom of the container. Now count the number of layers. A quick way to count the total number of blocks in the box is to multiply the number of

TABLE 7.4 Liter and Cubic Unit Equivalents

Liter Units	Cubic Units
1 milliliter (ml) = 1 cubic centimeter (cm^3)	
1 centiliter (cl) = 10 cubic centimeters (cm^3)	
1 deciliter (dl) = 100 cubic centimeters (cm^3)	
1 liter (L) = 1000 cm^3 or 1 cubic decimeter (dm^3)	
1 dekaliter (dal) = 10,000 cubic centimeters (cm^3)	
1 hectoliter (hl) = 100,000 cubic centimeters (cm^3)	
1 kiloliter (kl) = 1,000,000 cm^3 or 1 cubic meter (m^3)	

cubes in each layer by the total number of layers. Therefore, the formula for the volume of rectangular boxes—*volume equals length × width × height*—is simply an efficient means for counting the number of unit cubes required to fill the container. To compute the volume of rectangular boxes, it is generally more convenient to use cubic centimeters than milliliters, because the edge lengths can be directly related to the cubic unit (e.g., 4 cm × 5 cm × 10 cm = 200 cm^3, or 200 ml).

Construct a box 5 cm × 6 cm × 4 cm using squared paper and use white Cuisenaire rods, units from base-10 blocks, or Centicubes to verify the volume formula (volume = length × width × height). What would happen if the length were 4.5 cm? Could you easily pack it full of cubes? Can you show that the volume formula still works for boxes with fractional dimensions?

The volumes of nonrectangular space figures can only be estimated using centimeter cubes. The interiors of these space figures cannot be packed with cubes without leaving gaps. The development of formulas for the volume of cylinders and prisms is patterned after that of a rectangular box, a right rectangular prism. First, find the area of the base and multiply the result by the height. For example, the area of a right triangular prism (the bases form 90° angles with the sides) is the area of its triangle base times its height. This holds similarly for a right cylinder (see Figure 7.20).

Formulas are available for computing the volumes of many regular space figures. Although the development of these formulas is beyond the scope of this book, it is interesting to compare volumes computed using formulas with measurements made using displacement techniques or water and a graduated cylinder.

Demonstrate that the volume of nonright cylinders and prisms (the parallel bases are not perpendicular to the sides) can be calculated using the area of the base times height procedure. Draw a sketch to show your solution.

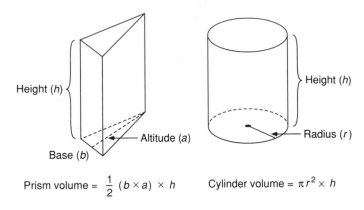

Prism volume = $\frac{1}{2}(b \times a) \times h$ Cylinder volume = $\pi r^2 \times h$

FIGURE 7.20 Prism and Cylinder

A paper **cone** or **pyramid** can be constructed with the same base and height as a cylinder or prism. As shown in Figure 7.21, it takes three cones or pyramids filled with rice to fill a corresponding cylinder or prism. Grade 5–6 children can construct paper models of a cone, a cylinder, and other corresponding space figures to motivate the volume formula.

Volume of a Cone or Pyramid $= \frac{1}{3} \times$ Area of base $\times$ Height

MEASURING TIME

Time can be interpreted as a quantity or a position in a sequence, depending on the context. Both applications are needed for everyday life. For example, in the results of a footrace, elapsed time (say, two hours and thirty minutes for a marathon) is a quantity measure. This measure is required to compare the results of *different* races. The finishing position (first, second, and so on), though directly related to elapsed time, is useful for comparing results within the *same* race. These two interpretations parallel the concepts of ordinal (position) and cardinal (quantity) numbers, which will be discussed in the next chapter.

The position-in-a-sequence interpretation of time develops early with most children (e.g., Wash your hands before you eat). A quantity understanding of time develops later as children's needs change (e.g., You can watch one hour of television tonight). Effective use of clocks requires the coordination of these two characteristic notions of time. The relation between the development of number and time is a good example of how concepts in one strand of mathematics may assist with the development of another.

School starting at 8:30 A.M. specifies a point in time. To place the minute hand, some children might count thirty minutes past the hour, but the result is a **point in**

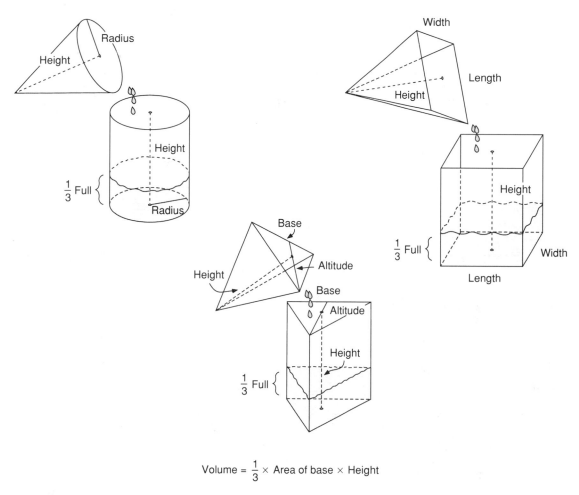

$$\text{Volume} = \frac{1}{3} \times \text{Area of base} \times \text{Height}$$

FIGURE 7.21 Volume of a Pyramid and a Cone

time. If a store is open from 7:00 A.M. to 11:00 P.M., the sixteen working hours are an example of **elapsed time.** To compute elapsed time, it is necessary to count the number of hours between the opening and closing times.

Both interpretations are required for effective participation in society. Getting to school on time or meeting friends at the playground depends on the ability to read a clock. Budgeting time for math, reading, science, and other subjects at school requires the ability to measure the quantity of elapsed time.

Piaget observed that children *conserve time* when they understand that two events can take the same amount of time even though the speed of the objects involved and distance traveled may vary. For example, suppose one child runs around a baseball diamond, and another child walks from second base to the home plate. If they both arrive at the plate at the same moment, we say that the two events took the same

| Card 1 | Card 2 | Card 3 | Card 4 | Card 5 |

FIGURE 7.22

amount of time. Some elementary-aged children may confuse distance and speed with elapsed time and report that, even though the events started and stopped at the same time, the elapsed time was different. For children to determine elapsed time, they must coordinate the speed of the two hands on the clock. They must understand that it takes the short hand the same amount of time to travel from 1:00 to 2:00 as it does for the long hand to travel once around the face. The long hand must travel faster than the short one in order for the two trips to take the same amount of time.

Comparison

Direct comparison activities help children learn to distinguish the concept of time from other associated attributes, such as distance and speed. Beginning activities should require children to place a series of events in temporal order. For example, given a shuffled set of pictures depicting a series of events in a story, a child could put them in order to match the narrative (see Figure 7.22).

Have children run races and keep track of the finishing order. Vary the type of race (hopping, three-legged, wheelbarrow) to allow for more potential winners.

Arbitrary Units

Arbitrary units of time, like dripping water, heartbeats, or pendulum swings, can be used to measure the duration of events. The number of time units (elapsed time) can be recorded and graphed.

For example, have children work in pairs. One child ties a shoe five times, and the other measures the elapsed time using a nonstandard unit such as the swings of a pendulum or the number of heartbeats. Reverse roles and repeat the experiment. Make a class record of the results by having the children show their elapsed time on a bar graph (see Figure 7.23). Discuss why some children took longer than others to complete the task. Is it fair to use different pendulum and heartbeat rates to time the event? How could children change things to make the experiment fair?

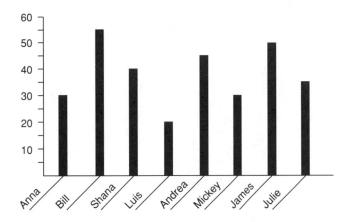

FIGURE 7.23 How Many Heartbeats to Tie a Shoe?

Standard Units

Operations on the clock face involving **seconds, minutes,** and **hours,** and the **days, weeks, months,** and **years** of the calendar can be introduced as standard units of time. It is common practice to open the school day with a calendar exercise and a discussion of the daily schedule. This provides an excellent opportunity to gradually introduce the clock face, day and month names, and notions of elapsed time.

A simple, effective method of helping children develop facility reading a clock is to place a large digital clock directly under the classroom circular wall clock. Seeing the two time displays side by side over a long period of time helps children see the relationship between the digits and the corresponding hand positions.

Initially, avoid the use of phrases such as "quarter-past" and "half-past" the hour. Children may confuse the idea of a quarter of a dollar (twenty-five cents) with a quarter of an hour (fifteen minutes). These phrases also depend on an understanding of fraction operations ($\frac{1}{2} \times 1$ hour = thirty minutes) that may not be adequately developed in the early grades. Though it is likely that students will need to understand phrases like these, such terms make more sense when introduced as an alternate to the additive form of reporting time—"two-forty-five" (2:45), "one-fifteen" (1:15), and "one-thirty" (1:30), for example. This method also encourages children to read the time from left to right.

Constructing the standard clock from a segmented timeline helps children to understand the relationships among time, speed, and distance on the circular face. Cut twelve 4 cm × 10 cm cardboard rectangles, write the numbers 1–12 in sequence in the upper right-hand corner of each, and fasten them together in a chain as shown in Figure 7.24a. Children record the hour on the timeline by positioning an arrow made of the same colored card under the appropriate number.

To construct a standard, one-hand clock face, form the timeline into a circle, glue it onto a circular piece of cardboard about 50 cm in diameter and attach the arrow in the center with a brass fastener. Children can practice recording the hour

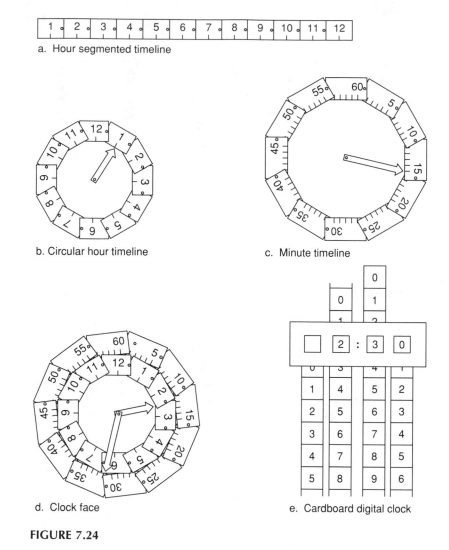

a. Hour segmented timeline

b. Circular hour timeline

c. Minute timeline

d. Clock face

e. Cardboard digital clock

FIGURE 7.24

by positioning the arrow to point at the appropriate number. Point out that, if the arrow points between two numbers, you always read the smaller number (see Figure 7.24b).

Make a new timeline from different colored cards with segments somewhat longer than the first (4 cm × 12 cm) so that it will form a concentric circle outside the hour timeline. Mark off five equal segments on each card and fasten them into a chain as before. Counting by five, label the right end of each card in sequence from 0 to 55. Use this scale to give children practice counting by fives to show fifteen minutes, thirty minutes, forty-five minutes, and the five-minute intervals in-between. Next, form the chain into a circle, glue it outside the hour scale, and attach a longer arrow that reaches the minute scale (see Figure 7.24c). Repeat the counting-by-five activity.

Using both arrows together allows children to read the hour and minute positions from the two concentric scales. If desired, the outside scale can be eliminated later by duplicating the minute segments on the hour scale.

Using the minute scale, have the students practice displaying fifteen minutes past the hour, thirty minutes past the hour, and forty-five minutes past the hour. The target times can be displayed using a large digital clock or a paper model using four sliding paper strips, as shown in Figure 7.24d,e. The long arrow can also be used to practice recording time to the nearest minute.

To read time from a clock face, have children count by fives, then by ones to reach the correct time to the nearest minute (e.g., 3:18 P.M.—5, 10, 15, 16, 17, 18). As the number of minutes rises above forty-five minutes, children often confuse the hour at which the arrow is pointing (e.g., 2:55 is often read as 3:55). Emphasize that, when the short hour arrow is between two numbers, one always uses the smaller hour value. The simultaneous display of an anolog clock face and a digital clock generally helps minimize this confusion until children can consistently read the correct time.

To provide K–2 children practice reading time in various forms, prepare several sets of cards showing various ways of displaying the same time using a clock face, digital clock, words (half-past four or four-thirty), and numerals (4:30). Mix them and distribute one card to each child. Have the children with no talking, form groups whose cards show the same time. Have each group explain why the cards show the same time. Mix the cards and repeat the process.

> Construct a clock face and digital clock like the ones shown in Figure 7.24d,e. Use both displays to show the beginning and ending time of an event. Calculate the elapsed time using each.

MEASURING TEMPERATURE

Measuring temperature is a comparison activity. The **Celsius** and **Fahrenheit** temperature scales use the freezing temperature (0°C, 32°F) and boiling temperature (100°C, 212°F) of water as starting points. All measures of temperature, then, are comparisons to these benchmarks.

To facilitate more accurate comparisons, the range of temperatures between freezing and boiling is divided into one hundred equal parts for the metric Celsius scale (°C) and 180 for the Fahrenheit scale (°F). Each unit is called one degree (not related to angle measure degrees).

Temperature can be measured by observing the expansion, or swelling, of heated objects. A thermometer is constructed by sealing mercury or alcohol in a glass vacuum tube. It is first immersed in ice water, and the level of the mercury is marked on the tube (0°C or 32°F). It is then immersed in boiling water, and again the level is marked (100°C or 212°F). Finally, this range is divided into the appropriate number of degrees (one hundred for Celsius and 180 for Fahrenheit).

Temperatures below zero are indicated by *negative* values (e.g., ⁻15°C). The temperature scale can be used to motivate the need for negative numbers with young children. This use of negative number values is generally accepted even by young students as an appropriate way to show temperatures below zero (see Figure 7.25). Further development of integers (negative and positive whole numbers) will be presented in chapter 13.

Note that a thermometer does not measure the *quantity of heat*, as objects of different sizes and densities require different amounts of heat to reach the same temperature (e.g., a heavy iron pan takes longer to heat up than a light aluminum pan because it absorbs more heat before passing it on to your finger). Quantities of heat energy are measured in **calories**, the heat energy needed to raise one gram of water 1°C. One kilogram of human fat contains approximately 7700 calories of energy potential. To lose one kilogram, your body must work hard enough to use up 7700 calories of energy. The body uses about one calorie per minute just to maintain

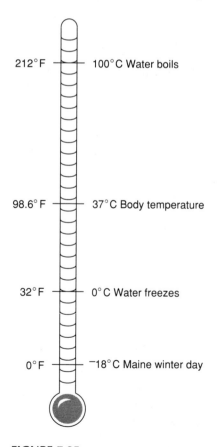

212°F — 100°C Water boils

98.6°F — 37°C Body temperature

32°F — 0°C Water freezes

0°F — ⁻18°C Maine winter day

FIGURE 7.25

internal functions. If you simply lay in bed all day you would use up about 1400 calories. Swimming requires about five calories per minute. It would take fifteen minutes to swim off a banana!

Comparison

To introduce comparison of temperatures, have children list objects that feel cool and warm when compared to their body temperatures. Take the class to the playground and find an area in the direct sunlight where the blacktop is next to a concrete sidewalk. Have the children straddle the two surfaces and carefully touch one with each hand (avoid this activity when the blacktop is too hot to touch). Ask if the surfaces are at the same temperature. Ask why the blacktop seems hotter. Does it get more sunlight?

To see the need for an objective temperature measurement instrument, have two people try the following experiment (take care not to involve students who are sensitive to heat).

1. Place three bowls of water on a table in front of the class—one quite hot, one quite cold, and one at room temperature.

2. Ask one person to place his or her right hand in the hot water (make sure the water is not too hot) and the second person to place his or her right hand in the cold water.

3. Ask each to describe the temperature of the water in the bowl.

4. After leaving their hands in the water for about ten seconds, have them simultaneously place the same hand in the center bowl and indicate whether it feels warm or cool.

5. Discuss why one thinks it is warm and the other thinks it is cool.

Standard Units

In order to ensure agreement of the results of measurements, we must use standard instruments and units. To introduce the process of reading a scale, use a large, cardboard demonstration model of a Celsius or Fahrenheit thermometer. Thermometer scales showing one mark for each degree are best for elementary school use. These models are available commercially or can easily be constructed by drawing the appropriate scale on a piece of heavy cardboard and inserting a piece of white ribbon, half of which is colored using a red felt marker, through a slit above and below the scale. Fasten the ends of the ribbon on the back of the cardboard, and pull it up and down to adjust the temperature reading.

Using a Celsius or Fahrenheit thermometer, have children measure the temperature of various objects, such as

☐ Their own temperatures before and immediately after a run.

☐ The classroom hamster, snake, or other pet (place the thermometer under a leg or belly for one minute).

☐ A basketball before and after it is filled (hold the bulb of the thermometer against the ball for one minute before and immediately after filling).

☐ The blacktop, sidewalk, and soil in the sun.

☐ The blacktop, sidewalk, and soil in the shade.

☐ The tire on the teacher's car just after arriving at school and one hour later.

☐ Inside and outside of a refrigerator at school or at home.

☐ The inside and outside temperature measured each hour throughout the school day.

Remind children that, when the thermometer shows a temperature *between* two marks on the scale, they should select the value of the closest mark. As there is not enough room to write all the temperature values on the scale, children will have to count the marks above the closest written value. After measuring and recording the temperatures, give each child an opportunity to duplicate one of the temperature readings on the cardboard thermometer model.

WORKING WITH MONEY

Activities with money can serve two instructional purposes. First, practical applications involving money abound in the child's world. Making change, counting coins, buying items at the store, and totaling the restaurant bills offer excellent opportunities for practicing skills learned at school.

Second, working with money can also provide a real-world application for using the base-10 place value system for counting, adding, and subtracting amounts. In addition to the increased use of the metric system and the availability of calculators, the convenient base-10 structure of our money system can significantly contribute to the emerging concept of place value numeration.

Initially, the penny is defined as the basic unit. Using only pennies and dimes, children can construct representations of amounts to 99 cents. Including the dollar allows children to represent values to 999 cents. Introduce children to the numerals for each coin (e.g., 1¢, 5¢, 10¢, 25¢, 50¢), then the amounts in between (12¢ or 38¢). The use of the nickel, quarter and fifty-cent piece is minimized until students gain facility using the penny, dime, and dollar (see Figure 7.26a). The decimal notation for coins can be introduced (e.g., 15¢ = $0.15) after amounts greater than $1.00 are learned. As shown in Figure 7.26b, money can be used later to introduce decimal fractions by defining the dollar as the basic unit (1.00 = one dollar; 0.10 = one dime; 0.01 = one penny).

Primary children can use play money to practice relating the number values of the coins and bills. Show a target coin (a dime). Have each child match the value using only one type of coin (two nickels or ten pennies) and record the results by drawing a sketch or writing the values. Using pictures of small sets of coins (one

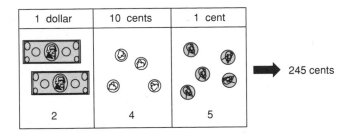

a. Money place value table

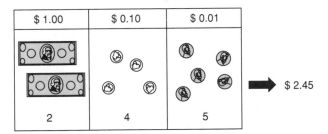

b. Decimal money place value table

FIGURE 7.26

dime and two pennies), have children find as many ways as possible to make the same amount using other coins (twelve·pennies, one nickel and seven pennies, two nickels and two pennies). Have them circle the set containing the fewest number of coins for each amount. Older children can do the same activity with larger amounts.

A classroom store offers a practical environment for simulated practice with money. Have children bring clean, empty cans and cartons from home. Label each item with a price and set up a learning center with shelves and a table. Using play money and working in pairs, one child makes a shopping list of items to buy; the second acts as the storekeeper and gives the customer the appropriate amount of change. The total cost of the items should be computed by both the customer and the storekeeper to make sure they agree on the change. The children then reverse roles and repeat the activity. The type and cost of the items can be adjusted to enable children at various grade levels to participate in a classroom store (e.g., 1¢ and 2¢ for kindergarten and amounts up to $1.00 for Grade 2).

Give Grade 3–6 children a real or simulated copy of a menu from a local restaurant and let them select their favorite meal and calculate the cost. Have them list two items that cost $1.00 or less. Older children can plan a meal for their whole family that must cost less than a specified amount (e.g., $25.00). Using play money and working in pairs, have one child pay the bill and the other count out the correct change. The menu items and costs can be varied according to grade level and eating habits of the class. Older students can select items from mail-order catalogs, calculate shipping charges and taxes, and complete an order form.

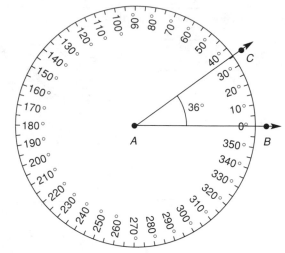

a. *m∠BAC* or *m∠A* = 36°

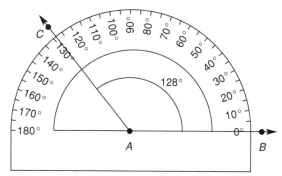

b. *m∠BAC* or *m∠A* = 128°

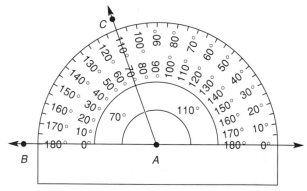

c. *m∠BAC* = 70° or 110° ?

FIGURE 7.27

MEASURING ANGLES

Angles are commonly measured using a tool called the **protractor** (see Figure 7.27a). This device is simply a circle (or half-circle) on which 360 equidistant points (or 180 for a half-circle) have been marked.

Place the center of a circle protractor on the vertex of an angle. Count the number of degrees measured by the angle. Notice the degree count can be read directly off the scale if one of the rays passes through the zero point on the protractor.

A standard, half-circle protractor has 180 divisions, or **degrees,** marked along its outer edge. The center is midway along the straight edge. To measure an angle, first place the straight edge along one of the rays with its center over the vertex of the angle. The angle measure is indicated by the point where the second ray crosses the semicircular scale (see Figure 7.27b).

Some protractor scales are marked in both left and right sequences. These are often confusing to students first learning to measure angles. The dual scale protractor is more convenient when measuring angles in a variety of positions, but users must be able to recognize which scale value is reasonable for the angle being measured. The angle measure shown in Figure 7.27c could be interpreted as 70° or 110°. Experience estimating angle measures and identifying acute and obtuse angles will help children observe that the angle being measured is less than 90° and, therefore, that the correct measure is $m\angle BAC = 70°$.

Use a protractor to determine which pairs of angles in Figure 7.28 have approximately the same measure. Write the angle measure of each.

CLASSROOM ACTIVITIES

As with other experiences suggested in this text, it is important to actually try out the following activities in a laboratory setting. Think about how you would introduce each activity as a classroom lesson. What materials would be necessary? How would they be distributed? How would you evaluate student progress? The references at the end of this chapter contain examples of additional instructional materials and suggested activities.

Body Measures (Comparison)

Children can practice comparing lengths by cutting pieces of string to match the length of various body parts and seriating the bundle on a graph. Working in pairs,

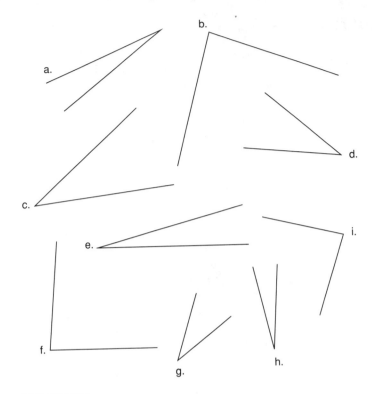

FIGURE 7.28

one student acts as the model and the other cuts pieces of string equal to the length of an arm, leg, handspan, foot, waist, and height. Students reverse roles and cut a second bundle of strings. Each piece of string is labelled with the appropriate body part. Have the children compare the different lengths (e.g., the length of both arms outstretched is about equal to one's height). Each child attaches the bundle of strings about five centimeters apart on a prepared bulletin board graph. Each string should begin at a common base-line and be in order from short to long (see Figure 7.29).

FIGURE 7.29 Body Bundle Wall Graph

Heart Throb (Arbitrary Unit)

Children can measure the duration of various events using heartbeats as units. Students practice finding their own pulses by gently pressing the forefinger to one side of the throat just under the jawbone. The children then work in pairs with one student completing a task (for example, write name backwards, tie shoe five times, write this week's spelling words, say the nines multiplication facts, and so on) while the other times the event by counting the number of heartbeats required to finish. They reverse roles and graph the results to compare elapsed time. Problems associated with using nonstandard units to measure the duration of events can be discussed. For example, does it really take longer to tie your shoe if your partner's heart beats faster?

Are You Fit? (Standard Unit)

Children can measure their height and weight in metric units to determine their height-to-weight ratio (the approximate normal ratio for elementary-aged children is about four centimeters height for each one kilogram of weight). Children work in pairs, and using a meter stick, measure each other's height in centimeters. Students also determine their weight using a bath scale calibrated in kilograms. Each height-to-weight ratio is computed by dividing weight in kilograms into height in centimeters (e.g., 150 cm ÷ 40 kg). Note that this ratio is only approximate. It is important to note that it is perfectly normal for results to be somewhat above or below the ratio of 4 to 1.

Example: Height = 146 centimeters (cm)
Weight = 37 kilograms (kg)
Height/Weight Ratio = $\frac{146}{37}$ = 3.95

With young children, it is fun to graph each student's height and weight at the beginning and end of the school year to chart growth.

PROBLEM-SOLVING EXPERIENCES INVOLVING MEASUREMENT

Orange-Peel Sandwich

Which is larger, the surface area of an orange or the area of one face of a slice of bread? Children can use the actual objects to *conduct an experiment,* then *draw a sketch* to compare their results. Think about what would happen if you made an orange-peel sandwich. Would the peel hang over the edge, or would it be possible to rearrange the peel to fit entirely on the face of a slice of bread without overlapping?

Problem-solving experiences like this introduce the idea of the approximate nature of measures (what to do with the holes in the flattened orange peel) and help students differentiate the area concept from that of other measures such as volume and length.

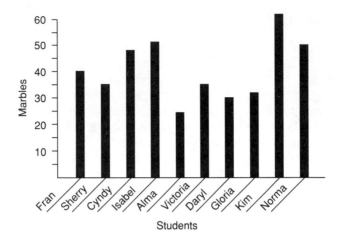

FIGURE 7.30 How Big Is Your Shoe?

Fat Foot

What is the volume of your foot? One way to find out is by counting the number of marbles your shoe will hold. Children can *conduct an experiment* to help solve this problem and compare their results by *constructing a graph*. When doing this experiment in a classroom, remind students to wear shoes without any openings. Count the number of marbles (or other uniform objects) that fit into the left shoe and record the results on a class graph. Students can make individual bars for a class graph by laying a strip of construction paper along the vertical scale and cutting it to the appropriate length. The strip is fastened to the graph above the student's name (see Figure 7.30). Do left and right shoes have the same volume? How would you find out? How could you make a more accurate measure of the volume of each foot? How could you make a graph to compare the volumes of left and right feet? What other objects' volumes could you measure using this method?

Packing Crates

How many different, rectangular boxes can be designed that will hold exactly sixty-four centimeter cubes? Children can *construct a model* to help solve this problem. Using sixty-four centimeter cubes (base-10 unit blocks or Centicubes), sheets of centimeter squared paper, and tape, form a rectangular base using some of the blocks and construct equal layers until all the cubes are used. If there are any holes in the top layer, start over. Construct a paper packing crate for each completed rectangular structure using centimeter squared paper and tape. This problem can be extended to other packing problems using 36, 48, 81, and 100 cubes. Older children can try to discover a rule for finding the dimensions of a packing crate (i.e., finding sets of factors for each pile of cubes—$64 = 4 \times 4 \times 4$).

Denominate Numbers

Many real-world problems require estimating and measuring the length, weight, area, volume, or temperature of an object. Often events must be temporally ordered, or elapsed time must be computed. The values associated with these measurements can then be operated upon to find totals, differences, averages, and ratios as necessary. Measures written with unit labels (e.g., 22.5 kilograms, $4.55, 6:00 P.M., 350 millimeters) are called **denominate numbers.** For example, to find the total height of three children we add their respective heights:

> Paul   — 132 centimeters
> Andria — 145 centimeters
> Jamala — 158 centimeters
> ———————————————
> 435 centimeters = 4 meters 35 centimeters

If the children lay end to end, they would form a line four meters thirty-five centimeters long. Though an individual's height is often represented in centimeters only, we generally regroup measures into larger units whenever possible (meters and centimeters). Although it is possible to write the previous answer using the intermediate unit decimeter (4m, 3dm, 5cm), it is not common practice.

Notice that the metric system makes it easy to convert between units. Since the system is based on powers of 10, operations with metric measures parallel those of the decimal number system and the denominations of U.S. money. Knowing the relative size of the units allows children to move the decimal point (multiply or divide by powers of 10) the appropriate number of places to the left or right to make conversions from one unit to another in the metric system. For example:

1. 325 centimeters = 3.25 meters. (To translate centimeters to meters, divide the number of centimeters by 100—the value is *reduced* by moving the decimal point two places to the left as the unit is proportionally *increased*.)
2. 2.7 kilograms = 2700 grams. (To translate kilograms to grams, multiply the number of kilograms by 1000—the value is *increased* by moving the decimal point three places to the right as the unit is proportionally *decreased*.)
3. 5.76 square kilometers = 5,760,000 square meters. (To translate square kilometers to square meters, multiply the number of square kilometers by 1,000,000—the value is *increased* by moving the decimal point six places to the right as the unit is proportionally *decreased*.)

Though a numeral may get smaller or larger, the overall size of the measure remains constant since the change in unit size exactly compensates for the change in the numeral. Writing decimal equivalents of actual measures may be misleading because of the way we interpret measurement error. If the weight of an object is measured as 2 kilograms, it means that the actual weight is between 1.5 kg and 2.5 kg—a measurement error of one-half kilogram. If we use the method above to convert 2 kilograms to 2000 grams, we might interpret the actual weight as between 1999.5 g and 2000.5 g—a measurement error of one-half gram. Though the amounts

2 kg and 2000 g are mathematically equivalent, they are interpreted differently as measures of real-world objects.

It may be necessary to convert between metric units when solving problems. However, it is inadvisable (and unnecessary for most practical purposes) to introduce exact conversions between metric and customary units. Approximate equivalents (e.g., meter and yard or liter and quart) may be introduced as benchmarks to help children compare the size of metric and customary units.

Solve the following exercises using metric tables as required.

1. 450 cm = ? m

2. 450 cm = ? dm

3. 1276 g = ? kg

4. 1.5 kg = ? g

5. 3.5 cm = ? mm

COMPUTER APPLICATIONS

Measurement Software Tools

With access to quality software tools, teachers are not as limited in the type of exercises available as they are with CAI materials. When employing software tools, however, the teacher becomes responsible for planning the content and sequence of the instructional activities. To help plan effective activities, the teacher should review the instructor's guide supplied with the software or supplementary resource books written for the specific software.

Geometric Supposer. A good example of a software tool that can be used to develop both measurement and geometric concepts is the *Geometric Supposer* series. Grade 4–6 students can use this tool to explore the area formulas of triangles and parallelograms (rectangles, squares, trapezoids, and so on). Given the formula for determining the area of a parallelogram (area = base × altitude), students can ask the *Supposer* to present a parallelogram, measure its base and altitude, and compute the product. The *Supposer* can then directly measure the area, and the two solutions can be compared (see Figure 7.31). The process can be repeated several times with randomly selected parallelograms to build confidence in the standard formula. Area formulas for other quadrilaterals can be similarly explored.

The system can also directly measure the perimeter and area of figures, allowing students to test invented formulas and other conjectures (e.g., the perimeter of a rectangle equals twice the sum of the length and width; or the sum of the interior angles of any quadrilateral equals 360°).

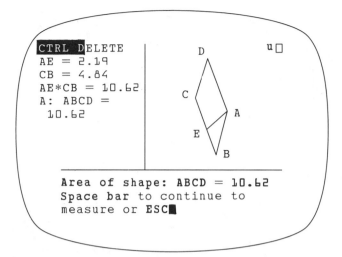

```
CTRL DELETE          D              u□
AE = 2.19
CB = 4.84
AE*CB = 10.62
A: ABCD =       C
   10.62
                             A
               E

                    B


Area of shape: ABCD = 10.62
Space bar to continue to
measure or ESC█
```

FIGURE 7.31 Area of a Parallelogram. *Note:* From "Geometric Supposer" [Computer Program] by Sunburst Communications, 1985. Copyright 1985 by Sunburst Communications. Reprinted by permission.

Students can carry out similar experiments on triangles using the *Geometric Supposer: Triangle*. For example, the *Supposer* can test the conjecture that the triangle formed by connecting the midpoints of two sides of any triangle equals one-quarter (0.25) of the area of the original triangle (see Figure 7.32).

Teachers have also successfully used the *Geometric Supposer* with an overhead projector display pad or large monitor when introducing measurement and geometric concepts to the entire class. If only one computer is available, it is also effective to set up a computer learning center with task cards and schedule pairs of students to work at the center throughout the day.

ADAPTING INSTRUCTION FOR CHILDREN WITH SPECIAL NEEDS

Children who have experienced protracted difficulty working with numbers and geometric figures may have difficulty understanding the measurement concepts and using measurement instruments. It is particularly important to allow extra time for these children to experience comparison activities before introducing arbitrary and standard units of measure. Direct comparison activities encourage children to focus on one measurement attribute at a time. The visual feedback provided by these experiences also helps special-needs children attain the conservation abilities (number, quantity, length, area, volume, and time) so important to the measurement process.

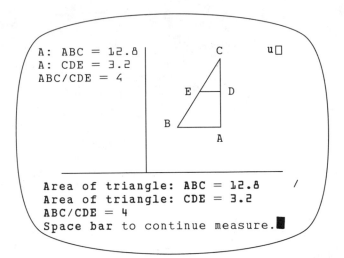

```
A: ABC = 12.8
A: CDE = 3.2
ABC/CDE = 4



Area of triangle: ABC = 12.8          /
Area of triangle: CDE = 3.2
ABC/CDE = 4
Space bar to continue measure.
```

FIGURE 7.32 Solving Area Problem Using *Geometric Supposer*. *Note:* From "Geometric Supposer" [Computer Program] by Sunburst Communications, 1985. Copyright 1985 by Sunburst Communications. Reprinted by permission.

Physically involving children in measurement activities is especially important for special-needs learners. Each measurement attribute can be related to a characteristic of their own bodies (length—pace; area—footprint; volume—handful; weight—body weight; time—heart rate; temperature—body temperature).

Use measurement instruments with large, clear scale markings. Initially, a meter (yard) stick can be constructed from strips of wood with no intermediate markings. When greater precision is required, a decimeter (foot), then a centimeter (inch) scale can be added. Instead of using a tape measure to measure distances longer than a meter (yard), allow special-needs children to use several meter sticks laid end to end. This technique helps the child understand the unit iteration process involved in measuring. Similarly, use a simple two-pan balance and single-unit mass pieces (marbles, one-gram, or one-ounce) to weigh objects instead of a beam balance or multiple-unit mass pieces (washers and marbles, 10-gram, or 8-ounce) until the child has a good understanding of the process.

Mathematically gifted children can be introduced to specialized measurement instruments like the micrometer, beam balance, and transit. Have children research the uses of such instruments in science and the workplace. Children can construct a *hypsometer* to indirectly measure the heights of tall objects. To use this device, the child first measures ten meters (or paces, feet, or yards) from the base of the object (see Figure 7.33).

Standing ten meters from the object, sight through the straw and pinch the string holding the bob. Where the string crosses the bottom *altitude scale*, read the height of the object above eye level (i.e., add the child's eye-level height to the value given

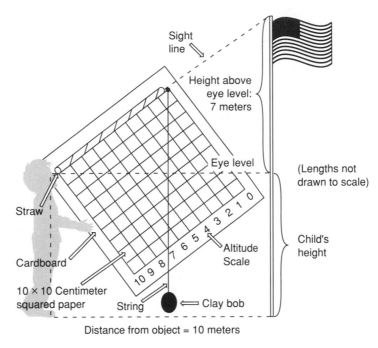

FIGURE 7.33 Constructing a Hypsometer

by the altitude scale to find the total height of the object). If the object is too close to sight through the straw and still have the string cross the altitude scale, walk twenty meters from the base of the object and multiply the scale value by two. Have students try to show why this device gives the correct height.

SUMMARY

Measurement is the process of comparing or assigning values to particular attributes of objects or events. Various measurement instruments are introduced at the elementary level including the ruler, balance, graduated cylinder, thermometer, clock, and the protractor. Common measurement attributes introduced at the elementary level include length, weight, area, volume, time, temperature, and angle. Piaget observed that the cognitive abilities of conservation, seriation, and transitivity are important for the development of meaningful measurement skills. There are three levels of measurement that can be used to introduce each measurement attribute: direct comparison, arbitrary (nonstandard) units, and standard (metric and customary) units. When learning measurement, students are encouraged to estimate measures prior to carrying out experiments. Estimations are followed with manipulative activities consistent with the three levels of measurement. Finally, problem-solving experiences are presented that require the application of newly learned concepts and skills. Other important measurement concepts are measurement error and ordinal and cardinal numbers. The Metric Conversion Act of 1975 allowed for the voluntary introduction of the SI metric system of measures in the United States. While some progress has been made in the transition to metric units, the United States remains the last industrial country in the world without a specific conversion plan. While children must therefore be familiar with both metric and customary units of measure, conversion between the two systems should remain informal. Measurement lessons often include the construction of individual or class graphs. Such graphs organize data for interpretation and provide a record of individual participation in an activity. Care must be taken to allow additional time for special-needs children to work with concrete experiences at each of the three levels of measurement to ensure adequate understanding of each attribute and procedure. The use of simple, uncluttered measurement instruments is also recommended for these children. Gifted and talented children can benefit from the study of specialized measurement instruments and their uses in science and the workplace.

COURSE ACTIVITIES

1. List the common measurement attributes that are taught at the elementary level. For each attribute, list several applications that children experience in everyday life. Work in a small group to compile as large a list as possible.
2. Choose a measurement attribute other than length. Develop two sets of comparison, arbitrary, and standard unit activities designed to introduce the concept, one appropriate for the primary level and the other for the intermediate grades. Write a lesson plan for each activity. If possible, implement the lessons in an actual classroom setting and compare your results with others.
3. Using the three levels of measurement (comparison, arbitrary units, and standard units), develop a series of three lessons to teach angle measurement (also see the section on angles in chapter 6). Compare your lesson plans with others in your class.
4. Review the measurement objectives listed in appendix I. Select a focus grade level and one measurement attribute. Using supplementary resource books, identify three activities for each objective selected. Write a descriptive title for each activity and specify whether it is an example of comparison, arbitrary, or standard unit measurement.
5. Create a bulletin board idea for introducing metric length, weight, and volume units for Grade 5–6 children. Include several questions so that the display actively engages the viewer in learning rather than being simply decorative.
6. Make a set of homemade metric measurement instruments. Construct a meter stick, pan balance, set of metric mass pieces, and a graduated liter container. Review one of the supplementary resource books for construction details. Test the instruments for ease of use and accuracy.
7. Organize a study group with your peers to help you become familiar with the metric system. Learn the prefixes, common units, and relationships among the units.
8. Include examples of measurement activities in your lesson idea file. Organize them according to level of measurement and appropriate grade level.

9. Read one of the *Arithmetic Teacher* articles listed in the reference section. Write a brief report summarizing the main ideas of the article and describe how the recommendations for instruction might apply to your own mathematics teaching.

MICROCOMPUTER SOFTWARE

Bumble Games   Practice locating positions on coordinate graphs (The Learning Company).

Clock   Practice reading time on a simulated clock face (Hartley Software).

Explorer Metros: A Metric Adventure   Problem-solving experiences involving metric length, mass, capacity, and temperature (Sunburst Communications).

Geometric Supposer: Points and Lines, Triangles, Quadrilaterals, and *Circles*   A versatile series of software tools for evaluating conjectures about area formulas and geometric relationships (Sunburst Communications).

Magic Cash Register   Practice making change for simulated purchases (Avant-Garde).

Metric and Problem Solving   Practice estimating length and logic games (MECC).

Micros for Micros: Estimation   Practice estimating length, number, and time (Lawrence Hall of Science, University of California Berkeley).

Money and Time Adventures of the Lollipop Dragon   Practice telling time using digital and traditional clocks, counting money, and making change (Society for Visual Communication).

Money Matters   Primary-aged children can learn to recognize money (U.S. or Canadian), count money, and make change (MECC).

REFERENCES AND READINGS

Baffington, A. (1973). *Meters, liters and grams*. New York: Random House.

Barnett, C., Judd, W., & Young, S. (1976). *Measure matters*. Palo Alto, CA: Creative Publications.

Bates, J. (1972). *Exploring metric measure—Primary teacher's source book*. New York: McGraw-Hill.

Bitter, G., Mikesell, J., & Maurdeff, K. (1976). *Activities handbook for teaching the metric system*. Boston: Allyn & Bacon.

Cavanaugh, M. (1977). *Metric madness*. Denver: Scott Resources.

Christopher, L. (1982). Graphs can jazz up the mathematics curriculum. *Arithmetic Teacher, 30*(1), 28–30.

Clements, D., & Battista, M. (1986). Geometry and geometric measurement. *Arithmetic Teacher, 33*(6), 29–32.

Elementary Science Study. (1966). *The balance book*. New York: McGraw-Hill.

Harrison, W. (1987). What lies behind measurement? *Arithmetic Teacher, 34*(7), 19–21.

Hart, K. (1984). Which comes first—Length, area, or volume? *Arithmetic Teacher, 31*(9), 16–18, 26–27.

Hildreth, D. (1983). The use of strategies in estimating measurements. *Arithmetic Teacher, 34*(7), 50–54.

Horak, V., & Horak, W. (1983). Teaching time with slit clocks. *Arithmetic Teacher, 30*(5), 8–12

Johnson, E. (1981). Bar graphs for first graders. *Arithmetic Teacher, 29*(4), 30–31.

Keyser, T., & Souviney, R. (1980). *Measurement and the child's environment*. Glenview, IL: Scott, Foresman.

Mullen, G. (1985). How do you measure up? *Arithmetic Teacher, 33*(2), 16–21.

Nelson, D., & Reys, R. (Eds.). (1976). *Measurement in school mathematics: 1976 Yearbook*. Reston, VA: National Council of Teachers of Mathematics.

Nibbelink, W. (1982). Graphing for any grade. *Arithmetic Teacher, 30*(3), 28–31.

Roper, A. (1977). *Metric recipes for the classroom.* Palo Alto, CA: Creative Publications.

Thompson, C., & Van de Walle, J. (1985). Learning about rulers and measuring. *Arithmetic Teacher, 2*(8), 8–12.

8

Developing Number Sense

You may fly to poetry and music, . . . and number will face you in your rhythms and your octaves.

—Alfred North Whitehead

Upon completing Chapter 8, the reader will be able to

1. Describe three fundamental principles of meaningful counting and give examples of appropriate instructional activities.

2. Describe the ordinal and cardinal notions of number and explain why both are important for mathematics learning.

3. Describe three basic cognitive abilities that Piaget claims are important to a meaningful understanding of whole number.

4. List three classification operations important to the concept of number and give an example of an appropriate instructional activity for each.

5. Describe the three properties of a mathematical equivalence relation and give examples of each.

6. Describe the greater-than and less-than relations and give an example of an appropriate instructional activity.

7. Describe the role of patterns in the development of the whole number concept and give examples of appropriate instructional activities.

8. Describe the concepts of place value and regrouping and give examples of appropriate classroom activities.

9. Describe the two ways fractional part can be defined and give an example of each.

10. Describe a model for introducing decimal fractions and give an example of an appropriate classroom activity.

All the children are sitting in rows participating in an oral exercise. The teacher moves through the aisles to the refrain —"One, two, buckle my shoe. Three, four, shut the door. Five, six, pick up sticks." Approaching the back of the room, the teacher overhears a child singing—"La, la, la la la. La, la, la la la. La, la, la la la." When asked what he is saying, he replies, "I know the tune, but I haven't got the words yet."

Most children learn the counting song long before they begin school. Though rote counting can provide a basis for future concept development, the ability to repeat the sequence of number names should not be interpreted as an indication of a rational understanding of number. Just as kindergartners are able to reproduce recognizable utterances of the "Pledge of Allegiance," it may be years before the concepts of patriotism and freedom expressed in this familiar passage take on a meaning.

Repeated recitation of the number names in sequence is a necessary though insufficient condition of the meaningful development of counting. Figure 8.1 shows the types of errors reported in children's oral counting (Driscoll, 1980; Gelman & Gallistel, 1978; Saxe, 1979). The loops above the objects trace the movement of the child's finger during counting.

Example 1 shows a **coordination error**, where more than one number name is assigned to a single object. Example 2 shows an **omission error**, where one or more objects are skipped. In both cases, the child accurately applies the sequence of number names but violates the **one-to-one correspondence** requirement. In

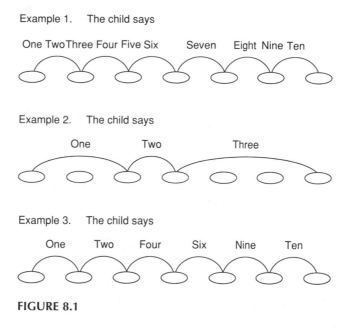

Example 1.   The child says

One TwoThree Four Five Six   Seven   Eight Nine Ten

Example 2.   The child says

One   Two   Three

Example 3.   The child says

One   Two   Four   Six   Nine   Ten

FIGURE 8.1

Example 3, the child assigns labels in a one-to-one fashion; however, the number names are not in the standard sequence.

In addition to rational counting, children must also understand the **cardinal principle** (the special significance applied to the last object counted). For example, when one is counting seven objects, the final number name indicates both that there are seven objects in the set and that the last object is the seventh in the sequence (Fuson & Hall, 1983). To apply the number concept to problems in the environment effectively, a child must therefore be able to coordinate the **ordinal** and **cardinal** notions of counting. Ordinal refers to numbers that define a *position*, such as a ZIP code. Cardinal refers to numbers that define a *quantity*, such as the U.S. census.

Measurement activities offer children opportunities to explore the ordinal and cardinal aspects of number. Ordinal values are the labels assigned to such ordered items as the first, second, and third place winners in a race. Similarly, by employing unit measures, students are able to seriate objects according to some attribute, such as length.

The concepts of ordinal and cardinal numbers can be applied to the meter stick. Here, an ordinal position, say thirtieth, corresponds to thirty centimeters, a measure of length (the distance from the zeroth to the thirtieth position). It is this convenient correspondence between the ordinal position names (tenth, hundredth) and the cardinal quantity names (10, 100) that facilitates the use of scales.

Difficulty reading scales can often be traced to some violation of this principle. For example, in Figure 8.2, a child asked to label the position indicated by *X* may give the answer 5. Because no zero starting position is specified, and some intermediate positions are skipped, some children may have difficulty completing the number sequence with the information provided. Experience measuring objects with various

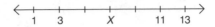

FIGURE 8.2 What Value Should Replace the *X*?

scales can help children learn to coordinate the ordinal and cardinal notions of number.

In summary, to develop **rational counting** a child should be able to:

1. Repeat the number names in the correct order.
2. Establish a one-to-one correspondence between the number names and the objects counted.
3. Understand that the last number named stands for the total number of items counted.

LEARNING THE NUMBER NAMES

When learning to count, children first copy the verbal patterns of adults and older children. Each word must eventually take on an invariant numerical meaning. Experience counting and comparing groups of physical objects helps children make the transition from an oral recitation of the counting song to a more coordinated understanding of the whole number concept.

Reviewing the class calendar during opening exercises each day affords an excellent opportunity to engage Grade K–3 students in relevant counting and numbering activities (Baratta-Lorton, 1976). Attendance can be easily recorded using an array of picture cards. On the first day of school, take a picture of each student. Mount each picture on a piece of sturdy cardboard with the child's name on the back. Display the cards with the names showing in an array on a bulletin board. Each morning have the students turn over their own cards so the picture faces out. The names left showing represent absent students.

As shown in Figure 8.3, the daily calendar can be expanded to provide additional number experiences. Children can keep track of the date, day of the week, number of school days completed (introduce a new number each day with a special celebration on each regrouping day), birthdays each month (include birthdays in July and August with September birthdays), daily weather, and lunch counts.

Learning the numbers between ten and twenty frequently causes difficulty because of a conflicting relationship between the number names and their underlying place values. In English, the words eleven and twelve have no clear linguistic relation to ten-and-one and ten-and-two. With thirteen through nineteen, a clear linguistic pattern exists based on ten (teen) and the digits (3–9). However, the place values are reversed for these number words, making it difficult for some children to interpret these values (see Table 8.1). For example, children may write 31 for thirteen.

To help avoid this reversal problem, first grade children should be encouraged to represent numbers to twenty using proportional place value materials and to

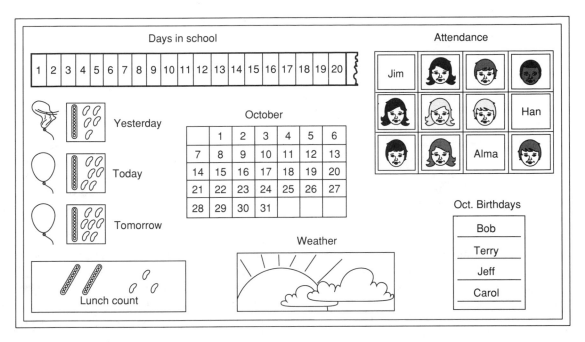

FIGURE 8.3 Bulletin Board Daily Calendar

record the numerals in place value tables labeled with the number words. This highly perceptual display provides a ready reference for coordinating the second decade of number symbols with their inconsistent linguistic equivalents.

It is interesting to note that, in most languages, the number names between ten and twenty do not parallel the base-10 place value system. It is generally thought that these words may have already been well-established in languages before the widespread acceptance of the Hindu–Arabic place value representation. Some languages, such as Spanish, however, are more consistent than English in this regard.

Other Number Readiness Concepts

In addition to rational counting, other concepts associated with the understanding of whole numbers include conservation of number, seriation, logical classification, and class inclusion.

TABLE 8.1 Reversed Place Values for Teens

Fourteen	Nineteen
1 4	1 9

FIGURE 8.4

As Piaget observed, **conservation of number** seems to be associated with the rational development of the concept of number. Conservation refers to holding invariant the number of objects in a set as they are moved about in space. For example, children frequently learn to associate a particular graphic pattern with values less than ten. A child might recognize the number five using the standard dice display but not recognize the same value when shown in a random pattern (Gelman & Gallistel, 1978). Small numbers can be displayed using various patterns to assist children in developing the understanding that the number associated with a set does not vary when the objects are rearranged. For the cardinal number of a set to be unique, the quantity of objects must remain constant regardless of the configuration of the objects or the order in which they are counted (see Figure 8.4).

The ability to order, or **seriate**, objects according to some attribute (length, weight, volume, time, number) is also important to the meaningful development of whole number. The ability to seriate helps students develop the notions of equality, inequality, one-more, and one-less.

Experience **logically classifying** objects provides needed organizing and counting opportunities for young children. Children first classify sets of objects according to color, size, shape, texture, or other physical attribute. Later, objects can be classified according to a number attribute of each object (e.g., number of wheels on each of a set of toy vehicles or chocolate chips in each cookie in a bag). Older children can classify sets of numbers according to whether each number is odd or even, prime or composite, divisible by 5 or not, and so on.

Class inclusion is a special type of classification reflecting the notion that every collection of objects is composed of a sequence of nested subsets, each set one larger than its predecessor. A set that is completely contained within another set is called an *included* set. For example, the set of six objects in Figure 8.5 is composed of six nested subsets. The first subset contains one object, the second contains two, and so on.

Piaget claimed that class inclusion embodies the coordination of the cardinal and ordinal notions of number. Class inclusion also leads to the development of the one-more and one-less concepts, and later, the operations of addition and subtraction.

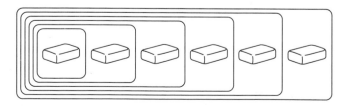

FIGURE 8.5

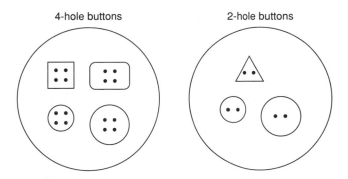

FIGURE 8.6

Introducing Classification

Grade K–3 children need ample opportunity to explore, observe, describe, measure, select, and group a rich assortment of objects. Classification experiences are fundamental to the development of a meaningful understanding of whole number. Initial experiences that involve the identification of tactile, visual, measurement, and geometric attributes are important for later use in classifying objects. Length, area, volume, weight, color, shape, geometric features, texture, and quantity attributes are introduced in kindergarten and extended throughout the primary years. Everyday objects and structured instructional materials can be employed.

Out of such classification experiences, number becomes simply another attribute for classifying objects. For example, when classifying buttons, children might form groups according to size, shape, texture, or number of holes (see Figure 8.6).

Children can engage in a range of classification experiences once they are able to identify the various attributes of objects. Three classification operations should be introduced (see Figure 8.7)

1. Union of disjoint sets—groupings with no common elements.
2. Intersection of two or more sets—groupings with common elements.
3. Included sets—groupings with all the elements of one set in common with the other.

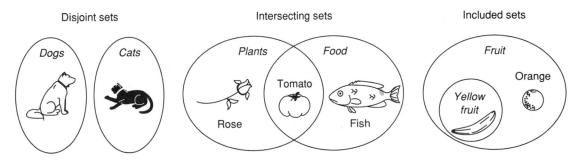

FIGURE 8.7

The combining, or **union**, of two or more disjoint sets (sets with no common elements) is a natural model for counting the total number of objects in two or more sets. Later, this procedure forms the basis for addition.

The logical connective *or* can be used to explore the union of sets. For example, in Figure 8.8a, a standard set of thirty-two attribute blocks (triangles, squares, diamonds, and circles in four colors and two sizes) is used to make a set of all the *red* things and a set of all *green* things.

The new set constructed by putting together these two groupings can be described as the set of objects that are red *or* green. Notice in Figure 8.8b that the number of elements in the new set is the *sum* of the numbers of red and green objects.

In Figure 8.8a, all the objects were in one set or the other, but not in both. Sometimes sets have members in common. For example, in Figure 8.9 the set of all teachers and the set of all parents include some teachers who are also parents.

Sets that have members in common are said to **intersect**. Activities involving intersection can be used to explore the features of the logical connective *and*. For example, using attribute blocks, all square objects can be grouped in one set and all red objects in another (see Figure 8.10).

Notice that two squares are also red. The objects placed in the intersection are at the same time square *and* red. To find the total number of elements in both sets (eight square blocks, eight red blocks) we must reduce the total (sixteen) by the number of objects counted twice (subtract two red squares), giving a total of fourteen

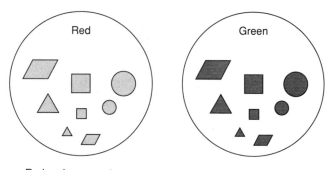

a. Red and green sets

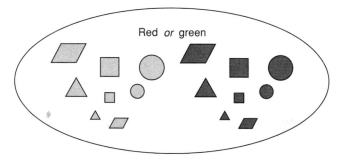

b. Union of red and green sets

FIGURE 8.8

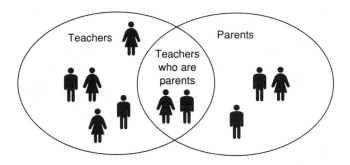

FIGURE 8.9 Intersecting Sets

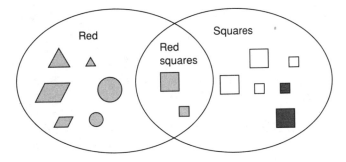

FIGURE 8.10

blocks. Two sets that have intersecting attributes cannot be used as a model for the addition operation.

The notion of intersection provides a useful transition to **class inclusion** situations where *all* items in one set are included in another set. For example, consider the set of all student teachers and the set of all adults at a school. Clearly, student teachers are included in the set of adults (see Figure 8.11).

Inclusion is the special case of intersection where the elements of one set are completely contained within another (often larger) set. As you will see in chapter 9, class inclusion holds special importance in the development of addition and subtraction.

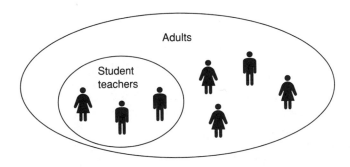

FIGURE 8.11

> Using a standard set of thirty-two attribute blocks, select attributes that show the union of two disjoint sets, the intersection of two sets, and set inclusion. Record the results by drawing a sketch of each grouping and writing a sentence using the logical connectives *or* and *and*.

Equivalence Relation

The everyday notion of equivalence involves the idea that two or more objects, sets, or ideas are alike in some way. In mathematics, an equivalence relation takes on a more specific meaning. The most commonly used equivalence relationship is the equals (=) relation. Three conditions must be true for a mathematical relationship to be called an equivalence relation:

1. *Reflexive Property*—an object must be equivalent to itself ($2 = 2$).
2. *Symmetric Property*—if an object is equivalent to a second object, the reverse must also be true (if $3 + 1 = 4$, then $4 = 3 + 1$).
3. *Transitive Property*—if one object is equivalent to a second object, and the second is equivalent to a third, then the first object must be equivalent to the third object (if $2 + 2 = 3 + 1$, and $3 + 1 = 4$, then $2 + 2 = 4$)

> Some relationships do not meet all three of the above conditions and are therefore not equivalence relations. For example, *is the brother of* meets only condition 3. Why doesn't *is the brother of* always meet condition 2? The relation *one more than* does not meet any of the three conditions. Is the relation *is the same weight as* an equivalence relation? Think of five other relationships and test them to see if they meet the conditions necessary to be equivalence relations.

Greater Than and Less Than

Greater than and *less than* relations do not satisfy the reflexive property, because a quantity cannot be larger (or smaller) than itself. They also do not satisfy the symmetric property—if a is larger than b, then b cannot be larger than a. The transitive property is satisfied—if $a > b$ and $b > c$ then $a > c$. The greater than and less than relations are therefore not equivalence relations.

Comparing the size of objects and numbers begins early as children learn to order the size of toys, pencils, and books. Most children can easily order two physical objects according to some perceptually salient attribute (length, weight, size,

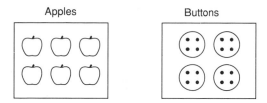

FIGURE 8.12 Are There More Apples or Buttons?

FIGURE 8.13

number). Seriating more than two objects is generally more difficult. Grade 1–3 children often have difficulty using the symbols for greater than ($>$) and less than ($<$). The problem may be related to a more general symbol reversal problem experienced by many children when first learning the alphabet and numerals (e.g., reversing the letters *d*, *b* and *p*, *q*, or the numerals 3, 5). To minimize this problem, avoid using the greater than and less than symbols ($<$, $>$) and focus on simply naming the larger or smaller sets of concrete objects. Shown the sets in Figure 8.12, children can circle the picture of the apples to show there are more apples than buttons.

Later, a hungry crocodile character can be introduced as a metaphor for inequalities. A hungry crocodile always eats the larger set. Notice his mouth forms the shapes of the standard inequality symbols (see Figure 8.13).

One More and One Less

Another representation of inequality involves the idea of **one more** and **one less**. These relations satisfy none of the three properties of equivalence relations. The process of constructing the counting numbers from a nested set of objects, each set one more than its predecessor, is fundamental to the development of a well-coordinated understanding of whole number. Children should have ample experience building and comparing stacks of blocks; comparing measures of water, rice, and other materials; choosing teams for classwork and sports; distributing lunches, milk, coats, hats, and mittens; and keeping track of the number of class fish, hamsters, or other classroom animals. All these experiences provide opportunities to work with sets using the one more and one less relations (see Figure 8.14).

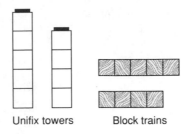

Unifix towers   Block trains

FIGURE 8.14

Number Patterns

Inventing, discovering, and describing patterns is an important component in the development of classification, number, and problem-solving skills. Patterns can be based on geometric attributes (shapes, regions, and angles), measurement attributes (color, texture, length, weight, volume, number), relational attributes (proportion, sequence, function), and affective attributes (values, likes, dislikes, familiarity, heritage, culture). Young children may line up toys in the order in which they want to play with them. Picture books may be stacked in the preferential order for reading. Structured blocks and construction materials provide opportunities for stacking and fitting objects together according to size or shape.

School work should build on these early experiences with patterns by expanding on and refining the attributes available for discriminating between, forming classes of, and ordering objects. Unifix Cubes (multicolored interlocking blocks) are an excellent medium for constructing patterns based on number and color. For example, children can construct trains of blocks that follow patterns they invent. They can then display their patterns, and class members can suggest the next car in the trains (see Figure 8.15a). Later, number patterns can be recorded using Unifix trains, as shown in Figure 8.15b.

Number sequences that are based on a Unifix block pattern can also be developed. For example, the sequence in Figure 8.15c is developed by adding two blocks to each preceding block element.

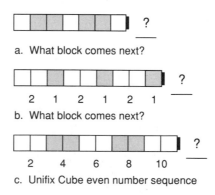

a. What block comes next?

2   1   2   1   2   1

b. What block comes next?

2   4   6   8   10

c. Unifix Cube even number sequence

FIGURE 8.15

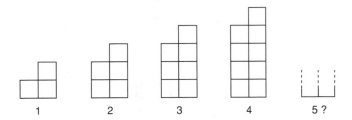

FIGURE 8.16 Block Sequence $(2x + 1)$

The three dots (...) at the end of each sequence are called an **ellipsis** and mean that the pattern continues indefinitely. Construct block patterns that show the following sequences:

1. 1, 3, 5, 7, 9, ...
2. 3, 6, 9, 12, 15, ...
3. 4, 7, 10, 13, 16, ...

Older students can construct more complicated number sequences using subtraction and multiplication. As shown in Figure 8.16, working out the rules used to generate number sequences is a useful way to introduce the function concept (see chapter 11).

HISTORY OF OUR NUMERATION SYSTEM

To broaden our discussion of how children develop a meaningful understanding of whole number, it is instructive to explore how our present numeration system came into being. As soon as society developed the ability to create a surplus of commodities, it became necessary to invent ways to keep track of them. **Tally systems** were probably the first means of monitoring the amount of valuable items. Shepherds in ancient times placed a pebble in a shallow hole in the evening for each sheep in the herd. The following morning, one pebble was removed for each sheep counted. Once all the sheep were counted, any pebbles left in the hole indicated the number of sheep lost to wolves during the night. Tally systems record numbers by assigning one counter to represent each object counted.

The Kilinge people in Papua New Guinea traditionally use a tally system when planning intervillage feasts. They prepare two identical bundles of sticks containing one twig for each day until the scheduled celebration. The village leaders agree to remove one stick each day. When the sticks are gone, it is time for the feast.

Like the handle of the successful gunslinger's revolver, tally systems lose their attractiveness as values get large. In order to record larger quantities accurately, **additive systems** were invented. These systems differ from tally systems in that specific symbols represent more than one object. For example, the ancient Egyptian

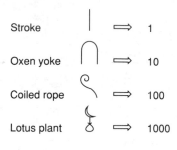

Stroke	$\Rightarrow$	1
Oxen yoke	$\Rightarrow$	10
Coiled rope	$\Rightarrow$	100
Lotus plant	$\Rightarrow$	1000

FIGURE 8.17 Egyptian Numerals

system uses a ten-to-one exchange rate. A stroke (|) represents one unit. When the count reaches ten (|||||||||), the set is exchanged for one oxen yoke (∩). Other symbols represent hundreds (coil of rope—◟), thousands (lotus plant—⚱), and larger place values (see Figure 8.17).

The total value represented is the sum of the symbols. The symbols can be placed in any order in additive systems (see Figure 8.18). There is no convenient way to represent zero.

Write the standard Hindu–Arabic values for the following ancient Egyptian numerals.

1. ∩∩∩||| → ?

2. ∩∩∩∩∩||| → ?

3. ◟◟◟∩∩||| → ?

4. ⚱◟◟|||| → ?

5. Write your current age in Egyptian numerals.

Roman numerals also constitute an additive system with two modifications. To reduce the number of symbols required to represent large values, intermediate exchange units for five, fifty, and five hundred were introduced (see Figure 8.19).

◟◟|||| = 24

||◟◟|| = 24

◟||||◟ = 24

FIGURE 8.18 Egyptian Non–Place Value System

$$I = 1$$
$$V = 5$$
$$X = 10$$
$$L = 50$$
$$C = 100$$
$$D = 500$$
$$M = 1000$$

FIGURE 8.19 Roman
Numeration System

A **subtraction rule** was also incorporated whereby symbols of smaller value were subtracted from a larger value positioned immediately to the right.

The benefits of the Roman system are apparent when you compare the number of symbols required to represent numbers such as 48 or 469 with the Egyptian system (see Figure 8.20).

Write the Hindu–Arabic values for the following Roman numerals.

1. XVII $\rightarrow$?

2. LIX $\rightarrow$?

3. CLXVI $\rightarrow$?

4. CMXXIV $\rightarrow$?

5. Write the current year in Roman numerals.

Tally and additive systems became inconvenient as merchants were required to keep track of larger and larger values. Complicated procedures were invented for carrying out basic computation operations.

For example, Figure 8.21 shows how ancient Egyptians multiplied large numbers by using powers of 10 and successive doubling, two operations easily accomplished using the Egyptian system. To compute 34 $\times$ 23, the largest factor (34) is first multiplied by 10, then successively doubled until the count column reaches the tens value of the second factor (in this case 20 in the factor 23). Egyptians multiplied by 10 by replacing each symbol with the next larger numeral (e.g., the one is replaced with a ten). Next, the largest factor (34) is multiplied by the next smaller power of 10 (in this case 1) and successively doubled until reaching the ones value of the second

Hindu–Arabic	Egyptian	Roman
48		XLVIII

FIGURE 8.20

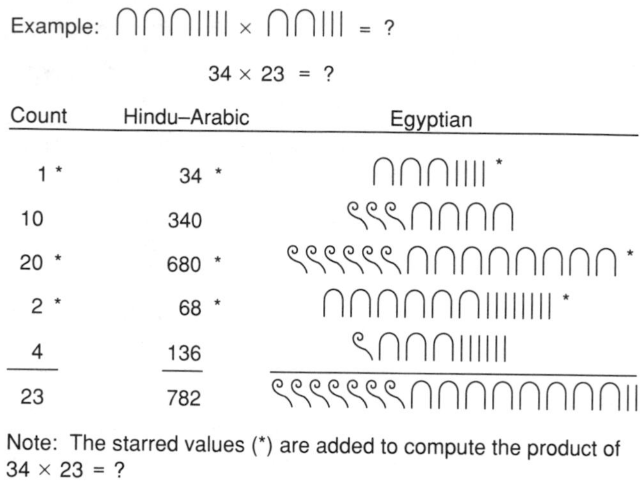

Example: ∩∩∩|||| × ∩∩||| = ?

34 × 23 = ?

Count	Hindu–Arabic	Egyptian
1 *	34 *	∩∩∩|||| *
10	340	ᓚᓚᓚ∩∩∩∩
20 *	680 *	ᓚᓚᓚᓚᓚ∩∩∩∩∩∩∩∩ *
2 *	68 *	∩∩∩∩∩∩|||||||| *
4	136	ᓚ∩∩∩||||||
23	782	ᓚᓚᓚᓚᓚᓚ∩∩∩∩∩∩∩∩∩||

Note: The starred values (*) are added to compute the product of 34 × 23 = ?

FIGURE 8.21 Egyptian Multiplication

factor (in this case 3 in the factor 23). The rows in the count column that add up to 23 are selected (1 + 20 + 2). Adding the starred (*) rows gives the product of (34 × 23).

Though Hindu–Arabic numerals are used in Figure 8.21 to make it easier to follow the example, it is possible to complete the exercise using only Egyptian symbols just as the Egyptians did thousands of years ago. The Romans adapted this system for their use, and in some parts of Eastern Europe it is still used today (with Hindu–Arabic numerals, of course).

To simplify the representation of large numbers and facilitate computation, **place value systems** were invented by several cultures around the world. The Chinese developed symbols for the numbers one through nine and different symbols for each power of 10. For example, the number 268 is represented either vertically or horizontally by writing each digit followed by the corresponding power of ten symbol (see Figure 8.22). There is no symbol for zero.

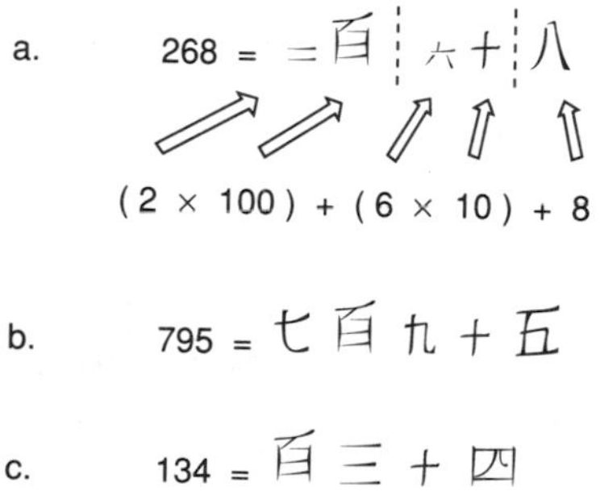

a.　　268 = 二百 六十 八

(2 × 100) + (6 × 10) + 8

b.　　795 = 七百九十五

c.　　134 = 百三十四

FIGURE 8.22 Chinese Numerals

> Write the counting numbers from 1 to 50 using Chinese symbols.

The Babylonians developed a **cuneiform** place value system based on powers of 60 with intermediate multiples of ten. Numbers were written by pressing vertical and horizontal wedge-shaped impressions into clay tablets. The vertical impressions represented the powers of 60, and the horizontal impressions each of the place values multiplied by ten (see Figure 8.23).

Again, the early Babylonian system had no way to represent zero. Without a zero place holder, a symbol can be interpreted as having several values, as shown in Figure 8.24.

The **zero** place holder was introduced by the Hindus, who lived in what is now called India, and was subsequently adapted by Arab traders. The system proved to be so useful that, by 1500 A.D., it had been adopted by people throughout most of Europe and Asia. If the exchange described in *A for Effort, Zero for Arab* (pp. 248–249) did not occur, it should have.

It is interesting to note that, at about the same time as the Hindu developments in numeration (600–800 A.D.), the Mayan culture in Central America independently developed a place value system for calendar computations. It had a modified base-20 structure and included a zero place holder pictured as a snail shell. Without the shell numeral, the system is simply additive (e.g., 15 + 4 = 19 in Figure 8.25). Numbers are written vertically with the symbols for each place value separated by a space. For example, when a shell numeral is included, the symbols in the place value above the shell are multiplied by 20 (the base). The next higher place value is designated as 18 × 20 = 360, the number of days in the Mayan year. Larger place values are determined by multiplying each subsequent place value by 20 (20 × 360 = 7200; 20 × 7200 = 14,400, and so on). As shown in Figure 8.25, the Hindu–Arabic value for a Mayan number can be calculated by adding the values of each place value.

36000	3600	600	60	10	1
◁	Y	◁	Y	◁	Y

FIGURE 8.23 Babylonian Cuneiform

◁   ◁   ◁   Y   Y = (3 × 10) + (2 × 1) = 32
or = (3 × 600) + (2 × 1) = 1802
or = (3 × 600) + (2 × 60) = 1920
or many other combinations

FIGURE 8.24 Multiple Cuneiform Values

A for Effort, Zero for Arab

Many years ago a Roman civil engineer, who was a high official in Alexandria, was approached by a young Arabian mathematician with an idea that the Easterner believed would be of much value to the Roman government in its road-building, navigating, tax-collecting, and census-taking activities. As the Arab explained in his manuscript, he had discovered a new type of notation for number writing that was inspired by some Hindu inscriptions. The Roman official presumably studied this manuscript very carefully for several hours, then wrote his reply:

Your courier brought your proposal at a time when my duties were light, so fortunately I have had the opportunity to study it carefully and am glad to be able to submit these detailed comments.

Your new notation may have a number of merits, as you claim, but it is doubtful whether it ever would be of any practical value to the Roman Empire. Even if authorized by the Emperor himself, as a proposal of this magnitude would have to be, it would be vigorously opposed by the populace, principally because those who had to use it would not sympathize with your radical ideas. Our scribes complain loudly that they have too many letters in the Roman Alphabet as it is, and now you propose these ten additional symbols of your number system, namely:

1, 2, 3, 4, 5, 6, 7, 8, 9, and your 0.

It is clear that your mark **1** has the same meaning as our mark **I**, but since our **I** is already a well-established character, why is there any need for yours?

Then you explain that last circle mark, like our letter **0**, as representing an empty column, or meaning nothing. If it means nothing, what is the purpose of writing it? I cannot see that it is serving any useful purpose; but to make sure, I asked my assistant to read this section, and he drew the same conclusion.

You say the number **01** means the same as just **1**. This is an intolerable ambiguity and could not be permitted in any Roman legal documents. Your notation has other ambiguities that seem even worse: You explain that the mark **1** means ONE, yet on the very same page you show it to mean TEN in **10**, and one HUNDRED in your **100**. If my official duties had not been light while reading this, I would have stopped here; you must realize that examiners will not pay much attention to material containing such obvious errors.

Further on, you claim that your system of enumeration is much simpler than the Roman Numerals. I regret to advise that I have examined this point very carefully and must conclude otherwise. For example, counting up to FIVE, you require five new symbols, whereas we Romans accomplish this with just two old ones, the mark **I** and the mark **V**. At first sight the combination of **IV** (meaning ONE before FIVE) for four may seem less direct than the old **IIII**, but note that this alert representation involves LESS EFFORT, and that gain is the conquering principle of the Empire.

Counting up to twenty (the typical counting range among the populace), you require ten symbols whereas we now need but three: the **I**, **V**, and **X**. Note particularly the pictorial suggestiveness of the **V** as half of the **X**. Moreover, it is pictorially evident that **XX** means ten-and-ten, and this seems much preferred over your **20**. These pictorial associations are very important to the lower classes.

You claim that your numbers as a whole are briefer than the Roman Numerals, but this is not made evident in your proofs. Even if true, it is doubtful that this would mean much to the welfare of the Empire, since numbers account for only a small fraction of the written records; and in any case, there are plenty of slaves with plenty of time to do this work.

When you attempt to show that you can manipulate these numbers much more readily than Roman Numerals, your explanations are particularly bad and obscure. For example, you show in one addition that **2** and **3** equals **5**. Yet you seem to get a different result in the case you write as:

$$
\begin{array}{r}
79 \\
+\,16 \\
\hline
95
\end{array}
$$

This indicates that **9** and **6** also equals **5**. How can this be? While it is not clear, it is evident that the other part is in error, for **7** and **1** equals **8**, not **9**.

Your so-called *repeating and dividing* tables also require much more explanation, and possibly correction of errors. I can see that your *Nine Times Table* gives sets that add up to nine, namely:

18, 27, 36, 45, 54, 63, 72, 81, and 90

But I see no such useful correlation in the *Seven Times Table*. Since we have SEVEN, not nine days in the Roman Week, it seems far more important that we have a system that gives more sensible combinations in this table.

All in all, I would advise you to forget this overly ambitious proposal, return to your sand piles, and leave the number reckoning to the official Census Takers and Tax Collectors. I am sure that they give these matters a great deal more thought than you or I can.

Author anonymous

Place value and the zero place holder are the key characteristics of the Hindu–Arabic system of numeration that enable us to represent and carry out operations with large numbers conveniently. As it has been throughout history, understanding place value is fundamental for children learning to compute with large values. The addition, subtraction, multiplication, and division algorithms are all based on the systematic regrouping of amounts represented in place value columns.

PLACE VALUE

It is possible to choose any arbitrary value to name the columns in a place value system. For example, Grade 1 children can be introduced to place value using beans and small paper cups. Give each child eight beans. To group by five, have the children count the beans one at a time, and when the count reaches five, put the five beans in a cup. After counting all eight beans, the answer is recorded as *one cup and three beans*. A group of thirteen beans gives the result, *two cups and three beans*. Have children practice with sets of up to twenty-five beans (see Figure 8.26). Later, children can group by 3, group by 8, and finally, group by 10.

It is convenient to name the place value columns using successive powers of

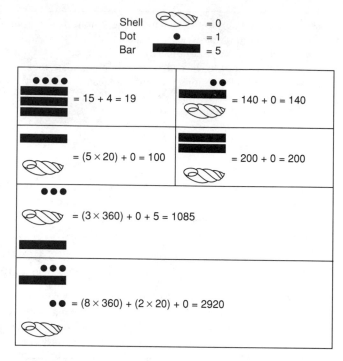

FIGURE 8.25 Mayan Numbers

a small number so that a consistent set of counting digits can be employed. For example, Table 8.2 shows the set of place values if we choose to group by powers of 2 (base 2).

For the expression 2^3, 2 is called the **base** and 3 is called the **exponent**. To compute 2^3, we multiply $(2 \times 2 \times 2 = 8)$. So, $2^3 = 8$.

To introduce primary children to the idea of place value, display on an overhead projector a pile of marbles, as shown in Figure 8.27a. Circle the largest possible group of marbles based on the base-2 place values (sixteen in this case). Of the remaining marbles, circle the next largest place value grouping.

In Figure 8.27b, there are fewer than eight marbles remaining, so continue to make groupings based on the smaller place values (in this case, four and two) until

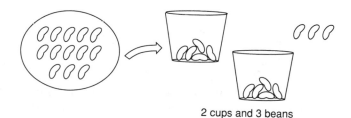

2 cups and 3 beans

FIGURE 8.26 Grouping by Five

TABLE 8.2   Powers of 2 Place Values

2^5	2^4	2^3	2^2	2^1	2^0
32	16	8	4	2	1

all the marbles are circled. A record of the number of marbles can be written in a base-2 place value table (see Figure 8.27c).

The number of marbles is $10110_{\text{base-2}}$ when represented by base-2 groupings. Notice that zero place holders are inserted in the first and fourth columns from right

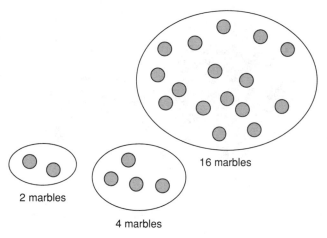

a. Twenty-two marbles

b. Marbles grouped by powers of 2

64	32	16	8	4	2	1
		1		1	1	

c. Base-2 place value table

FIGURE 8.27

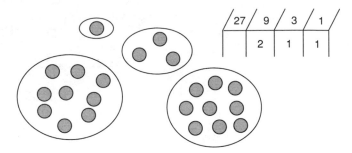

a. Base-3 place value table

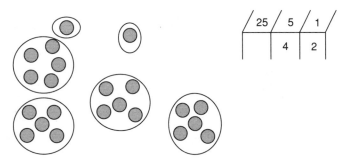

b. Base-5 place value table

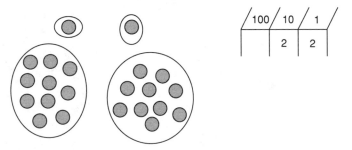

c. Base-10 place value table

FIGURE 8.28

to left in order to ensure that the three 1 digits are interpreted correctly when written outside the place value grid.

We can change the groupings by introducing a new base. The place value grids in Figure 8.28 show how the same set of twenty-two marbles would be represented using base-3, base-5, and base-10 groupings.

How old are you in base-2, base-3, base-5, and base-10? Does your age change depending on the base in which it is written?

FIGURE 8.29 Ten-Ones Regrouped to One-Ten

Regrouping

In order to perform computations on large numbers, systematic procedures called **algorithms** have been developed. These algorithms are based on the notion of place value **regrouping**. When counting objects, unique value names are available for amounts zero through nine. To represent ten objects, however, we group the ten units together (regroup them) and rename them *1-ten* (see Figure 8.29). Similarly, when the count reaches one hundred, 10-tens are exchanged for 1-hundred.

Several types of commercial and student-constructed materials are available to represent this regrouping process. The most perceptually salient materials are those that show the proportional number base relationship among place values. While commercial base-10 blocks are convenient, student-constructed **bean sticks** offer added benefits, in that the users personally construct the materials and are therefore familiar with the proportional relationships among the place values (see Figure 8.30).

Other teacher-made place value materials shown in Figure 8.31 include **sticks and frames, clothespin hangers,** and the **pocket chart**. A set of place value frames can be constructed from two 10 cm x 10 cm squares of 1 cm thick corrugated cardboard and a supply of plastic coffee stirrers. Using a nail, punch nine holes in one square just large enough to hold one coffee stirrer. Using a leather punch, make nine holes in a second piece of cardboard, each just large enough to hold ten coffee stirrers. This model can be used to show numbers to ninety-nine. A place value clothespin hanger is constructed from a wooden clothes hanger and a supply of clothespins (two colors). Twist the hook perpendicular to the hanger so it can be supported on the edge of a table or desk. Let one color of clothespin represent

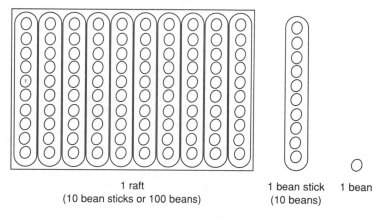

1 raft
(10 bean sticks or 100 beans)

1 bean stick
(10 beans)

1 bean

FIGURE 8.30 Bean Sticks

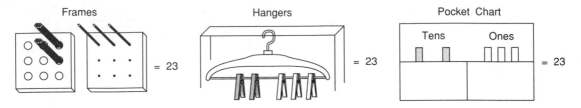

FIGURE 8.31 Place Value Materials

one unit and another 1-ten. The unit pins are placed on the right side of the hook and the ten-pins on the left to represent values to ninety-nine. A place value pocket chart can be easily constructed from tagboard for demonstrations of regrouping or for individual use. Strips of construction paper to represent ones and tens, individual and bundled coffee stirrers, clothespins, bean sticks, or other place value materials can be placed in the pockets to display values and regrouping.

Commercial base-10 blocks are available that readily model place values to the thousands place (see Figure 8.32).

Nonproportional materials are introduced *after* children have had successful experiences with the perceptually salient materials. The *abacus* provides a place value model where each column represents a successive power of 10. Notice, however, that a bead in the tens column looks just like a ones-bead. The value of each bead is indicated by its position, hence the label *place value* (see Figure 8.33a). Using nonproportional materials requires the child to remember the underlying number base relationship.

Chip trading activities offer an abacus-type representation for place value. Each place value is represented by plastic chips of a different color. Chip trading activities help children make the transition from the place value representation using proportional materials to the standard symbolic notation (see Figure 8.33b).

With young children, begin regrouping using a three-to-one exchange rate (base-3). The smaller exchange value affords easier counting and more frequent regrouping. A simple *Bankers Game* involves rolling a die and depositing the amount indicated in a cash till. This activity provides regrouping practice in a motivating game

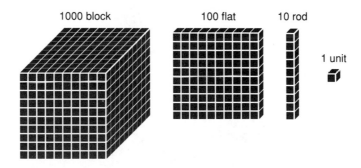

FIGURE 8.32 Commercial Base-10 Blocks

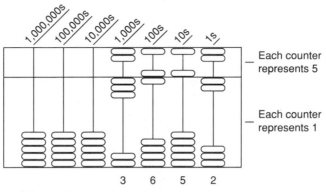

a. Chinese abacus

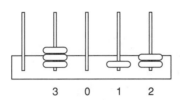

b. Chip trading abacus

FIGURE 8.33

environment. After students have had experience with a three-to-one exchange rate, introduce a five-to-one, then a 10-to-one rate of exchange (see Figure 8.34). Such experiences using nonproportional materials can be a useful extension of place value work with bean sticks and base-10 blocks. Nonproportional chip trading materials and the abacus can also be used by older students experiencing difficulty making the transition between proportional and symbolic representations of regrouping.

Concurrent with activities involving proportional (base-10 blocks) and transitional (chip trading) materials, students can record the results of concrete exercises using symbolic place value tables drawn on centimeter squared paper (see Figure 8.35). *Note:* It is good practice to include a small *s* after each shaded place value label (i.e., 100s, 10s, 1s) to minimize the possibility of children adding these labels to addends in exercises.

○○○ = ●   ○○○○○○○○○○ = ●

3-to-1 exchange   10-to-1 exchange

3 yellow = 1 blue   10 yellow = 1 blue
3 blue = 1 green   10 blue = 1 green
3 green = 1 red   10 green = 1 red
3 red = 1 black   10 red = 1 black

FIGURE 8.34 Bankers Game

1000s	100s	10s	1s	
		2	1	Exercise 1
	1	4	3	Exercise 2
				Exercise 3

FIGURE 8.35 Squared-Paper Place Value Table

Bean sticks, chip trading, and squared-paper place value tables can be employed to provide dynamic instructional support as discussed in chapters 2 and 3. For example, the number base relationship of the place value system is more compellingly displayed using proportional materials than when using chip trading or standard symbols in place value tables. Comparing the relative size of numbers and modeling regrouping actions during addition and subtraction is facilitated by the use of perceptually salient base-10 materials. Gradually introducing abacus-type chip trading and symbolic representation requires the student to internalize the underlying place value relationship involved in representing numbers.

RATIONAL NUMBERS

Introducing Common Fractions

The notion of simple fractional parts should be well-established prior to the introduction of decimal fractions. Simple fractions are used frequently by children when opportunities for equal sharing arise. Every mother learns that, in order to reduce sibling rivalry when two children share a cookie, one child divides it in half and the other chooses first. If this happens without a major disagreement, the children are well on their way to discovering the concept of *one-half*.

The idea of fractional part can be represented in two ways:

☐ Unit partitioning
☐ Set partitioning

As shown in Figure 8.36a, pies and rectangles can serve as concrete models for representing the **unit partitioning** concept of fractional **partitioning of a whole**.

Figure 8.36b shows how equal sharing of groups of objects (e.g., peanuts, popcorn, counters) can be used to introduce the idea of fractional **partitioning of a set**.

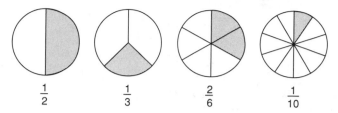

$$\frac{1}{2} \qquad \frac{1}{3} \qquad \frac{2}{6} \qquad \frac{1}{10}$$

a. Fractional parts of a whole

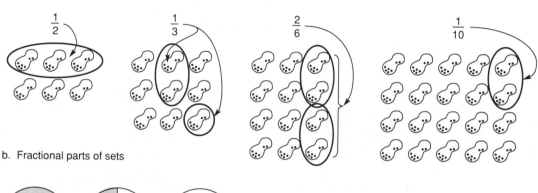

b. Fractional parts of sets

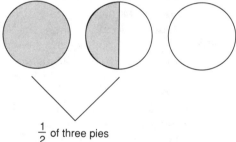

$\frac{1}{2}$ of three pies

c. Unit and set partitioning

FIGURE 8.36

Coordinating both fractional part models (partitioning a whole unit and partitioning a set) is necessary when the total number of objects in a set cannot be evenly divided (see Figure 8.36c).

Another meaning for the fraction $\frac{a}{b}$ is $a \div b$. This interpretation provides an easy method for calculating the decimal equivalent of common fractions (e.g., $\frac{1}{2} = 1 \div 2 = 0.5$).

Introducing Decimal Fractions

Physical models and procedures similar to those used for whole numbers can be used to introduce **decimal fractions** and the **decimal point**. Base-10 blocks can

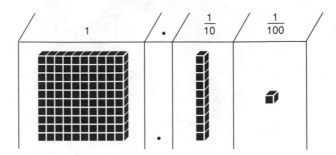

FIGURE 8.37 Decimal Representation Using Base-10 Materials

be used as a proportional display of decimal place values by assigning to the flat block the value of one unit (see Figure 8.37). Tenths and hundredths are therefore represented by the rod (0.1) and small cube (0.01).

Money can also be used at all grade levels to represent decimal fractions. Children can play chip trading exchange games using pennies, dimes, and dollars until the place value relationship among the coins is well-understood. Avoid confusion by not using nickels, quarters, and fifty-cent pieces during these experiences. Money is another example of a nonproportional place value model like chip trading and the abacus. Initially, children use the penny as the unit. Small amounts of money can be conveniently represented in this manner. When the amount approaches a hundred pennies, it becomes more convenient to use the dollar as the unit, with dimes and pennies represented as fractional parts of the unit dollar (a penny = 0.01, and a dime = 0.10).

The decimal point is simply a standard convention used to separate the whole number place values from the fractional place values (see Figure 8.38).

Different symbols are used for the decimal point in countries around the world. Many countries in Europe use a comma ($342.84 = $342,84) while New Zealand uses a raised dot ($342.84 = $342·84). The SI metric system also specifies the raised dot as the preferred decimal point indicator to avoid confusion in international trade. With the increased use of calculators, computers, and the metric system, the convenience of using decimal representation over common fractions is becoming more apparent.

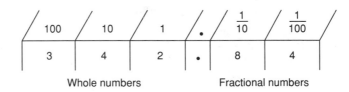

FIGURE 8.38

Rounding

Developing a sense of the relative magnitude of numbers (number sense) is an important problem-solving skill. It becomes especially important when working with numbers greater than 1000 and less than 0.1. One way to introduce estimation is through simple **rounding** exercises.

The easiest way to gauge the relative magnitude of a number is to focus on the leading digit (mentally set all the remaining digits to zero). For example, the number 4368 could be thought of as about 4000. Using this procedure, the number 4986 would *also* be about 4000. This **rounding down** procedure can also be used with numbers containing decimal fractions. We commonly write the rounded value with the same number of decimal places as the original number (see Table 8.3).

Children can practice rounding down by mentally computing grocery bills using **front end estimation**. They first add the dollar values only (round down to the nearest dollar) and adjust the sum if necessary by combining the cents into approximately one dollar amounts. For example:

	Register Tape		*Front-End Estimate*	
Meat	3.56			
Carrots	0.87			
Apples	1.22	Rounded-down total—	$7.00	
Batteries	2.57	Adjustment for cents—	+ $3.00	
Milk	0.89	Total Estimate—	$10.00	
Cheese	1.25			

Prepare overhead slides of several register tapes using appropriately sized values (three to five items of $0.01 − $1.99 for primary level; five to ten items under $5.00 for Grades 4–6). Have the class mentally calculate the sums using front-end estimation and record their answers. Have children with different answers explain their procedures. Discuss which estimates are reasonable and which might be misleading. For example, when checking to see if you have enough money to buy the groceries, estimates may be sufficient. However, to calculate the sales tax due, it is necessary to calculate the exact total of the nonfood items.

In some cases, a more accurate approximation could be derived if we **rounded up** to the next highest value of the leading digit (the third item in Table 8.3 is rounded up to 3,000,000.000). Of course, if we consistently rounded up, other values would be correspondingly inflated (the first item in Table 8.3 is rounded up to 400).

TABLE 8.3 Rounding Down

327.24	≈	300.00
0.421	≈	0.400
2,989,046.009	≈	2,000,000.000

TABLE 8.4   Rounding Off

Original Value		Rounded Value	
2,682	≈	3,000	round up
624,600	≈	600,000	round down
5,555.55	≈	6,000.00	round up
4,300	≈	4,000	round down

Customarily, when the key digit (the one following the leading digit) is five or more, the number is rounded up. Otherwise the value is rounded down. We call this process **rounding off**. In the examples shown in Table 8.4, the key digit is underscored.

A number line can be used to help children visualize that rounding off the number 628 to the nearest hundred involves determining if it is *closer* to 600 or 700. One way to gauge quickly which is closer is to compare its position to 650, the value halfway between 600 and 700. Since 628 is to the left of 650 on the number line, it is closest to 600 (see Figure 8.39).

Each of these rounding procedures finds applications in everyday problems. A short-cut method to calculate the approximate cost of a number of items is to round down to the leading digit. The sum will always be lower than the actual amount but this method gives a quick, ball-park estimate. Carpenters round up their measures in order to ensure each board will be long enough. It is less wasteful to cut a bit off a board than to replace it because it is too short. When computing sales tax, registers are set to round off to the nearest cent. When computing interest, banks carry out all computations to five or more decimal places for greater accuracy before rounding off.

It is sometimes necessary to round to a place value other than the leading digit. For example, if you wish to purchase a new guitar that costs $127.83 (including tax), you might go to the bank and withdraw enough money for the purchase, rounded up to the nearest $10 ($130.00). It is possible to round up, round down, or round off to the any place value represented in a number. Each value in Table 8.5 is rounded to the value shown in bold.

Students should have experience working with all three rounding procedures. Initially, rounding down offers a simple procedure to help students develop a sense of the relative magnitudes of numbers. Assessing the appropriateness of answers to long division exercises and other problems depends on a well-developed sense of the magnitude of numbers. Rounding procedures can also be employed to improve measurement estimation skills. Rounding up and rounding off have many useful applications when measuring length, time, and money.

FIGURE 8.39   Rounding Off on the Number Line

TABLE 8.5

Place Value			Rounding Down	Rounding Off	Rounding Up
Nearest 10	4367	≈	4360	4370	4370
Nearest 1000	47,256	≈	47,000	47,000	48,000
Nearest 0.01	24.5384	≈	24.5300	24.5400	24.5400

Round the following numbers to the correct amount for bold place value using each of the three rounding procedures.

		Round down	Round off	Round up
1.	34,6**5**1	?	?	?
2.	127.4**4**	?	?	?
3.	9**4**4,500	?	?	?
4.	**9**50,345	?	?	?
5.	0.00**4**3	?	?	?

CLASSROOM ACTIVITIES

Anchor Numbers

To help Grade K–1 children develop the ability to estimate and compare numbers readily, introduce the idea of benchmark, or **anchor**, numbers. Most children naturally develop the ability to recognize the cardinal value of sets containing 1, 2, 3, 5, and 10 objects. Children should be encouraged to identify one or more real-world anchor situations to which new sets of objects can be compared (1 → me; 2 → eyes; 3 → triangle; 5 → toes on one foot; 10 → fingers on two hands). Anchors for other small values should be encouraged (4 → tires on a car).

The next perceptual anchor number is twenty. To help children immediately recognize sets containing about twenty objects, Carlow (1986) suggests presenting pictures of several sets of twenty objects displayed in different configurations and densities and asking children to compare target sets to these anchor displays (see Figure 8.40). As children become more skillful, larger anchor numbers can be introduced (e.g., 50 and 100).

Bean Sticks

Activities involving the construction of multidigit numbers using proportional materials are important for the development of place value. K–3 children enjoy constructing bean sticks and using them to display numbers. The materials needed are

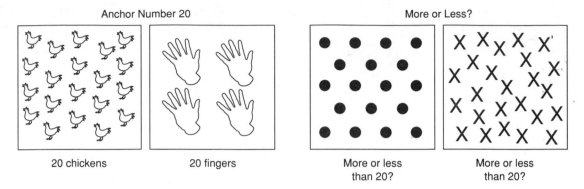

Anchor Number 20   More or Less?

20 chickens   20 fingers   More or less than 20?   More or less than 20?

FIGURE 8.40 Anchor Numbers

tongue depressors (Popsicle sticks), pinto beans (split peas), and white glue. Each child makes a set of fifteen bean sticks by gluing exactly ten beans onto each stick. **Hundred-rafts** can be constructed by gluing two extra Popsicle sticks across the backs of ten additional bean sticks.

Children can use bean sticks to display the month and day, daily attendance, cumulative days in school, number of students buying lunch, and so on. Once a value has been displayed using bean sticks, the numerals can be recorded in place value tables (see Figure 8.41).

Children can use their personal set of bean sticks to help them count, compare, and regroup numbers. Later, these materials are used to introduce whole number operations.

Using the Abacus

An open-end abacus can easily be constructed by Grade 3–6 children using wooden or plastic counters, a small block of wood, and long finishing nails. Holes should be drilled in each counter so they will stack on the nails in the block.

Each nail represents a place value and should be just tall enough to hold nine counters. This encourages regrouping when ten counters are collected in any place value. Separate colors can be used for each place value to further distinguish between place values when regrouping (see Figure 8.33b).

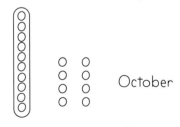

October

FIGURE 8.41 18 October

1000	100	10	1

Banker's Game

a. Squared-paper place value table

1000	100	10	1
	2	0	7

10	1	$\frac{1}{10}$	$\frac{1}{100}$
1	4 •	0	3

b. Zero place holder in place value table

FIGURE 8.42

Squared-Paper Place Value Tables

A useful device to help children accurately record the results of their concrete counting and numbering experiences is squared-paper place value tables. Construct a page of place value tables on squared centimeter (1 cm^2) graph paper. Reproduce enough of these record sheets to accompany counting and numbering exercises (see Figure 8.42a).

The tables help children keep the place values organized when recording results of activities such as the Bankers Game. It is especially helpful when an interior place value contains a zero. Decimal fractions are similarly represented (see Figure 8.42b).

Geoboard Fractions

Make an activity sheet for each child like the one shown in Figure 8.43 and an overhead transparency. Give each child a geoboard and a supply of rubber bands. Children must imagine how to divide each figure into the correct number of fractional parts and shade the fraction indicated. Have them use the geoboard to display each figure and the parts as necessary (see Figure 8.43). After completing the activity, children can invent similar problems for their peers to solve. Many other geoboard fraction activities have been developed for classroom use (Zawojewski, 1986).

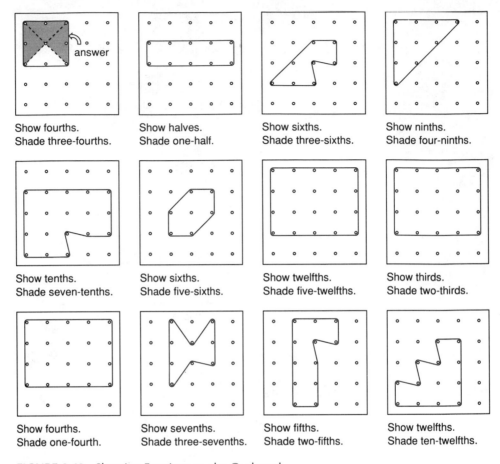

FIGURE 8.43 Showing Fractions on the Geoboard

Fraction Table

To provide practice comparing fractions, construct the fraction table in Figure 8.44. The strips show a number line scale of 0–2; each is divided into one of the following fractional parts: $1, \frac{1}{2}, \frac{1}{3}, \frac{1}{4}, \frac{1}{5}, \frac{1}{6}, \frac{1}{8}, \frac{1}{9}, \frac{1}{10}, \frac{1}{12}$.

To compare the fractions $\frac{2}{3}$ and $\frac{3}{5}$, children can represent each value by laying a strip of paper on each scale and cutting it the appropriate length. The fractions can then be compared by placing the two strips next to each other (see Figure 8.45). The fraction table can be used to display addition, subtraction, multiplication, and division of fractions as well (Wiebe, 1985).

PROBLEM-SOLVING EXPERIENCES INVOLVING NUMBER

Patterns are the systematic relationships observed among objects and ideas. In chapter 6, the idea of geometric patterns was introduced. A paper pattern is used by tailors to help them fashion identical shirts. A tiled floor continues in a systematic tessellated pattern (one without gaps between the tiles) to the edge of the room (see Figure 8.46).

Searching for patterns is often a useful strategy when solving problems. Different number patterns result from a series of systematic measures of counts, lengths, times, and so forth. Two types of number patterns, or **sequences**, are commonly introduced at the elementary level.

☐ *Arithmetic sequences*
☐ *Geometric sequences*

A sequence is a list of numbers in which order matters. Arithmetic sequences grow by adding the same value to each element in succession. For example, the following sequences are the result of successively adding two to preceding values.

☐ *Odd numbers:* 1, 3, 5, 7, 9, . . .
☐ *Even numbers:* 0, 2, 4, 6, 8, . . .

Many problems encountered in everyday life depend on understanding underlying **number patterns**. For example, a simple table setting of a number of knives, forks, and spoons grows in an arithmetic pattern. One diner requires three pieces of flatware, two need six, three need nine, and so on:

Add three: 0, 3, 6, 9, 12, . . .

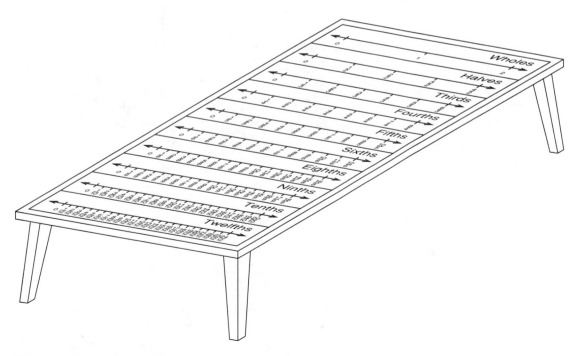

FIGURE 8.44 Fraction Table

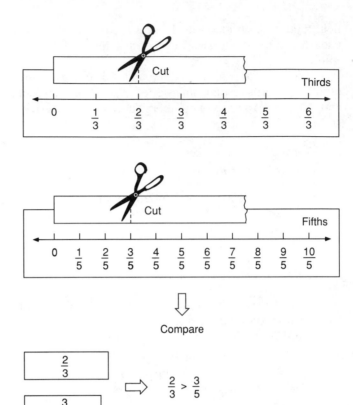

FIGURE 8.45 Comparing Fractions

Other examples of arithmetic sequences include:

1. The number of bricks needed, row-by-row, to build a wall.
2. The number of shoelaces in a collection of pairs of shoes.
3. The number of socks in a drawer (you hope).

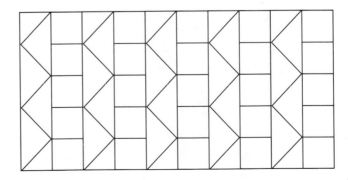

FIGURE 8.46 Tessellated Tile Pattern

4. The number of tires in a fleet of cars.
5. Each row in the multiplication table.

Suppose a wall has rows twenty-five bricks long. There would be twenty-five bricks in a wall one brick high, fifty for a wall two bricks high, seventy-five for three rows, and so on. Write the number sequence for this situation and the other four above.

Geometric sequences grow faster than arithmetic sequences, since each successive value is *multiplied* by the same number:

Multiply by 2: 1, 2, 4, 8, 16, . . .
Multiply by 5: 1, 5, 25, 125, . . .

To show how rapidly geometric sequences grow, repeatedly fold a piece of paper in half and keep track of the number of layers (see Table 8.6).

If you continue the pattern in Table 8.6, you will discover that seven folds gives one-hundred twenty-eight layers (about the maximum number of folds possible even with a large sheet of paper).

Tradition passes down the story that the Hindu inventor of chess made the following request when offered by his patron a gift of his choice:

I require but one grain of rice on the first square of chess board. On the next square, place two grains and on the next four. Continue the pattern for all 64 squares. This modest gift of rice will adequately compensate me for my small achievement.

The inventor of chess certainly deserved a most generous gift. Using a calculator, try to compute the amount of rice requested. Upon which square of the chessboard does the calculator give up?

TABLE 8.6 Folding Paper

Folds	Layers
0	1
1	2
2	4
3	8
.	.
.	.
.	.

TABLE 8.7 Planting Gardens

Edge Length (l)	Area (l^2)
0	0
1	1
2	4
3	9
4	16
5	25
.	.
.	.
.	.

Other number patterns resulting from quadratic or higher-order polynomial equations grow faster than arithmetic sequences but not as fast as geometric sequences. For example, a square garden grows in area as the edges increase in length. A square garden one meter on an edge has an area of one square meter (1 m^2). When the edges are increased to two meters, the resulting area is four square meters. Three meter edges give a nine square meter garden.

A pattern becomes apparent as we continue to lengthen the edges. Table 8.7 summarizes the relationship between edge length (l) and area of square gardens (l^2), or l to the *power* of 2.

> Write the rule that gives the area of a square garden based on the edge length. How large is a square garden ten meters on an edge? twenty-five meters on an edge? fifty meters on an edge?

CALCULATOR AND COMPUTER APPLICATIONS

Calculator Games

Ninety-nine. Strategy games can help maintain student motivation while they practice desired number skills (Moursund, 1981). To play the game *Ninety-nine*, students use place value to develop a winning strategy. Students play in pairs (or one student can compete against the class) sharing one calculator.

Player 1 secretly enters any two-digit number into the calculator and gives Player 2 the following clues:

1. The number of 9s in the secret number.
2. The value of one digit that is not a 9.

TABLE 8.8 Sample Ninety-nine Game

Player 2 Guesses	Calculator Display	Player 1 Gives Hints
	27 (secret number)	1. No nines 2. One digit is 2
7	27 + 7 = **34**	1. No nines 2. One digit is 4
63	34 + 63 = **97**	1. One nine 2. One digit is 7
20	97 + 20 = 117 117 − 20 = **97**	Overflow
2	97 + 2 = **99**	**End**

Player 2 guesses a number that, when added to the secret number in the calculator, will equal 99. In response to clues from Player 1, Player 2 continues to guess values to be added to the cumulative sum in the calculator until reaching 99. If the total goes over 99, Player 1 says "overflow" and subtracts the current guess before resuming. The object is for Player 2 to reach exactly 99 in the fewest number of guesses (see Table 8.8).

Ninety-nine seems deceptively simple until you play the game. With experience, it is possible to work out a strategy using the place value concept and the addition facts to arrive at 99 using a maximum of three guesses.

> Play Ninety-nine with a partner until you work out an expert strategy. Extend the goal to 3 digit values to a maximum of 999. Will the same strategy work? Using an optimal strategy, what is the minimum number of moves that ensure you will reach 999?

Number Pattern Software

The Pond. This computer simulation allows primary children to predict the pattern of the path of a frog as it jumps on lily pads across a pond. The complexity of the pattern increases with experience. The program encourages guess-and-test, working backwards, and systematic listing strategies. For example, after looking at a graphic display of part of the pond, the student might predict a jumping pattern of *up 3, left 2*. The frog then hops across the lily pads following the selected pattern. If the path is incorrect, the frog gets wet (see Figure 8.47).

```
You have 35 moves left
Press ↑, ↓, ←, →
to move your frog.

Press RETURN when you know
         the pattern.
```

FIGURE 8.47 The Pond. *Note:* From "The Pond" [Computer program] by Sunburst Communications, 1984. Copyright 1984 by Sunburst Communications. Reprinted by permission.

ADAPTING INSTRUCTION FOR CHILDREN WITH SPECIAL NEEDS

Persistent difficulties with arithmetic are often due to a poor understanding of place value and regrouping. In the early grades, children experiencing persistent problems reversing digits or sequencing numbers should be given opportunities to seriate and classify concrete objects and record counts using proportional base-10 materials. Such students may need to rely on perceptually salient materials for a longer period than other children before the gradual transition to more symbolic representations can be accomplished.

In kindergarten, extra help learning to write the number symbols may be necessary for some children. Children can copy numerals in a shallow tray filled with sand so mistakes can be easily erased. To help children develop a mental image of each numeral, make a set of large digits using sandpaper for them to reproduce in the sand tray without looking. Have pairs of children trace numerals on each other's back with a finger and have the partner reproduce it in a sand tray or on the board using a damp sponge. When counting above nineteen, many children have difficulty with the transition between decades (e.g., 29 to 30). Display a number line from one to one hundred or use a hundreds chart for children to refer to when counting. When the count reaches the end of a decade, have the child point to the next value and say its name.

When place value is introduced, generally in Grade 1, some children need extra help relating the physical representations of multidigit numbers with the symbols.

After children have displayed a value using base-10 materials or an abacus, have them write the symbol using a different color for each place value. When using chip trading materials, it is convenient to use the the same colors (yellow—1s; blue—10s; green—100s; red—1000s). It may be helpful later for these children to continue using colored crayons when learning to add and subtract multidigit values as well.

Children in upper grades experiencing consistent difficulty with computation may need to be reintroduced to place value and regrouping with materials such as chip trading, an abacus, and base-10 blocks. Several computer simulations of base-10 materials are also available for classroom use (*Base Ten on Basic Arithmetic*, MECC; *Math Ideas with Base Ten Blocks*, Cuisenaire; *Place Value Place*, InterLearn).

Gifted and talented children can extend their understanding of place value by playing the *Pico-Centro-Nada*. This game involves using logical thinking and a knowledge of place value to guess a target number consisting of three or more digits. The game can also be played by less mature children using two-digit target numbers. To play, a leader secretly selects a target number (say, 704), and the class takes turns guessing. After each guess, the leader gives one of the following clues:

1. Pico—at least one digit is correct and in the correct position.
2. Centro—at least one digit is correct but is in the wrong position.
3. Nada—no digits are correct.

A sample game for the secret target number 704 might proceed as shown in Table 8.9.

With practice, any three-digit target number can be found in eight guesses or fewer. Two-digit numbers can be found more easily using the same logic. A number of commercial games (e.g., *Mastermind*) are based on the same principles.

TABLE 8.9 Game of Pico-Centro-Nada

Guess	Clue
978	Centro—at least one digit is correct but in the wrong position.
999	Nada—no digit is correct (9 is not a digit in the target number).
888	Nada—no digit is correct (8 is not a digit in the target number).
987	Centro—7 is a correct digit but is in the wrong place (it must be in the hundreds position due to the first and fourth guess).
123	Nada—no digit is correct (1, 2, and 3 are not in the target number).
456	Centro—at least one digit is correct but is in the wrong position.
944	Pico—at least one digit is correct (4 is in tens or ones position).
956	Nada—no digit is correct (5 and 6 not in target number).
990	Centro—at least one digit correct but in wrong position (0 must be in the ten's position).
704	Pico-Pico-Pico—target number guessed in 10 tries!

SUMMARY

Number is one of the fundamental concepts of mathematics. Though most children learn to count early in life, meaningful counting requires the consistent one-to-one coordination of each number word with its associated object. Piaget suggests that an understanding of conservation of number, seriation, and classification are also important to the meaningful development of the ordinal (position) and cardinal (quantity) notions of number. A thorough understanding of Hindu–Arabic place value notation is necessary for work with number operations involving larger values and decimal fractions. Perceptually salient instructional materials, such as bean sticks, frames and sticks, base-10 blocks, Cuisenaire Rods, chip trading materials, the abacus, and Unifix Cubes are available to help children develop a meaningful understanding of place value and regrouping. Later these materials can be used to introduce whole number and decimal algorithms. Establishing anchor numbers, estimation and rounding skills, and work with number patterns help children develop a sense of the relative magnitude of values. Base-10 blocks, money, and metric measures can be used to introduce decimal fractions. Common fractions ($\frac{1}{2}$, $\frac{2}{3}$, etc.) are introduced using unit partitioning and set partitioning. Both models can be used later to introduce equivalent fractions and fraction operations. Calculator number games and computer simulations can provide practice with counting, place value, and number patterns. Studying the history of our current numeration system can be an enlightening activity for many children, especially gifted students. Special-needs children often require additional experiences counting and classifying objects and later with structured base-10 materials to develop a meaningful understanding of number. Special help in writing the number symbols can be facilitated by using sand trays, textured numerals, tracing numerals on students' backs, and writing multidigit numbers with colored crayons.

COURSE ACTIVITIES

1. Construct a display showing how the numbers 1, 15, 72, 256, and 1111 are represented using examples of each of the three types of numeration systems—tally, additive, and place value.

2. List the advantages and disadvantages of tally, additive, and place value numeration systems. What unique features of the Hindu–Arabic numeration system have made it so popular throughout the world?

3. Construct a set of attribute materials (four colors, four shapes, and two sizes) and review a series of commercially available activities that introduce ordering, simple classification, intersection, and class in-

clusion. Try several of these activities with a partner and, if possible, with a group of children. Write a brief description of any problems encountered in completing these activities.

4. Describe five situations that require students to seriate two or more objects or ideas. How could you incorporate these situations into lessons to help children coordinate the ordinal and cardinal notions of number? Write a lesson plan for a seriation task and implement the lesson using peers as students or with a small group of children in an appropriate classroom. Write a summary of the experi-

ence and discuss with others in your class any problems encountered .

5. Develop a series of three lessons using money, metric units, and/or the calculator to introduce the tenths and hundredths place values. Implement your activities with your peers as students or in an appropriate classroom. Keep a log of your experience and review it with others in your class.

6. Construct a set of bean sticks using Popsicle sticks and beans or other proportional base-10 material. Use these materials and a set of dice to play the Bankers Game in a small group. Work out procedures for a *Tax Game*, where players start with one hundred counters and pay taxes on each roll of the dice until nothing is left.

7. Design a bulletin board display showing the historical development of our current numeration system. Include examples of tally, additive, and place value systems. Engage the viewer in some activity or problem-solving experience that could later be incorporated into a classroom lesson.

8. How many different ways are there to make change for $1.11? One possible way is with the five coins (50¢, 50¢, 5¢, 5¢, and 1¢). This is a class size problem, so work with a group and partition the problem into subtasks. What is the smallest possible number of coins needed to make $1.11 using only pennies, dimes, and silver dollars? How is this result related to the place value concept?

9. List five examples of arithmetic, polynomial, and geometric sequences found in real-world situations. Make sketches or concrete models to show each sequence. This is a class size problem, so you may wish to work in small groups and share results.

10. Include examples of counting, place value, fraction, and decimal number activities in your idea file. Specify an objective, appropriate grade level, materials, and procedures.

11. Review the article by Ian Beattie listed in the references. Construct the *Number Namer* described in the article and design a lesson in which it is used to introduce place value to a group of Grade 1 students. How could this tool be used in conjunction with other instructional materials? List several ways the Number Namer could be used in the classroom.

MICROCOMPUTER SOFTWARE

Base Ten on Basic Arithmetic Multiplication and place value practice (MECC).

Gertrude's Puzzles Problem-solving experiences involving reasoning skills and classification tasks (The Learning Company).

Gertrude's Secrets Practice identifying attributes and classifying objects (The Learning Company).

Math Ideas with Base Ten Blocks Practice with place-value and number concepts using graphic representations of base-10 materials (Cuisenaire Company of America, Inc.).

Moptown Practice with classifying objects and logical thinking (Special Delivery Software).

The Pond Offers practice predicting number patterns using a simulated frog jumping on lily pads in a pond (Sunburst Communications).

Place Value Allows the user to represent numbers with base-10 blocks, assign numerals and number words to base-10 displays, compare numbers, and play a game in which children use regrouping to win (Educational Materials and Equipment).

Place Value Place Two programs employing a multi-column graphic calculator with which users can explore place-value concepts involving various bases, addition, and subtraction in both tool and game formats (InterLearn Inc.).

Shark Estimation Games Four estimation activities simulating a hunt for a hidden shark involving integer and decimal values on the number line and coordinate plane (InterLearn Inc.).

REFERENCES AND READINGS

Baratta-Lorton, M. (1976). *Mathematics their way.* Palo Alto, CA: Addison-Wesley.

Baroody, A. (1986). The value of informal approaches to mathematics instruction and remediation. *Arithmetic Teacher, 33*(5), 14–18.

Beattie, I. The Number Namer: An aid to understanding place value. *Arithmetic Teacher, 33*(5), 24–28.

Carlow, C. (1986). Estimation and mental computation: Early focus on number awareness. *Arithmetic Teacher, 34*(4), 16–17.

Driscoll, M. (1980). Research within reach: Elementary school mathematics. St. Louis: CAMREL.

Fuson, K., & Hall, J. (1983). The acquisition of early number word meanings. In H. Ginsburg (Ed.), *The development of mathematical thinking* (pp. 50–107). New York: Academic Press.

Gelman, R., & Gallistel, C. (1978). *The child's understanding of number.* Cambridge: Harvard University Press.

Harrison, M., & Harrison, B. (1986). Developing numeration concepts and skills. *Arithmetic Teacher, 33*(6), 18–21.

Hopkins, M. (1987). Number facts—or fantasy? *Arithmetic Teacher, 44*(7), 38–42.

Jacobs, H. (1970). *Mathematics: A human endeavor.* San Francisco: W. H. Freeman.

Labinowicz, E. (1980). *The Piaget primer.* Menlo Park, CA: Addison-Wesley.

Menninger, K. (1969). *Number words and number symbols: A cultural history of numbers.* Cambridge: The M.I.T. Press.

Moursund, D. (1981). *Calculators in the classroom: With applications for elementary and middle school teachers.* New York: John Wiley.

Reys, R. (1985). Estimation. *Arithmetic Teacher, 32*(6), 37–41.

Saxe, J. (1979). Developmental relations between notational counting and number conservation. *Child Development, 50*(1), 180–187.

Schoen, J. (Ed.). (1986) *Estimation and mental computation: 1986 yearbook.* Reston, VA: National Council of Teachers of Mathematics.

Schoen, J. (1987). Estimation and mental computation: Front-end estimation. *Arithmetic Teacher, 34*(6), 28–29.

Scott, L. (1960). *Rhymes for fingers and flannelboards.* St. Louis: Webster.

Souviney, R., Keyser, T., & Sarver, A. (1978). *Mathmatters: Developing computational skills with developmental activity sequences.* Glenview, IL: Scott Foresman.

Trafton, P., & Zawojewski, J. (1985). Estimation and mental computation: Rounding wisely. *Arithmetic Teacher, 32*(6), 36–37.

Van de Walle, J., & Thompson, C. (1985). Let's do it: Estimate how much. *Arithmetic Teacher, 32*(9), 4–8.

Wiebe, J. (1985) Discovering fractions on a fraction table. *Arithmetic Teacher, 33*(4), 49–51.

Zawojewski, J. (1986). Ideas: Making equal-sized pieces. *Arithmetic Teacher, 34*(4), 18–25.

9

Addition and Subtraction of Whole Numbers

"Can you do addition?" The White Queen asked, "What's one and one and one and one and one and one and one and one and one and one?" "I don't know," said Alice, "I lost count." "She can't do addition," the Red Queen interrupted.

—Lewis Carroll

Upon completing Chapter 9, the reader will be able to

1. Describe two developmental concepts associated with the meaningful understanding of addition and subtraction.

2. Describe an instructional hierarchy for introducing whole number addition and subtraction operations.

3. Describe two models for introducing the concept of addition and give an appropriate classroom lesson for each.

4. Describe the four basic properties of the addition operation and give an example of the instructional application of each.

5. Describe the inverse relationship between addition and subtraction and give an example of an appropriate classroom lesson.

6. Describe several techniques to help children memorize the addition and subtraction facts.

7. Describe how a number line can be used as a model for addition and subtraction and specify its limitations.

8. Describe three models of subtraction and give an example of each.

9. List five instructional materials that can be used to introduce the addition and subtraction algorithms and give sample lessons for each.

10. Describe examples of developmental activity sequences to introduce the multi-digit addition and subtraction algorithms.

11. Describe five reasons to integrate mental computation into the mathematics curriculum.

12. Describe how rounding off can be used as a technique for estimating sums and differences and give an example of an appropriate classroom lesson.

13. Describe three ways to introduce addition and subtraction story problems to primary and upper-grade children.

14. Describe nonroutine problems involving addition and subtraction and give a sample lesson.

15. Describe how preprogrammed calculators and computers can be used to teach addition and subtraction.

16. Describe six criteria for selecting computer-supported drill-and-practice materials and give an example of quality software for drill and practice, concept development, and problem solving.

17. Describe alternative algorithms for addition and subtraction and discuss their application to the instruction of special-needs children.

Children learn addition and subtraction by building on previously developed notions of geometry, measurement, and numeration. The formal introduction of arithmetic begins in kindergarten and extends at least through Grade 4 for the more complex exercises. Addition and subtraction are introduced in kindergarten and Grade 1. Multiplication and division are generally introduced in Grades 2 and 3. In practice, the mathematics curriculum periodically spirals back to operations introduced previously to review and extend algorithms for work with larger values and rational numbers. This cycle continues through Grade 6 and, for many children, into middle school.

The availability of inexpensive calculators and microcomputers raises the question whether learning to calculate with large numbers quickly and accurately should remain a primary goal for elementary school children. For example, the Mathematics Framework for California Public Schools (1985) recommends that elementary instruction emphasize practice in choosing whether to use mental computation, a calculator, or a written algorithm to solve a problem. The ability to apply the appropriate technology and to estimate results prior to solving a problem are likely to become increasingly valuable mathematics skills. However, this trend does not diminish the importance of gaining a good conceptual understanding of each of the number operations.

Teaching addition and subtraction of whole numbers follows a hierarchy of systematic steps including early counting experiences, conceptual development, the memorization of basic facts, and work with symbolic algorithms. An instructional hierarchy can be used as an overall guide when planning the pace, sequence, and content of computation lessons. Referring to a sequence of concepts and skills will help you organize the ideas presented in this chapter into effective lessons. Figure 9.1 shows possible hierarchies for addition and for subtraction.

INTRODUCING THE ADDITION CONCEPT

There are two common models of addition:

☐ *Union of sets* addition
☐ *Measurement* addition

Union of Sets Addition

A typical display using the **union of sets** model for addition is shown in Figure 9.2. Putting sets A and B together (the *union* of A and B) represents the symbolic sum 4 + 3. The answer 7 is represented by set C. When asked how many balls in set C, many five- and six-year-olds report that there are none; they claim that all the balls are in sets A and B. At first this reply may seem to be the result of some miscommunication. In fact, many children of this age group have difficulty with the **class inclusion** concept of keeping the whole and its parts in mind at the same time. When two sets of concrete objects are counted as shown in Figure 9.3 and subsequently added, the original values are no longer apparent.

Using objects of different shapes or colors makes it easier for the child to reconstruct the addends (see Figure 9.4).

An **operation** is an action that changes one or more objects into a new entity. Addition is an operation that combines two numbers (addends) to create a unique third value called the **sum**. Since addition operates simultaneously on two numbers, it is called a **binary** operation.

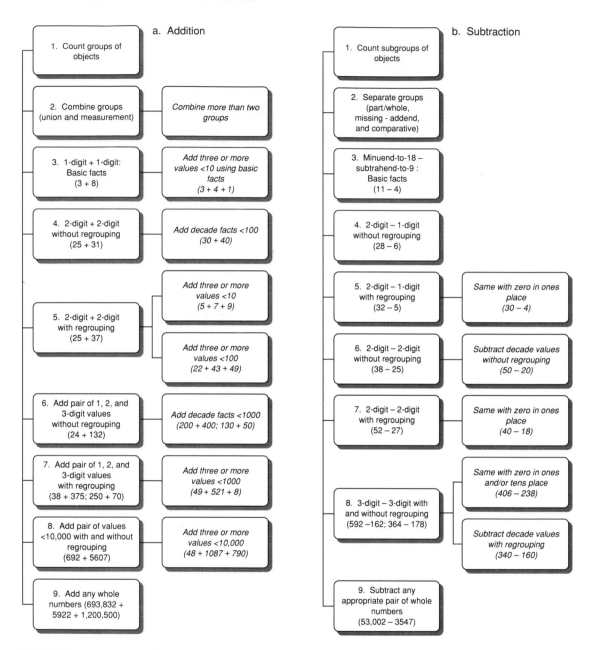

FIGURE 9.1   Instructional Hierarchies

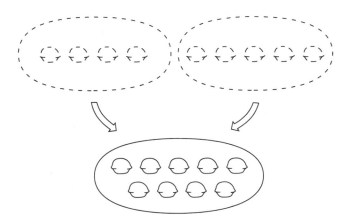

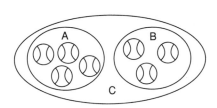

FIGURE 9.2 Union of Sets A and B Is Set C

FIGURE 9.3 Four Walnuts Plus Five Walnuts Equals ___ Walnuts?

Measurement Addition

To demonstrate how **measurement** is used to model addition, join two objects and find their total length. For example, have children construct two stacks, or **towers**, of Unifix cubes, using a different color for each addend. Join the two towers into a single tower. Finally, a third tower of another color is constructed that equals the height of the combined addend towers.

Figure 9.5 shows how the measures of the addends and the sum are displayed. **Unifix boats** (proportioned plastic trays that hold the Unifix towers 1–10) can also be used to relate each tower with its number symbol. Children can then invent story problems that are represented by the towers.

Reversibility

Piaget noted that young children who readily agree that $3 + 4 = 7$ may have difficulty recognizing that $7 = 3 + 4$. Children need to develop an ability to **reverse** physical and mental actions in order to verify the results of experiments. To practice reversing sums, ask children to produce as many number pairs as possible that add up to a given value. For example, using a regular deck of cards, have each child find all the pairs that have the sum **8**.

4   +   3   =   7

FIGURE 9.4 Four Triangles Plus Three Squares Equals ___ Blocks?

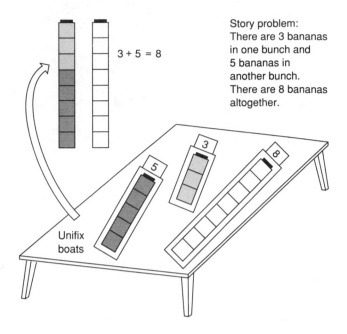

Story problem:
There are 3 bananas
in one bunch and
5 bananas in
another bunch.
There are 8 bananas
altogether.

3 + 5 = 8

Unifix
boats

FIGURE 9.5 Unifix Towers

How many pairs of whole numbers have sums equal to 10, 11, 12, 13, . . . ?
Look for a pattern to help determine the number of possible added pairs
for sums of 50 and 100.

The same activity can be done using Hainstock blocks. These small, plastic boxes each contain a specific number of balls (1–10) in two sections separated by a divider with a hole through which the balls can pass (see Figure 9.6). Children use the blocks to find the pairs of values whose sums equal the total number of balls. Lima beans or buttons painted on one side can serve the same purpose. These materials allow work with zero as well (e.g., 0 + 8 = 8).

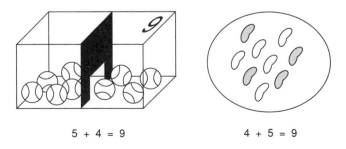

5 + 4 = 9   4 + 5 = 9

FIGURE 9.6 Hainstock Nine Block

Older children can explore number combinations by listing the number of different ways to make change using two or more coins. For example, how many ways can you make change for ten cents? twenty-five cents? fifty cents? Such activities encourage practice with basic sums while focusing on a larger problem goal.

Memorizing the Addition Facts

One of the most demanding tasks that confront children in the primary grades is committing to memory the one hundred basic addition facts (the single-digit sums from 0 to 9). This chapter introduces methods to help children develop a firm understanding of the addition concept while minimizing the memory requirements associated with the recall of the basic addition facts.

There are advantages to a model for addition that allows the direct comparison of addends and sums. Unifix cubes, discussed previously, and **Cuisenaire rods** are two readily available commercial materials used to introduce the sums to 18. Unifix cubes are an example of a **discrete** model for counting, as each number value is constructed from individual units.

Cuisenaire rods, however, represent a **continuous** model for counting; each number value is represented by a separate rod that cannot be physically divided into units. The set of ten multicolored rods represents the following whole number values:

white	=	1
red	=	2
light green	=	3
purple	=	4
yellow	=	5
dark green	=	6
black	=	7
brown	=	8
blue	=	9
orange	=	10

Cuisenaire rod **trains** are formed by placing two or more rods end to end. By matching the train's length with a single rod or with an orange 10-rod plus an additional rod it is possible to display all of the addition facts (see Figure 9.7). You can also display sums of three or more addends by constructing trains with additional "cars" and matching their lengths with orange 10-rods and one additional rod.

The results of such activities can then be recorded symbolically in place value tables. Sums greater than nine require students to coordinate their counting with place value notation. Simple place value tables can be constructed on centimeter squared paper, as shown in Figure 9.8. *Note:* As mentioned in Chapter 8, placing a small *s* after each shaded place value label minimizes the possibility of children including these labels as values in addition, subtraction, multiplication, and division exercises.

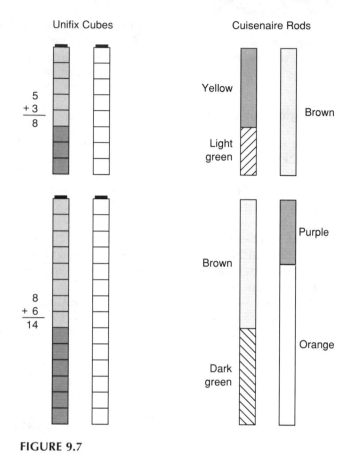

FIGURE 9.7

Two-Entry Addition Table

Once children thoroughly understand the basic concept of addition and have had practice constructing and recording sums to 10 and then to 18, the **two-entry addition table** can be introduced. The table, published in Greece by Pythagoras after his visit to Babylon and Egypt in the fifth century B.C., is normally laid out with the first addend on the vertical axis and the second addend across the top (see Figure 9.9). The table includes all combinations of sums of the values 0–9.

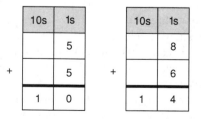

FIGURE 9.8 Place Value Table

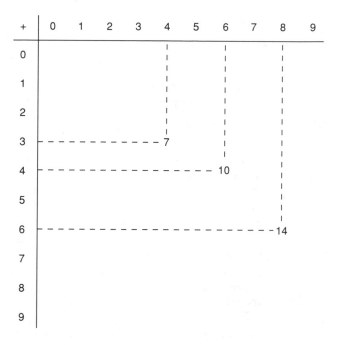

FIGURE 9.9 Two-Entry Addition Table

The addition table can be used as a reference by students as they solve multidigit addition exercises. Many students use mental arithmetic procedures, such as counting on to ten and adding the remaining value ($8 + 5 = (8 + 2) + 3 = 10 + 3$), or using doubles and adding the remaining value ($7 + 9 = (7 + 7) + 2 = 14 + 2 = 16$), to compute basic sums quickly. These methods slow the addition process only slightly and are appropriate strategies for rapid recall of the addition facts.

Addition Properties

Addition has several properties that can simplify mental and written computation procedures.

1. Additive Identity—adding 0 to any number gives the original value for an answer ($4 + 0 = 4$). To demonstrate the additive identity property, have the children put five blocks in one hand and none in the other. Ask how many blocks are in each hand. Have them put both hands together and determine the total number of blocks.

2. Commutative Property—reversing the order of (commuting) two addends does not affect the result ($4 + 3 = 3 + 4$). To demonstrate the commutative property, have children make a Cuisenaire train using two different rods ($4 + 3$). Find the sum by matching the length of the train with another rod (7), or an orange 10-rod and one additional rod if the total is larger than 10. Record the results by drawing a sketch. Have the children reverse the cars in the original train and again find the sum (3

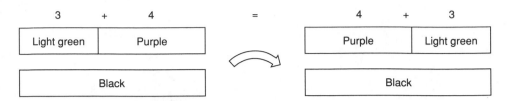

FIGURE 9.10 Commutative Property

+ 4). The answer is a 7-rod in both cases. Repeat the activity several times using different pairs of rods (see Figure 9.10).

3. Associative Property—the way in which three or more addends are grouped does not affect the result $((2 + 3) + 5 = 2 + (3 + 5))$. To demonstrate the associative property, have each child make a three-car train (2-, 3-, and 5-rods) using Cuisenaire rods. Replace the first two rods (2 + 3) with one rod equal to their sum (5). Then find the total of this sum and the remaining rod (5 + 5). Repeat the process by replacing the second two rods (3 + 5) with one rod equal to their sum (8) and then finding the total of this sum and the remaining rod (2 + 8). The answer is a 10-rod in both cases. Repeat the activity for several sets of rods (see Figure 9.11).

4. Closure Property—a set of numbers is closed under addition if the sum of any pair of values is also a member of the set (e.g., in the set of whole numbers, all pairs of values have sums that are also whole numbers: 9 + 1 = 10; 2 + 5 = 7; and so on). To demonstrate the closure property, have children pick any pair of Cuisenaire rods and find their sum by matching their combined length with another rod, or a 10-rod and one additional rod. For any pair, a sum can always be shown using other Cuisenaire rods (see Figure 9.12). Would this be true for any length train if we had enough orange 10-rods to show the answer?

The mathematical terms themselves are not as important for children to learn as how the properties can be used to save time and effort when making mental and written calculations. If technical terms such as additive identity and commutative property are introduced at all, it is generally only after the student has a firm grasp of the concepts and their applications.

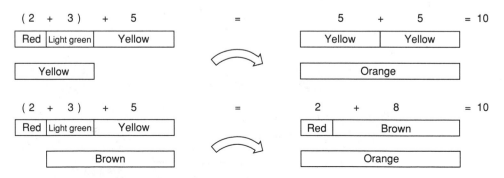

FIGURE 9.11 Associative Property

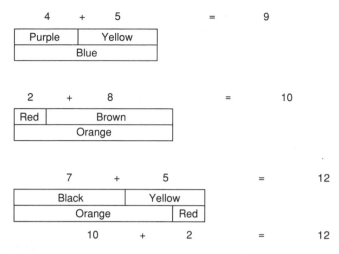

FIGURE 9.12 Closure Property

Applications of Number Properties

The *commutative property* can significantly reduce the memory requirements associated with learning the basic addition (and multiplication) facts. For example, knowing $6 + 9 = 15$ means a student does not have to memorize $9 + 6 = 15$. Knowledge of the commutative property reduces from one hundred to fifty-five, the total number of addition facts that must be memorized. The forty-five addition facts thus eliminated can be quickly remembered by reversing the order of each associated commuted fact.

Applying the *additive identity* and the *one-more pattern* further reduces the addition table memorization requirements. Adding 0 to any number gives a sum equal to the original value ($4 + 0 = 4$). The one-more pattern allows students to compute quickly sums involving the number 1 ($7 + 1 = 8$).

> How many of the one hundred sums remain to be memorized if those involving the commutative property, the additive identity, and the one-more rule are eliminated?

The *associative property* allows several additional computation shortcuts. When adding long columns of values, it is often useful to add digits that increase the subtotal by exactly 10. For example:

$$
\begin{array}{r}
{}^{2}13 \\
27 \\
39 \\
21 \\
+16 \\
\hline
116
\end{array}
$$

When adding the ones column *think*:

$3 + 7 = 10$

$9 + 1 = 10$

$$
\begin{array}{r}
6 \\
\hline
26
\end{array}
$$

Put 6 in the ones column and regroup 2 tens.

Another simplifying technique involves *adding doubles*. To add $(7 + 8)$, think of $(7 + 7) + 1$; or for $(7 + 9)$, think of $(7 + 7) + 2$. Formally, this shortcut is an application of the associative property. The doubling approach provides a structure to help students memorize the number facts and promotes a strategy for checking facts.

$$6 + 8 = 6 + (6 + 2) = (6 + 6) + 2 = 12 + 2 = 14$$

Adding to 10 and *using doubles* are two common techniques used by children and adults when doing everyday arithmetic. List other techniques you can identify that can be used to assist with the recall of number facts or to ease mental computation.

The applications of the *closure property* are somewhat more obscure. This property helps us understand why fractions, negative numbers, and later, irrational numbers must be invented in order to solve problems that have no whole number answer. For example, problems such as $(4 - 6 = ?)$ could not be answered without the existence of negative integers. Problems such as $(4 \div 6 = ?)$ would be meaningless without fractions. Inventing new classes of numbers in order to solve problems is an important application of the closure property.

Addition Table Patterns

Knowledge of the commutative property may help students recall some of the addition facts. The commutative property can be displayed within the addition table. If a line is drawn along the diagonal (from upper left to lower right in Figure 9.13), the numbers above the line are the exact reflection of those below the line. If the table were folded along the diagonal, all the sums in the shaded area would touch an identical sum in the unshaded area. The reason the sums in the addition table are displayed in such a symmetrical pattern is due to the commutative nature of sums. The arrows in the figure connect some of the corresponding commutative pairs.

Other interesting patterns can be found as well. Notice in Figure 9.14 that all the sums are identical on the diagonals running from the lower left to upper right. By listing the paired addends associated with each sum along these diagonals, all possible single-digit combinations can be found.

Find patterns such as the odd numbers, even numbers, whole numbers, multiples of 3, multiples of 4, and multiples of 5 in the addition table. Circle each number in the pattern and connect them with a line. Can you find any patterns that are connected by curved lines? Expand the addition table if necessary (e.g., include addends for 11–20) on a piece of squared paper.

+	0	1	2	3	4	5	6	7	8	9
0	0	1	2	3	4	5	6	7	8	9
1	1	2	3	4	5	6	7	8	9	10
2	2	3	4	5	6	7	8	9	10	11
3	3	4	5	6	7	8	9	10	11	12
4	4	5	6	7	8	9	10	11	12	13
5	5	6	7	8	9	10	11	12	13	14
6	6	7	8	9	10	11	12	13	14	15
7	7	8	9	10	11	12	13	14	15	16
8	8	9	10	11	12	13	14	15	16	17
9	9	10	11	12	13	14	15	16	17	18

FIGURE 9.13 Commutative Pairs on Addition Table

+	0	1	2	3	4	5	6	7	8	9
0	0	1	2	3	(4)	5	6	7	8	9
1	1	2	3	(4)	5	6	7	8	9	10
2	2	3	(4)	5	6	7	8	9	10	11
3	3	(4)	5	6	7	8	9	10	11	12
4	(4)	5	6	7	8	9	10	11	12	13
5	5	6	7	8	9	10	11	12	13	14
6	6	7	8	9	10	11	12	13	14	15
7	7	8	9	10	11	12	13	14	15	16
8	8	9	10	11	12	13	14	15	16	17
9	9	10	11	12	13	14	15	16	17	18

$$0 + 4 = 4$$
$$1 + 3 = 4$$
$$2 + 2 = 4$$
$$3 + 1 = 4$$
$$4 + 0 = 4$$

FIGURE 9.14 Sums Equal to 4

Language

Several English words are commonly used when referring to addition:

- ☐ *add* —Add three and four.
- ☐ *plus* —Three plus four is how much?
- ☐ *sum* —Find the sum of three and four.
- ☐ *together* —Together, they had how much?
- ☐ *altogether* —How much altogether?
- ☐ *combined* —What is their combined worth?
- ☐ *total* —Total of three and four.

The fact that so many phrases are used to indicate when values should be added makes it inefficient to rely on only key words to help identify the appropriate operation in story problems. The issue is further complicated when more than one operation is called for in the problem description. Initially, story problems can be presented using alternate phrases that call for the application of a single operation. Without actually computing an answer, students practice identifying the operation needed to solve the problem. Once students can successfully identify when to invoke each operation, two-step problems can be gradually introduced.

INTRODUCING THE SUBTRACTION CONCEPT

The concept of subtraction is not new to most children when they enter school—people have been taking things away from them for years. It is no surprise that, when a big sister eats part of their popcorn, less is left for them. In school, this general notion of *take away* is transformed into a procedure that carefully attends to the specific values involved. In the case of the stolen popcorn, school subtraction requires that children attend to the exact number of original kernels, called the **minuend**, the number eaten, called the **subtrahend**, and the number remaining, called the **difference**.

There are three common models of subtraction:

- ☐ *take away* subtraction
- ☐ *comparative* subtraction
- ☐ *missing addend* subtraction

Take Away Subtraction

Like addition, subtraction is generally introduced using sets. Instead of putting two or more sets together as we would to show addition, sets are *partitioned* to show subtraction. **Take away**, or **part-whole**, subtraction involves removing objects from a set and counting the number remaining. For example, Figure 9.15 shows a flock of 10 birds with 3 flying away (10 − 3 = 7).

In the problem, "ten birds take away three birds," the subtrahend 3 does *not* represent three new birds, but three of the birds in the original flock of ten. This

FIGURE 9.15 Subtraction

often overlooked distinction is the cause of confusion for many students developing the subtraction algorithm.

Children must understand that the original set (minuend) is *exactly* made up by the subtrahend and the difference—another example of class inclusion. If reunited, these two sets would again equal the original amount. Piaget observed that class inclusion is related to the meaningful development of subtraction as well as addition. Figure 9.16a shows the take away model for $7 - 3 = ?$ using a Unifix tower.

Some children are not able to reverse the process of subtraction by reconstructing the original problem in their minds. Errors such as subtracting the top digit from

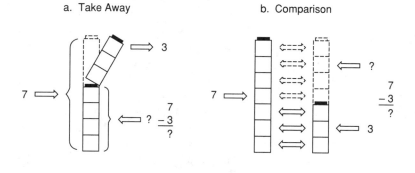

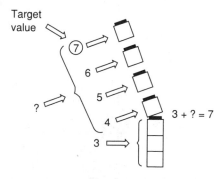

FIGURE 9.16 Three Models of Subtraction

the bottom digit to avoid regrouping ($12 - 8 = 16$) are likely to result in such cases. To minimize this problem, children should thoroughly understand the relationship among the minuend, subtrahend, and difference before being introduced to the subtraction algorithm. The mental ability to reverse the addition process allows children to think about subtraction without having to develop an entirely new concept. One useful application of the inverse relation is the admonition to check subtraction by adding upward:

$$
\begin{array}{r} 35 \\ -18 \\ \hline 17 \end{array}
\quad \rightarrow \quad
\begin{array}{r} 35 \\ +18 \\ \hline 17 \end{array}
$$

To check subtraction, the sum of the difference and the subtrahend should equal the minuend.

Comparative Subtraction

A second type of subtraction, **comparative subtraction**, uses one-to-one correspondence to find the difference. For example, in Figure 9.16b, the height of two Unifix towers is compared by matching the cubes in the shorter tower one to one with those in the taller tower. The number of cubes required to complete the one to one correspondence between the two towers is the difference. This type of subtraction situation models problems such as finding how many more school lunches were sold during one week than another.

Missing Addend Subtraction

A third type of subtraction, **missing addend subtraction**, involves counting on from a given amount (first addend) to achieve a target value (sum). The child keeps track of the number of counts (missing addend) between the given number and the target value. This problem cannot be solved using one-to-one correspondence, because the target value does not exist except in the mind of the solver. It is a process of finding the missing addend. For example, if a notebook cost $7 and the buyer had only $3, the problem can be written as a missing addend number sentence:

$$3 + ? = 7$$

To determine the value of the missing addend, the child keeps track of the counts from 3 to 7. Figure 9.16c shows this missing addend exercise using Unifix towers. The child places cubes one at a time on the given tower (3) until reaching the target value (7). The child must count the number of Unifix cubes required (missing addend) to reach the target value.

Subtraction problems are frequently presented in the missing addend format as early as Grade 1. Though missing addend equations do show the inverse relationship between addition and subtraction, the representation is often confusing to children. Missing addend problems are presented in several textbooks as an alternate subtraction algorithm. Since it is difficult to extend the procedure to multidigit exercises, the value of spending the time to ensure proficiency with this procedure is questionable. It is important, however, for children to have experience with all three types of subtraction situations to help them better understand when to apply subtraction in solving problems.

> Write a story problem to depict each of the three types of subtraction. Exchange the problems and try to identify the type of subtraction each represents.

Using the Number Line

Elementary textbooks frequently use the **number line** as a way to visualize numbers and operations. Often, number lines appear when children are first introduced to systematic counting exercises (see Figure 9.17).

The number line is convenient for coordinating the one-to-one correspondence between the number names and symbols. Problems can occur, however, when the number line model is introduced prematurely.

> To experience the way a child might view an initial introduction to sums on the number line, look at the letter line in Figure 9.18. Try to compute the sums $C + D$ and $B + F$ without reference to a standard number line.

How did you solve the problem in Figure 9.18? You may have found yourself trying to relate the letter line to the more familiar number line. If you matched the letter A to 0, B to 1, C to 3, and so on, and computed the answer using your knowledge of number facts, you should arrive at the solution given in Figure 9.19.

If you matched the letters to the values on the number line beginning with 1 instead of 0, you would get a different answer. Children often experience difficulty knowing where to start counting on the number line when computing sums, differences, and products. Counting on a number line focuses on an ordinal interpretation of number. For example, what will be the last year of this century? The year 1999 is

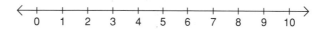

FIGURE 9.17 Number Line

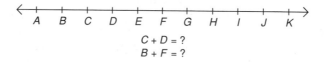

FIGURE 9.18 Letter Line

Example 1: 2 + 3 = 5

⇓ ⇓ ⇓

C + D = F

Example 2: 1 + 5 = 6

⇓ ⇓ ⇓

B + F = G

FIGURE 9.19 Number Line Sums Mapped to Letter Line Sums

incorrect. The first century of one hundred years had no year labeled 0. Therefore, the last year of the first century must have been the year 100. The last year of the twentieth century, therefore, will be the year 2000. Many people may celebrate the arrival of the next century one year early! Number lines showing whole numbers must have a zero point, however.

Addition and subtraction on the number line require a careful coordination of the ordinal (position) and cardinal (amount) notions of number. The cardinal information on the number line is represented by the distance between two points. In order for the ordinal number 1st and the cardinal number 1 to occur at the same point on the number line, a **zero point** must be defined. The cardinal value 1 is one unit of distance along the line. By definition, the end point of this segment is labeled ordinal position one, or 1st. The ability to coordinate the two complementary notions of number is fundamental to the efficient use of the number line as a model for addition and subtraction.

Children who are unable to **conserve length** (maintain invariance of distance through a transformation) may have difficulty using the number line. For example, in the problem 2 + 3 = 5, nonconservers may report that the segment representing 3 somehow gets *bigger* as it slides along the number line from its position starting at 0 to its position starting at 2 (see Figure 9.20). The use of a number line as a model for number operations is likely to be more effective when introduced after students attain conservation of length.

Subtraction should be introduced using multiple representations. Proportional materials such as bean sticks and base-10 blocks are used to help refine the informal notion of take away into a systematic procedure involving place value and regrouping. The number line can also be useful when adding or subtracting small numbers. However, it has limited application for large numbers, because it offers no convenient way to represent place value regrouping.

Nonproportional materials like chip trading counters or the abacus can be used to extend the subtraction concept with larger numbers. As with addition, the concrete actions should be recorded using place value tables. Eventually, symbolic exercises can be introduced and solved using the standard algorithm.

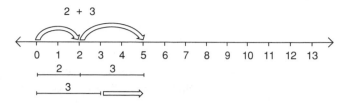

FIGURE 9.20 2 + 3 = 5

Memorizing the Subtraction Facts

Students can construct a subtraction matrix for exercises with minuends to 18 and subtrahends to 9. The result is identical to the addition table described previously. To use the subtraction table in Figure 9.21, find the number inside the circle (minuend) and the value in a triangle along the vertical axis (subtrahend). The difference is in the square along the horizontal axis.

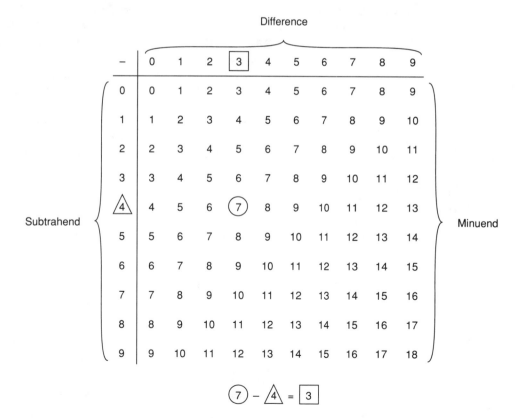

FIGURE 9.21 Subtraction Matrix

Related Facts

Before learning multidigit subtraction, students need to memorize the subtraction facts. To help with this task, it is useful to introduce the notion of addition and subtraction **families of related facts**. For example, using the values 3, 4, and 7, a family of four number sentences can be constructed.

$$3 + 4 = 7 \qquad 7 - 4 = 3$$
$$4 + 3 = 7 \qquad 7 - 3 = 4$$

Practice Games

Mastery of the addition facts should facilitate the recall of the related subtraction facts. Games involving flash cards, dice, and spinners can be effective in encouraging children to practice the recall of the subtraction facts. For example, construct two dice (dodecahedrons) by taping together twelve identical pentagons to form a soccer ball-shaped space figure (these dice are also available commercially). The dice can be used to generate subtraction exercises for drill activities. Number each face of one die with the values 0–9 (repeat 8 and 9 on two faces) and each face of the other die using the values 7–18. Some subtraction facts cannot be generated with these dice (e.g., $5 - 3 = ?$). To give practice with these smaller facts, substitute for the second die an additional die numbered with the values 0–11.

Playing in pairs, students roll the dice in turn and place them next to each other, with the larger value on the left, to form a subtraction problem. Each time a player calls out a correct answer, his or her uniquely colored counter is placed over the same value on the difference axis across the top of the two-entry table. When all the differences (0–9) are covered by counters, the player who has covered the most differences wins. Be creative and make up other games and activities to interest children in practicing their addition and subtraction facts.

Addition and Subtraction Slide Rule

Constructing a slide rule is an interesting way to use the number line as a simple addition and subtraction calculator. Each child needs two strips of heavy tagboard (or two identical rulers) marked off at equal intervals (1 cm works well). As shown in Figure 9.22, the top number line corresponds to the minuend, and the bottom to the subtrahend. Line up the two scales so the minuend 12 is directly above the subtrahend 7.

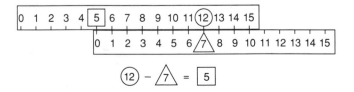

FIGURE 9.22 Subtraction/Addition Slide Rule

The difference 5 is the point on the minuend scale directly above the zero on the subtrahend scale. Notice that this configuration can be used to generate a whole family of subtraction exercises, all of which have the difference 5:

$$8 - 3 = 5$$
$$15 - 10 = 5$$
$$5 - 0 = 5$$

and so on.

By reversing the process, you can use the same slide rule to compute sums:

$$5 + 3 = 8$$
$$5 + 10 = 15$$
$$5 + 0 = 5$$

and so on.

Language

Students frequently confuse the minuend and subtrahend when interpreting the word *from* in subtraction problems. English is read from left to right. English descriptions of situations calling for addition generally read from left to right (e.g., How much is three plus two?). Subtraction, however, can require right-to-left or left-to-right processing.

- ☐ *from*—How much is three from five?
- ☐ *take away*—Take away six from eight.
- ☐ *remain*—If you have ten carrots, how many remain if a cow eats two?
- ☐ *subtract*—Subtract nine from twelve.
- ☐ *reduce*—Reduce the number eight by three.
- ☐ *left*—If there are five fish, and two die, how many are left?
- ☐ *how many*—If there are twenty in your class, and nine are boys, how many are girls?

Initially, children may simply write the problem *three from five* backwards (3 − 5 = ?) yet calculate the correct result. Later, however, this inaccurate procedure will cause problems when students are learning to do multidigit subtraction exercises that require regrouping. For example, the error found in the following worked exercise is quite common.

$$\begin{array}{r} 43 \\ -17 \\ \hline 34 \end{array}$$

When confronted with a regrouping situation, this student took the path of least resistance and simply computed the difference between the digits in each column *without regard to order*. Analyzing error patterns can be a useful aid when trying to determine what children are thinking when they carry out a procedure incorrectly.

> Review the following examples of subtraction errors and write a description of the underlying conceptual problem associated with each. What preliminary concepts are needed and how would you explain the correct procedure to a child?
>
> | | | | |
> |---|---|---|---|
> | 57 | 23 | 109 | 316 |
> | − 29 | − 12 | − 78 | − 118 |
> | 26 | 111 | 171 | 208 |

Consistent use of left-to-right constructions such as *five subtract three* or *five take away three* minimizes initial confusion. When the use of the term *from* is introduced, careful monitoring of value order must be maintained.

THE ADDITION ALGORITHM

Using the Place Value Model

The algorithms for addition, subtraction, multiplication, and division can be developed using a range of instructional materials (base-10 blocks, chip trading, Cuisenaire rods, place value charts). The development of number operations is facilitated by initially embodying each concept in a perceptually salient model, such as proportional base-10 materials. As the learner gains confidence through experience and can subsequently bear more responsibility for coordinating task components, externally imposed supports can be reduced. This process can be accomplished by systematically decreasing the use of perceptually rich materials and simultaneously increasing the use of graphic and symbolic displays.

It is also important to ensure that manipulations at all levels converge in a direct and obvious manner on the intended symbolic algorithm. Some concrete models, though convenient for introducing examples involving small values, do not lead directly to the standard algorithm. For example, any model that does not clearly embody the idea of place value and regrouping will be of limited use for larger sums and differences. Though no concrete model exactly matches the symbolic manipulations involved in the standard addition and subtraction algorithms, it is best to carefully review the range of materials and procedures available and choose those that best match your teaching style and the needs of your students.

Multidigit addition can be introduced after students have developed a thorough understanding of the addition concept and can display the basic addition facts either from memory or through the rapid use of shortcut mental computation. Premature implementation may seriously delay progress for those students still struggling with the recall of the basic number facts (Driscoll, 1980). If, because of scheduling or district policy, it is impossible to delay the introduction of multidigit sums, children can be allowed to use the addition table or a calculator as a tool until no longer needed. Practice sessions focusing on visual patterns, mnemonics, and drills should be continued for children experiencing difficulty memorizing the basic addition facts.

Note that little benefit is likely to derive from introducing multidigit exercises with the hope that they will provide the necessary practice for memorizing the basic facts. Because the memory task is embedded in the more complex multidigit addition process, it is unlikely that these children will make much progress in memorizing the facts *or* in learning the addition algorithm.

Regrouping Ones and Tens

Fundamental to multidigit addition is the notion of **regrouping**. Adding two numbers must give a sum represented in standard place value form. For example, 16 + 8 can be represented as shown in Figure 9.23. Without employing a place value table, it is possible for children to confuse the first representation with the number 114 (one hundred fourteen). To minimize confusion, each place value may contain only one digit (0–9). Ten units of any place value must be regrouped into one unit of the column representing the next larger place value.

When introducing two-digit addition without regrouping, care must be taken to ensure that children do not view each column (ones and tens) as a separate exercise. In the problem 14 + 15 = 29, for example, children may think the 9 digit represents a larger value than the 2.

Since children need to know all the basic facts with sums to 18 to be introduced to two-digit addition with regrouping, addition without regrouping is generally introduced first. Practice in adding the decade numbers with sums to 90 should also be introduced at this time (e.g., 20 + 40).

Children should be encouraged to focus on the place value relationship when adding two-digit values, including the decade numbers, without regrouping to ease the transition to exercises involving regrouping. Initial lessons incorporating such proportional place value materials as bean sticks or base-10 blocks can be very helpful. A tagboard place value mat, as shown in Figure 9.24, can be used to encourage the parallel development of concrete manipulative procedures and the symbolic algorithm.

Modeling the sequence of actions pictured in Figure 9.25 helps children carry out the steps of the standard addition algorithm using base-10 materials. First, the beans in the ones column representing 4-ones and 2-ones are moved down and collected in the answer box. Finally, the bean sticks representing 1-ten and 1-ten are moved down and collected in the answer box. The procedure works equally well in demonstrating

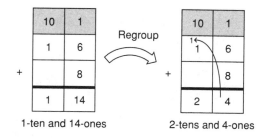

1-ten and 14-ones   2-tens and 4-ones

FIGURE 9.23 Regrouping 10-Ones for 1-Ten

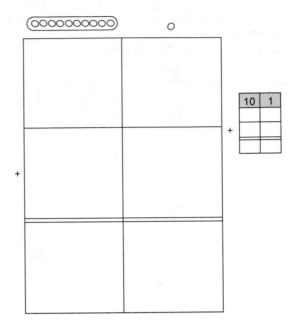

FIGURE 9.24 Addition Place Value Mat

addition with and without regrouping (later, when regrouping is necessary, ten beans (10-ones) are regrouped into one bean stick (1-ten) and moved to the tens column).

Using a set of proportional base-10 materials, carry out the concrete actions necessary to complete the following addition exercises. Record the results in place value tables.

1. 12 + 3 = ?

2. 11 + 5 = ?

3. 13 + 14 = ?

4. 32 + 43 = ?

5. 27 + 11 = ?

Once they have demonstrated the ability to solve exercises concretely, children record the actions and results using squared-paper (cm² graph paper) place value tables. The columns in the tables assist with the placement of digits in the correct place value column.

For students displaying difficulty in consistently recording results using place value tables, it may be useful to introduce intermediary activities using nonproportional chip trading or abacus-type materials. Using the same place value mat, colored counters are substituted for the proportional ones and tens. As shown in Figure 9.26,

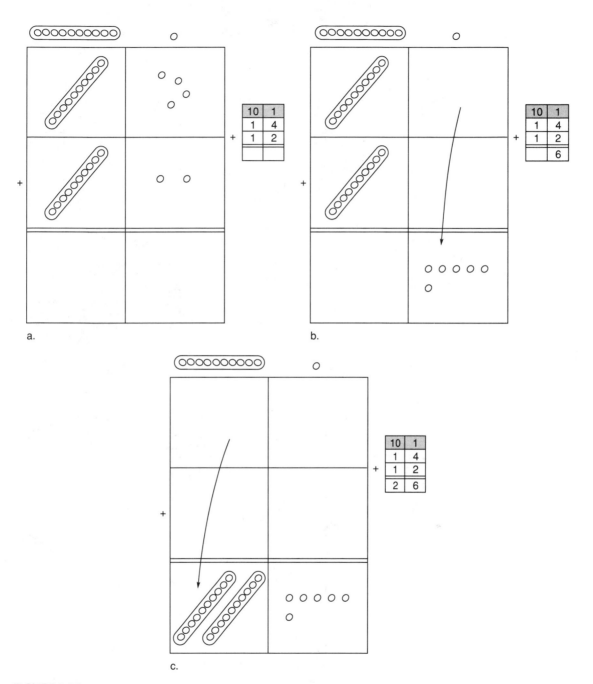

FIGURE 9.25

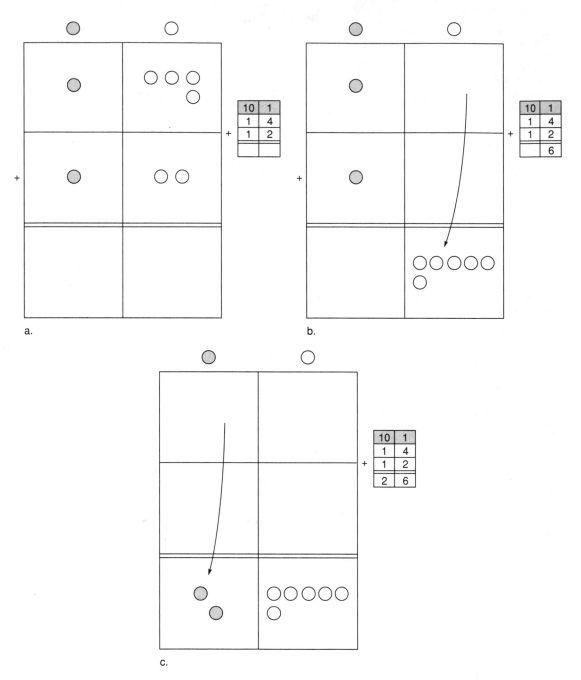

FIGURE 9.26

the regrouping process is somewhat less obvious using these materials, since only the column's position (and perhaps color) distinguishes a tens counter from a ones counter.

> Using the nonproportional counters, carry out the actions necessary to solve the following addition exercises.
>
> **1.** $12 + 3 = ?$
> **2.** $11 + 5 = ?$
> **3.** $13 + 14 = ?$
> **4.** $32 + 43 = ?$
> **5.** $27 + 11 = ?$

The same procedure is used to introduce sums of decade values. Since the ones column contains 0, collect the 2-tens and 3-tens in the answer box. Record the symbolic results in the place value table to the right.

> Using proportional base-10 materials, carry out the necessary actions to solve the following addition exercises. Record the results in place value tables.
>
> **1.** $30 + 40 = ?$
> **2.** $50 + 20 = ?$
> **3.** $10 + 20 = ?$
> **4.** $20 + 40 = ?$
> **5.** $60 + 20 = ?$

Two-place addition *with regrouping* is introduced after the place value relationship and basic facts are well established. The sequence of bean stick actions shown in Figure 9.27 demonstrates how the regrouping process is accomplished. First, the beans in the ones column are collected in the answer box. Since the number of beans exceeds nine, ten beans are regrouped into one bean stick and moved into the tens column (the symbolic record for this process is written in the place value table to the right). Finally, the regrouped bean stick and the two original bean sticks are collected in the answer box.

For children who need additional practice with regrouping, nonproportional place value materials can be substituted for bean sticks. The regrouping process parallels that for proportional materials, though the exchange relationship is indicated only by the position of the counters and perhaps their color (see Figure 9.28).

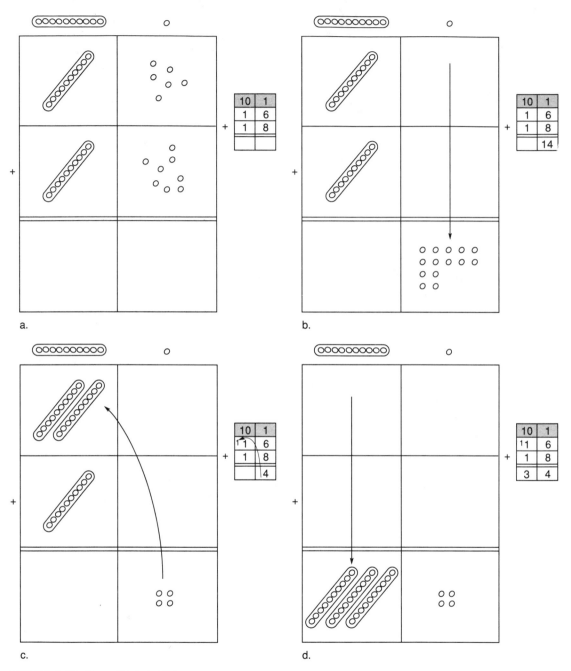

FIGURE 9.27 Bean Stick Addition

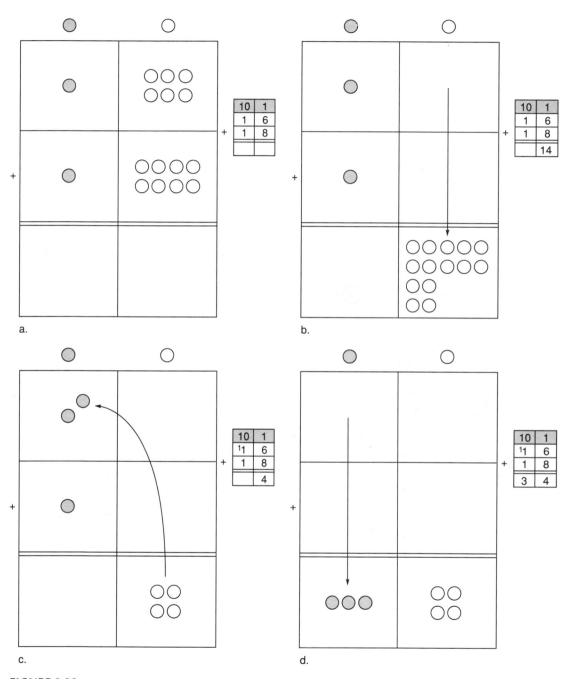

FIGURE 9.28

Using proportional and nonproportional base-10 materials, carry out the necessary actions to solve the following addition exercises. Record the results in place value tables.

1. $8 + 14 = ?$
2. $13 + 18 = ?$
3. $25 + 17 = ?$
4. $38 + 15 = ?$
5. $25 + 15 = ?$

As children become adept at manipulating the proportional materials, nonproportional objects and symbols are introduced. Gradually, the responsibility for coordinating the place value concept and regrouping process is shifted from the teacher to the students themselves. The process is one of gradually reducing external supports such as small group work with place value mats, proportional and nonproportional materials, and place value tables. As a result, the child's responsibility for coordinating task components increases. Eventually, problems are presented without place value tables to ensure that the child can automatically carry out the algorithm. The eventual goal is for the individual to assume full responsibility for coordinating all concepts and processes necessary to carry out the task at the symbolic level of representation. The process is an example of Vygotsky's notion of dynamic support for learning.

The following sequence can be used when extending an algorithm to work with larger values.

1. Introduce the procedure using proportional and, when appropriate, nonproportional concrete materials.
2. Have children record actions and results of concrete manipulations using place value tables.
3. When consistently successful with item 2, assign symbolic exercises in place value tables allowing the use of concrete materials for checking results as necessary.
4. When consistently successful with item 3, assign practice exercises using standard symbolic displays until procedures are established as a routine algorithm.

Extending the Addition Algorithm

Exercises involving more than two addends, or sums involving values with three or more digits, can be solved using the same procedures. Construct an addition mat for 3 addends, as shown in Figure 9.29. Display the exercise using bean sticks or base-10 materials. Collect the values in the ones column in the answer box. As there are more than 9-ones, regroup 10-ones into 1-ten and move it to the tens column. Record the results in the place value table to the right. Finally, combine the 1 regrouped ten and the 3 original tens in the answer box and record the results.

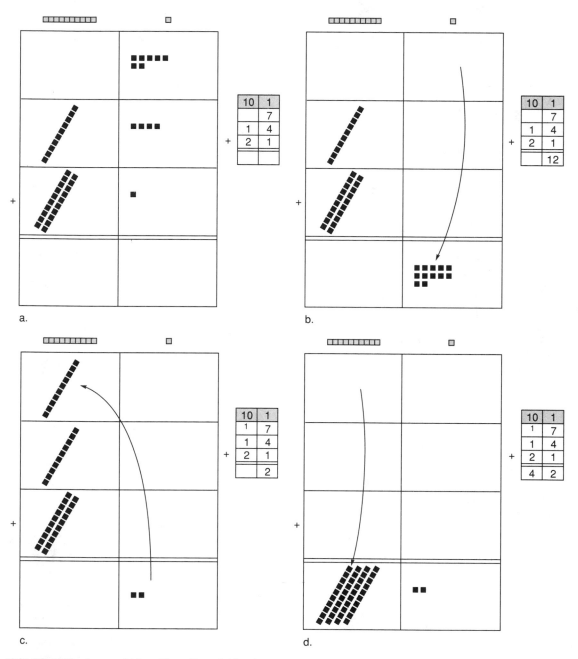

FIGURE 9.29   Sums of More Than Two Addends

Using proportional base-10 materials, carry out the necessary actions to solve the following addition exercises. Record the results in place value tables.

1. $8 + 10 + 17 = ?$
2. $14 + 7 + 5 = ?$
3. $12 + 10 + 9 = ?$
4. $15 + 26 + 12 = ?$
5. $13 + 18 + 44 = ?$

To add numbers with sums greater than 99, the hundreds column is introduced. As shown in Figure 9.30, 1-hundred can be represented by a *raft* of bean sticks or a *flat* of ten 10-rods using base-10 materials.

To add decade numbers with sums greater than 90, construct a three-column addition mat as shown in Figure 9.31. Since there are no blocks in the ones column, first collect the 5-tens rods (or bean sticks) and 6-tens rods in the answer box. Since there are more than nine rods, regroup 10-tens rods into a 1-hundreds flat (or bean stick raft) and move it to the hundreds column. Finally, the regrouped hundred is moved to the answer box. The symbolic record of this process is written in the place value table to the right.

The sequence of actions involved in adding multiples of 100 without regrouping follows a similar pattern. Since the ones and tens columns both contain 0, the 3-

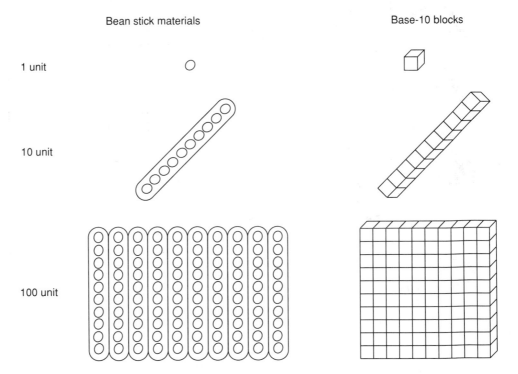

FIGURE 9.30

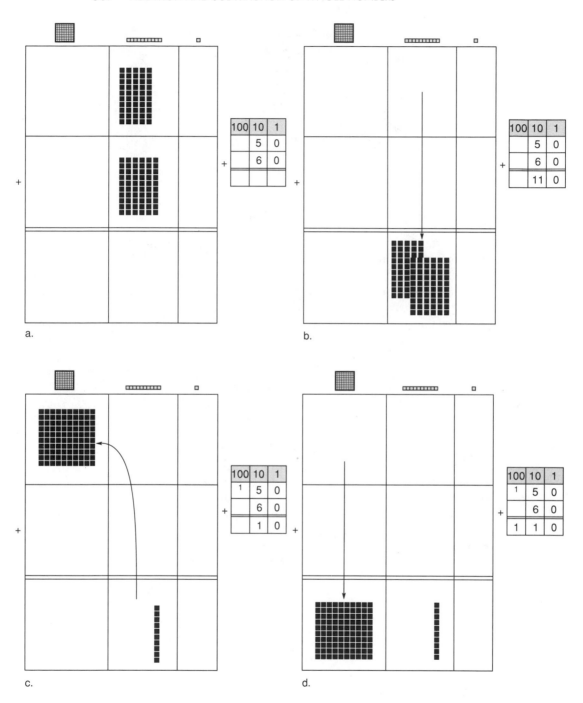

FIGURE 9.31

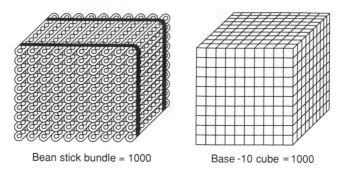

Bean stick bundle = 1000   Base -10 cube = 1000

FIGURE 9.32 Thousand-Unit Bean Stick Bundle and Base-10 Cube

hundreds and 5-hundreds are collected in the answer box and the symbolic manipulations recorded in a place value table.

Sums larger than 999 are normally introduced in Grade 2 or 3. The thousands column can be represented by a *bundle* of ten rafts or by the *cube* equal to ten flats using base-10 materials (see Figure 9.32).

Since the amount of physical materials required increases dramatically for modeling sums over 999, teachers generally use an abacus or other nonproportional model (e.g., chip trading) for operations with large values. Children who have had adequate experience with appropriate concrete models will normally prefer to use symbols and place value tables for such computations. Children who are experiencing difficulty can be encouraged to use concrete models as tools for checking their answers and to record their results in place value tables.

Exercises involving two or more values with sums to 9,999 can be displayed using proportional materials. Nonproportional chip trading or other abacus-type materials are more convenient for work with large numbers.

Figure 9.33 shows the sequence of actions needed to add two values where regrouping is required for both the ones and tens columns. First, the ones are collected in the answer box. Since there are more than 9-ones, 10-ones are regrouped into 1-ten and it is moved to the tens column. Next the tens are collected, 10-tens are regrouped into 1-hundred, and it is moved to the hundreds column. Finally, the hundreds are collected in the answer box. At each step, the symbolic record of the manipulation is written in the place value table to the right.

Using proportional and nonproportional base-10 materials, carry out the necessary actions to solve the following addition exercises. Record the results in place value tables.

1. 127 + 134 = ?

2. 225 + 196 = ?

3. 167 + 316 = ?

4. 308 + 156 = ?

5. 463 + 175 = ?

As children gain facility with the concrete actions and are able to record the results of their experiments consistently, the teacher can then introduce symbolic problems presented in place value tables and, later, in standard format with no place value labels. Concrete materials can be used as a tool to check answers (see Figure 9.34).

THE SUBTRACTION ALGORITHM

Children should exhibit quick recall of the basic subtraction facts before moving to multidigit subtraction. If it becomes necessary, because of district policy or scheduling constraints, to introduce subtraction with regrouping prematurely, students can be allowed to use the subtraction tables or a calculator as a tool for solving multidigit exercises. As with addition, the teacher should avoid using multidigit examples as a context for drilling basic subtraction facts because some children may focus on the memory task at the expense of the multidigit algorithm.

Regrouping Ones and Tens

Subtraction of multidigit numbers often requires regrouping. Regrouping in addition involves trading ten of a smaller unit for one of the next larger place value. In subtraction, the process is reversed. It is useful to develop a concrete representation that is analogous in form and procedure to the symbolic subtraction algorithm. For example, using base-10 materials, first position the minuend in the proper position on the place value mat as shown in Figure 9.35a.

The problem immediately arises of how to represent the subtrahend. If five unit blocks are placed under the minuend, there will be more than eight blocks on the mat. The subtrahend 5 represents five blocks in the original set of 8. The subtrahend simply serves as a reminder of how many blocks to remove from the minuend. Rather than using five new blocks, it is only necessary to use some indicator of the amount to be removed from each column. As shown in Figure 9.35b, tally marks can be used initially. Later, standard digits can be substituted.

The appropriate number of blocks can be moved from the minuend to the subtrahend and those remaining moved to the answer space below the double lines. The result can also be recorded symbolically on squared paper place value tables, as shown in Figure 9.35c.

Subtraction of decade values without regrouping is used to introduce subtraction of two-digit numbers. Figure 9.36 shows the sequence of actions required to display the exercise 60−40. If there are 0-ones in the ones column, and we remove 0-ones, no ones remain. Next, as there are 6-tens, and we need to subtract 4-tens, move 4-tens to the subtrahend and the remaining 2-tens to the answer box. Record the actions as they occur in the place value table to the right.

Subtraction of other two-digit numbers without regrouping can then be introduced using the same procedures.

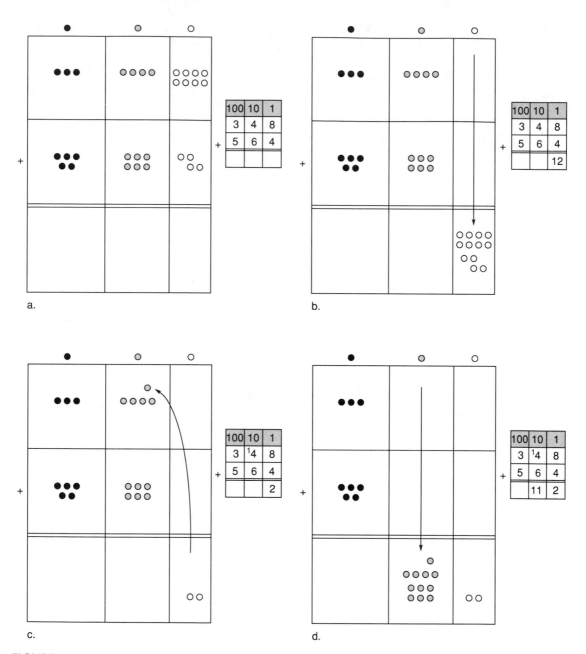

a.

b.

c.

d.

FIGURE 9.33

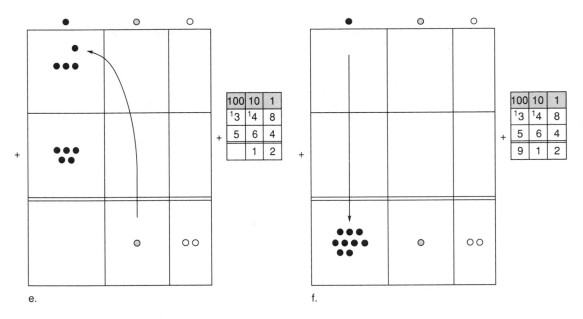

e.   f.

FIGURE 9.33   *continued*

Using proportional base-10 materials, carry out the necessary actions to complete the following subtraction exercises. Record the actions and results in place value tables.

1. 30 − 20 = ?

2. 40 − 10 = ?

3. 32 − 11 = ?

4. 45 − 13 = ?

5. 83 − 32 = ?

When introducing subtraction of two-digit values with regrouping, it is helpful to use exercises with one-digit differences initially (e.g., 24 − 16). Subtraction is normally carried out beginning with the ones column and progresses from right to left. When insufficient blocks are available in the ones column of the minuend to subtract the

Place value table

1000	100	10	1
	4	3	4	7

Place value table

1000	100	10	1
4	3	4	7
4	9	6	

+

Standard algorithm

4347
+496

FIGURE 9.34

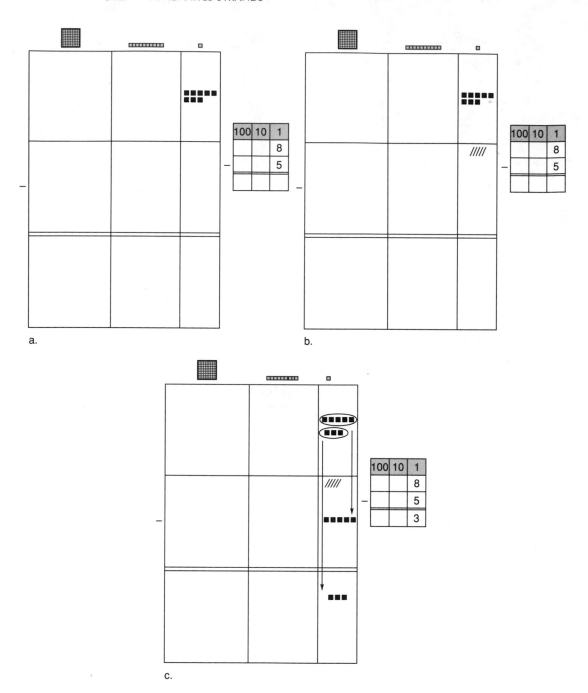

a.

b.

c.

FIGURE 9.35

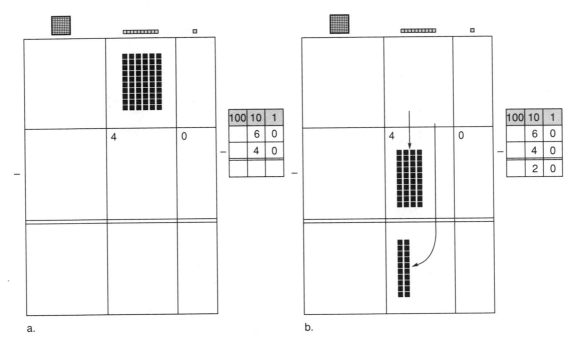

a.

b.

FIGURE 9.36

value indicated in the ones column of the subtrahend, 1-ten in the minuend must be regrouped and moved to the ones column where it is needed (note that the total value of the minuend does not change after regrouping). The difference can then be readily calculated.

Figure 9.37 shows the steps involved in completing the two-digit subtraction problem 25 − 18 using base-10 materials. Since there are insufficient ones blocks to subtract 8 as indicated in the subtrahend, 1-ten must be regrouped and moved to the ones column where it is needed (point out that the total value of the minuend remains the same before and after regrouping the 1-ten). Now, 8-ones can be subtracted from the 15-ones in the minuend by moving them to the subtrahend. The remaining 7-ones are then moved to the answer box. Next, subtract 1-ten from the remaining 1-ten in the minuend by moving it to the subtrahend, leaving 0-tens to move to the answer box. Use symbols to record the actions as they occur in the place value table to the right.

Children should practice using concrete representations of subtraction exercises with minuends to 99. A mixture of regrouping and nonregrouping exercises should be presented. The results of each exercise are recorded using place value tables. Because there are not enough ones in the subtrahend for the exercise shown in Figure 9.38, we regroup 1-ten into the ones column. Next, move 9-ones to the subtrahend and the remaining 7-ones to the answer box. Finally, move 3 of the remaining 7-tens to the subtrahend and the rest to the answer box.

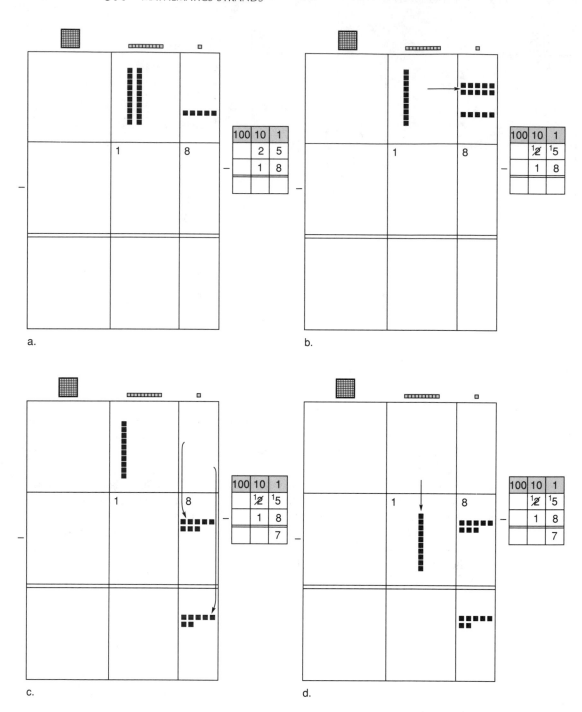

a.

b.

c.

d.

FIGURE 9.37

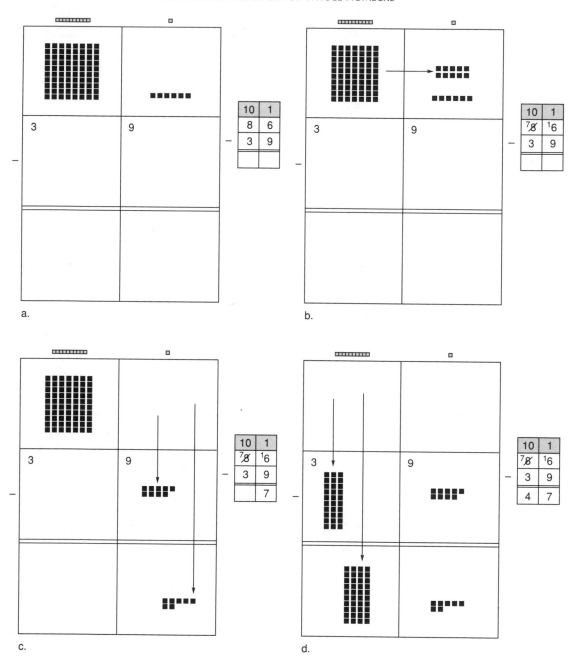

FIGURE 9.38 Subtraction Using Base-10 Materials

Using proportional base-10 materials, carry out the necessary actions to complete the following subtraction exercises. Record the actions and results in place value tables.

1. $32 - 24 = ?$

2. $27 - 11 = ?$

3. $32 - 15 = ?$

4. $45 - 15 = ?$

5. $83 - 37 = ?$

Minuends to 1000 and Larger

Exercises involving multiples of 100 and decade numbers larger than 100 can be used to introduce subtraction involving three-digit numbers. Figure 9.39 shows the sequence of actions involved in displaying the exercise $500 - 200$. Since there are zeros in the ones and tens columns of the minuend and subtrahend, the difference in each case is 0. To subtract 2-hundreds from the 5-hundreds in the minuend, move 2-hundreds to the subtrahend and the remaining 3-hundreds to the answer box. Record the actions as they occur in the place value table.

Exercises involving decade values larger than 100 can be introduced next. For the example $230 - 110$, the ones column for the minuend and subtrahend contains 0, so the difference for this column is 0. To subtract 1-ten from 3-tens, move 1-ten to the subtrahend and the remaining 2-tens to the answer box. Following the same procedure for the hundreds column gives a difference of 1-hundred. The actions should be recorded in the place value tables as they occur.

Carry out the necessary actions using base-10 materials to complete the following subtraction exercises. Record the results in place value tables.

1. $300 - 100 = ?$

2. $500 - 300 = ?$

3. $800 - 500 = ?$

4. $250 - 200 = ?$

5. $260 - 40 = ?$

Exercises with minuends to 999, then 9999 and larger, with and without regrouping, are introduced next. For values of this size, concrete representation using base-10 materials can be tedious. Students can complete exercises using nonproportional chip trading or other abacus-type materials and record their results in place value tables. Incorrect answers can be checked with base-10 materials as necessary.

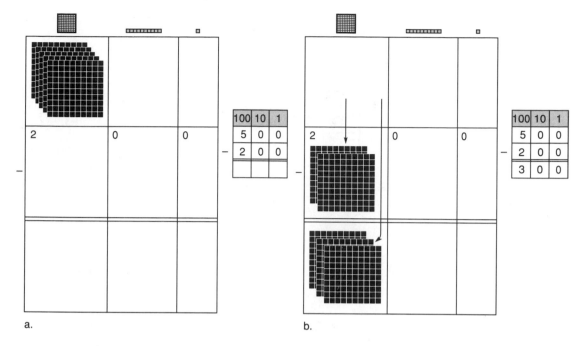

FIGURE 9.39

Figure 9.40 shows the sequence of actions needed to solve the subtraction exercise 437 − 159. First, we regroup 1-ten into 10-ones in order to subtract 9-ones. Move 9-ones to the subtrahend and the remaining 8-ones to the answer box. Since we cannot subtract 5-tens from the remaining 2-tens, we regroup 1-hundred into 10-tens giving us 12-tens in the tens column. Move 5-tens to the subtrahend and the remaining 7-tens to the answer box. Finally, move 1-hundred from the remaining 3-hundreds to the subtrahend and the rest to the answer box. Record the actions in the place value table as they occur. Notice that a *working space* has been added above the minuend in the place value table to make it easy to show the results of each regrouping.

Using nonproportional chip trading or abacus-type materials, carry out the necessary actions to complete the following subtraction exercises. Record the results in place value tables.

1. 236 − 15 = ?

2. 132 − 56 = ?

3. 1322 − 215 = ?

4. 114 − 28 = ?

5. 1223 − 234 = ?

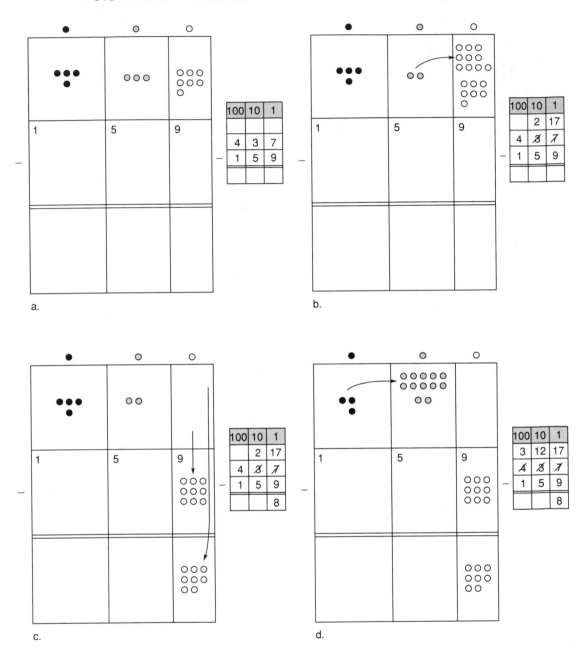

FIGURE 9.40

Regrouping Across Zero Digits

Subtraction exercises involving minuends with one or more zero digits require special attention (e.g., 205 − 127 or 2004 − 235). Children often have difficulty extending the regrouping process to problems with zero digits in the minuend. When

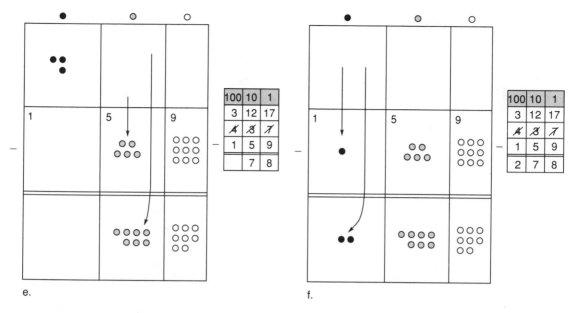

e.

f.

FIGURE 9.40   *continued*

regrouping is called for and a zero digit appears in the next larger place value, we must look farther to the left until we find a value that can be regrouped. Figure 9.41 shows the sequence of actions necessary to solve the exercise 403 − 156. Since the value in the ones column is not large enough to subtract 6-ones, we look to the tens column to regroup 1-ten. Since there are 0-tens in the tens column, we must look to the hundreds column. We regroup 1-hundred into 10-tens, and subsequently, 1 of these tens into 10-ones. This leaves 3-hundreds, 9-tens and 13-ones in the minuend. The subtraction can now be carried out as with the previous exercises.

Figure 9.42 shows a frequent regrouping mistake. Children may forget to deduct from the 10-tens the 1-ten regrouped to the ones column. This results in incorrectly leaving 10-tens in the tens column and 13-ones in the ones column. This obviously leads to an incorrect difference in the tens column.

The situation is further complicated when the minuend contains two or more adjacent zero digits. As shown in Figure 9.43, the process of regrouping one unit into ten of the next smaller units must be repeated for each zero involved.

To minimize this problem, it is helpful to refer to the 4-hundreds and 0-tens in the number 403 as 40 tens (an application of the distributive property). Then, when regrouping 1-ten, we are left with 39-tens, or 3-hundreds and 9-tens, plus the 13-ones in the ones column. Similarly, for 2001, if we borrow 1-ten from 200-tens, we are left with 199-tens, or 1-thousand, 9-hundreds, and 9-tens, plus the 11-ones in the ones column. This procedure can also be applied when working exercises using only symbols:

$$
\begin{array}{r}
4\ \ 9\ 15 \\
5\ 0\ 5 \\
-\ 1\ 3\ 8 \\
\hline
3\ 6\ 7
\end{array}
$$

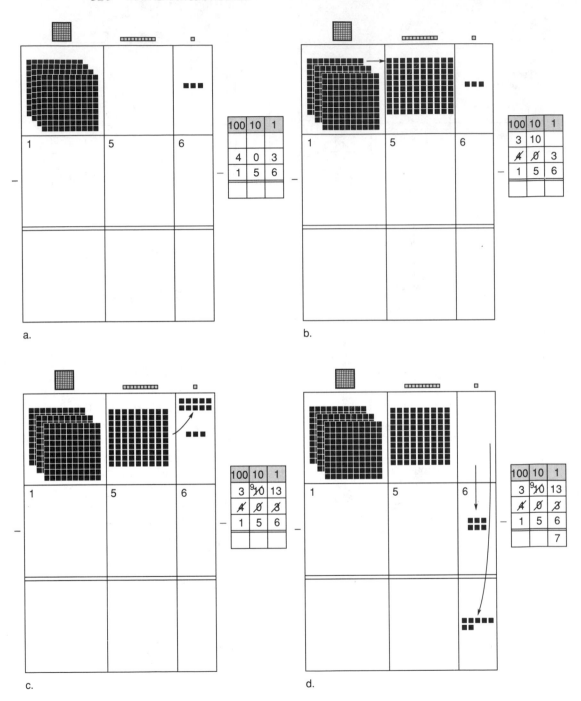

a.

b.

c.

d.

FIGURE 9.41

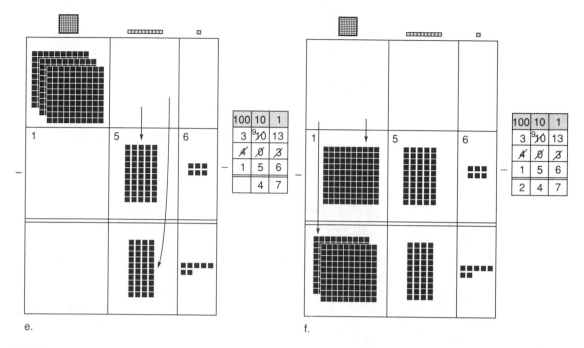

e.

f.

FIGURE 9.41 *continued*

100	10	1
3	(10)	13
~~4~~	~~Ø~~	~~Ø~~
− 1	5	6
2	(5)	7

Circled values in tens column
indicate result of improper regrouping.

FIGURE 9.42 Incorrect Regrouping

1000	100	10	1
1	⁹10	⁹10	11
~~2~~	~~Ø~~	~~Ø~~	~~1~~
−	4	3	6
1	5	6	5

FIGURE 9.43 Four-
Digit Subtraction

The use of proportional base-10 materials is important for the development of this standard **decomposition** subtraction algorithm. Nonproportional chip trading or other abacus-type materials and place value tables serve to effect a transition from the concrete manipulations needed for sound concept development to the systematic use of symbolic algorithms.

CLASSROOM ACTIVITIES

Mental Computation

In the early 1800s, a Vermont boy astounded audiences around the world with his mental computation abilities. Zerah Colburn could accomplish such feats as instantly multiplying two 4-digit numbers in his head. Coincidently, during this same period, considerable stress was placed on mental computations in the elementary and secondary curricula. The contemporary learning theory held that mental exercises of this type developed mental discipline and logical thinking. The mental discipline theory fell into disrepute in the early 1900s, the result of which has been the almost exclusive dependence on written calculations in schools, even to the present day.

Colburn was born with this exceptional ability (he could never explain the procedures he used or how he learned his skill); it also would be unrealistic to expect most children to develop such profound mental computation skills. However, it would be inefficient for students to resort to a written algorithm to calculate exercises like $300 + 500$. Hence, there has recently been a resurgence of interest in providing a more balanced approach to learning arithmetic that includes both written and mental computational components (Trafton, 1978). Reys (1985) suggests several ways that mental computation can enhance student thinking and problem-solving abilities.

1. It promotes understanding of number properties and place value.

> ***Example:*** Without computing an answer, determine the number of digits each answer will contain.
> a. $483 + 917 = ?$
> b. $24 \times 9 = ?$
> c. $1210 - 240 = ?$
> d. $884,396 + 6 = ?$

By employing knowledge of partial sums, products, and place value, each of these exercises can be solved without actually calculating an answer.

2. It encourages analysis of problem situations prior to selecting and carrying out a strategy or algorithm.

> ***Example:*** Andrea bought four items at the store. How much did she spend?

By rearranging prices given in Table 9.1, the answer can be quickly calculated mentally: $(60 + 40) + (18 + 12) = 100 + 30 = 130$, or $1.30.

TABLE 9.1 List of Purchased Items

Items	Cost
Comb	60 cents
Apple	18 cents
Juice	40 cents
Gum	12 cents

3. It encourages flexible application of different forms of numbers.

Example: Without paper and pencil, complete the following exercises.
a. $0.5 \times 150 = ? \rightarrow$ one half of 150 is 75.
b. 10% of $90 = ? \rightarrow$ one tenth of 90 is 9.

4. It encourages the development of alternative algorithms.

Example: Without paper and pencil, complete the following exercises.
a. $46 + 19 = ? \rightarrow (46 + 20) - 1 = 66 - 1 = 65$
b. $7 + 9 = ? \rightarrow (7 + 7) + 2 = 14 + 2 = 16$

5. It promotes the development of estimation skills.

Example: Without calculating the answer, estimate the solution to the following exercises.
a. $20 \times 34 = ? \rightarrow$ the answer is about halfway between $20 \times 30 = 600$ and $20 \times 40 = 800$, or about 700.
b. $1\frac{5}{8} \times 24 = ? \rightarrow$ The answer is a little more than $1\frac{1}{2} \times 24 = (1 \times 24) + (\frac{1}{2} \times 24) = 24 + 12 = 36$.

Mental computation activities can be systematically introduced throughout the school year. Repeated practice will encourage students to attempt mental solutions first and resort to written algorithms or the calculator only when required.

Developmental Activity Sequence

In Chapter 3, a series of two or more activities addressing the same objective at various levels of representation was called a developmental activity sequence (DAS). The implementation of sequenced activities of this type enables the teacher to meet the needs of a wider range of children.

For example, suppose a Grade 4 teacher plans to introduce three-digit addition with regrouping across two place values. Some children are likely to have mastered all the preliminary skills necessary to carry out the manipulation with speed and accuracy. At the other extreme, there will probably be at least one child who is still having difficulty remembering some of the basic number facts. Most will likely fall somewhere in between. The teacher can introduce sequential regrouping to the whole class using the three models shown in Figures 9.44, 9.45, and 9.46.

After a large group demonstration of how to use each material (perhaps using an overhead projector), several exercises should be assigned. Working in small groups, all children begin solving exercises using base-10 materials. Each must show competence at the concrete level before moving on to chip trading materials or place value tables. Students are likely to move through the exercises at different rates. In the time allowed, perhaps 5 periods, all the children will complete a number of practice exercises, though not necessarily at the symbolic level. At each level, encourage children to record their concrete actions and results in place value tables.

Similar sequences of activities can be used to introduce subtraction with regrouping. The procedures employed in each representation should correspond as closely as possible to the symbolic manipulations of the subtraction algorithm. The actions and results of each concrete activity are recorded as they occur in place value tables.

One benefit of the DAS approach is that it provides the means to meet individual needs without having to organize the class into ability groups. A wide range of students can work toward the same objective, though at different rates and levels of abstraction. The DAS can be an effective way to allow for individual differences while avoiding many of the negative aspects associated with achievement-level ability grouping.

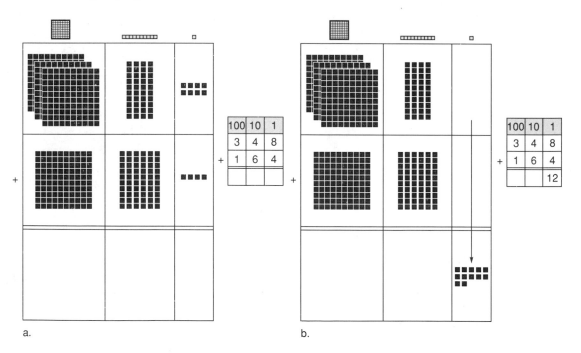

a.   b.

FIGURE 9.44 Addition Using Base-10 Materials

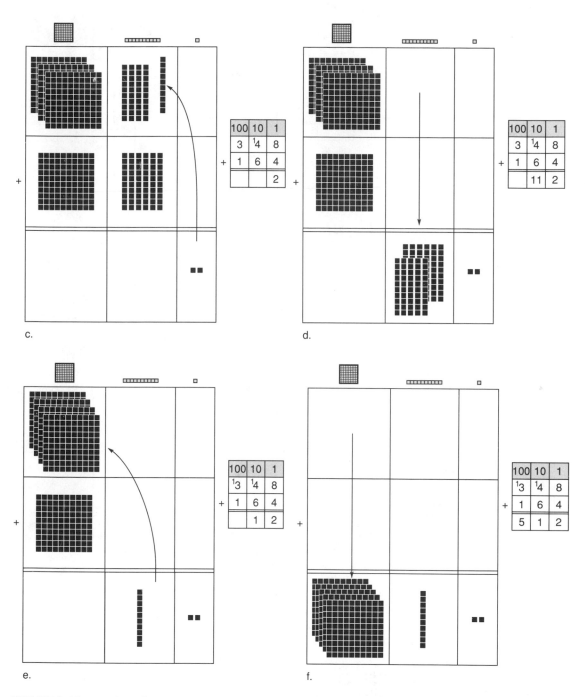

c.

d.

e.

f.

FIGURE 9.44 *continued*

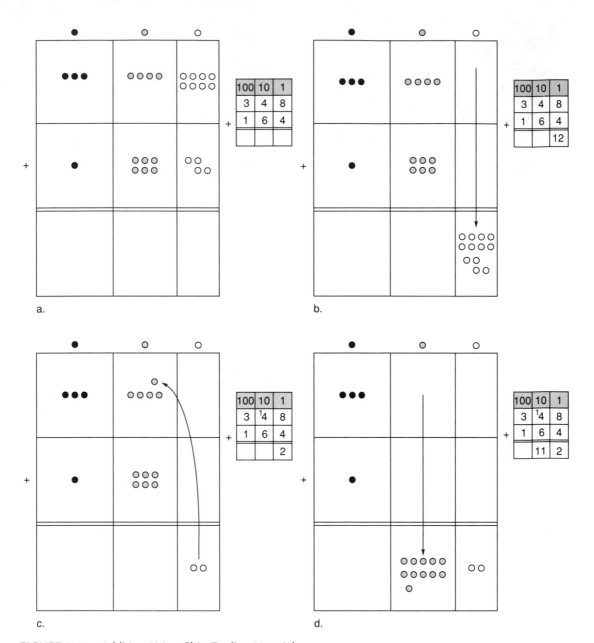

FIGURE 9.45 Addition Using Chip Trading Materials

PROBLEM-SOLVING EXPERIENCES INVOLVING ADDITION AND SUBTRACTION

Picture Problems

Addition and subtraction situations can be posed by using pictures of various sets of coins. What different amounts of money can be made using three pennies, one dime,

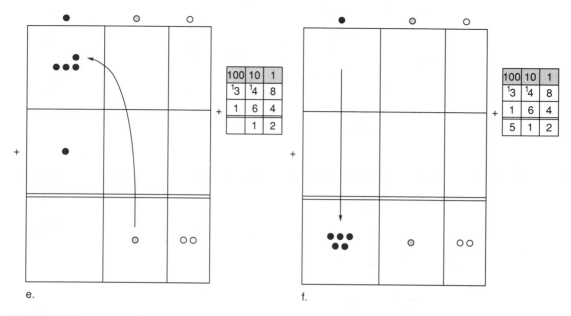

e.   f.

FIGURE 9.45 *continued*

and one quarter (e.g., 1¢, 2¢, 3¢, 10¢, and so on)? Again, each result is recorded in a place value table. Solution strategies include *conduct an experiment*, *draw a sketch*, and *make a systematic list*. Use fewer coins for K–1 children and a larger initial set of coins for older children (see Figure 9.47).

Headlines

A convenient way to introduce word problems involves presenting number sentence **headlines** and having students write the story. This activity encourages children to develop the language skills needed to describe and solve written problems. Since children select the structure of the story that goes with each number sentence, the memory requirements and language level are more likely to match individual needs (Case, 1982). For example:

$$\text{Headline:} \qquad \begin{array}{r} 5 \\ + 7 \\ \hline 12 \end{array} \quad \text{or} \quad 5 + 7 = 12$$

STORY 1. I ate 5 peanuts. Then I ate 7 more. I ate 12 altogether.

100	10	1
¹3	¹4	8
1	6	4
5	1	2

100	10	1
¹2	¹7	8
5	5	5
8	3	3

100	10	1
¹4	¹0	9
2	9	7
7	0	6

FIGURE 9.46 Addition Using Place Value Tables

Coins available

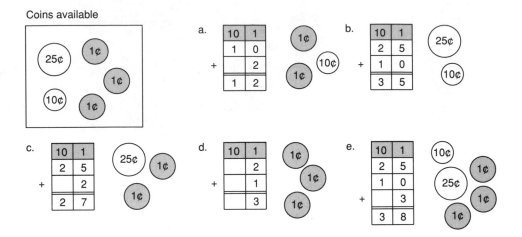

FIGURE 9.47 Making Change

STORY 2. Anna walks 5 blocks to school. I walk 7 blocks. Together we walk 12 blocks.

STORY 3. How much is 5 plus 7? The answer is 12.

The process can also be reversed, whereby story problems are presented and children are asked to write the number sentence headlines. The headline can be written as an **open** or **closed sentence**. An open number sentence has an open frame, or **variable**, that must be replaced by a value in order to complete the sentence.

$$\text{Open sentence:} \quad \begin{array}{r} 8 \\ +\,7 \\ \hline \triangle \end{array} \quad \text{or} \quad 8 + 7 = \triangle$$

$$\text{Number sentence:} \quad \begin{array}{r} 8 \\ -\,7 \\ \hline 1 \end{array} \quad \text{or} \quad 8 - 7 = 1$$

Children are encouraged to write open sentences first and then to compute the answers. Initial efforts in finding headlines for story problems are easier if the children can also choose the values included in the story. This can be accomplished by simply substituting blanks for the values in word problems found in textbooks. For example:

On our vacation, Luis bicycled _____ kilometers the first day, _____ kilometers the second day, and _____ kilometers the third day. How far did he pedal altogether?

Children fill in the blanks with values they can manage. A possible open headline might read $12 k + 18 k + 10 k = \triangle k$. Word problems presented in such a manner give children practice coordinating their language skills with the mathematical operations involved. Appropriate problems are introduced at every grade level. As multiplication and division are introduced, problems involving these operations can be included.

Multistep Story Problems

Solving everyday problems often requires employing more than one operation or step. Children can begin working with multistep story problems by selecting the appropriate sequence of operations without actually computing the results. Graphic presentations of problem situations can be used by primary level children to practice selecting and sequencing required operations. For example, in Figure 9.48 the post office worker must go through several steps to determine the total amount of postage and calculate the change owed. Children can write a story about the sequence of operations *without* actually calculating the change.

Story problems can also be presented so that children can initially focus on the appropriate order of operations without having to calculate the final result. For example:

Jenny had $5.00. She bought 5 apples for 35 cents each. Put a circle around the operation you would do first to find out how much change she should get. Put a square around the operation you would do second.

Add (+)   *Subtract (−)*   *Multiply (×)*   *Divide (÷)*

Estimating Sums by Rounding

Rounding skills can be directly applied as an estimation strategy when solving problems. Children who develop a sense of the relative magnitude of numbers, or

FIGURE 9.48

number sense, are more likely to recognize inappropriate answers when using a calculator and avoid unproductive solution strategies.

For example, when asked to share 45 pencils with 3 friends, a child might incorrectly decide to simply add the values, giving the answer 48. An understanding of the problem situation and of the magnitude of the values involved should preclude an answer this large and, therefore, disqualify addition as a solution procedure.

Rounding off to the *leading* place value is a convenient way to simplify the values in a problem to estimate reasonable results.

$$557 \rightarrow 600$$
$$550 \rightarrow 600$$
$$537 \rightarrow 500$$
$$37 \rightarrow 40$$

Students can practice estimating the sum of two or more numbers by first rounding to the most significant digit (largest place value). Such practice gives children opportunities to develop benchmark notions of magnitude that serve to guide decision making during problem-solving episodes. Teachers encourage the use of rounding and estimation by reducing the emphasis on children always producing exact answers. For example, using the country populations given in Table 9.2, what is the total number of people living in North America?

Students first round each population value to the leading digit. The total population can then be estimated by mentally adding the rounded values. Ask students if the resulting estimated total is reasonable for this problem. Why? In what problem situations are estimated results inappropriate (e.g., determining the number of people on an airplane to make sure it's not too heavy to take off)?

TABLE 9.2   Countries of North America

Country	Population (1983 Est.)
Belize	150,000
Canada	24,880,000
Costa Rica	2,620,000
Cuba	9,890,000
El Salvador	4,690,000
Guatamala	7,710,000
Mexico	75,700,000
Nicaragua	2,800,000
Panama	2,060,000
United States	234,250,000

Estimating Differences by Rounding

Rounding numbers can also be used to estimate differences. Mentally rounding the minuend and subtrahend to the leading digit (place value farthest to the left) before subtracting gives an approximation of the difference. This procedure offers a quick way to gauge the reasonableness of answers. A rough estimate may also indicate whether the correct operation was chosen during the solution of a problem. If the estimate for the answer seems unreasonable, children will be encouraged to review the problem and evaluate the selection of operation.

To estimate differences, round off to the leading digit of the minuend and subtrahend, then mentally compute the difference.

$$
\begin{array}{rcr}
557 & \rightarrow & 600 \\
-\,135 & \rightarrow & -\,100 \\
\hline
\text{Actual} \quad 422 & \rightarrow & 500 \quad \text{Estimate}
\end{array}
$$

The estimate will be off by no more than the value of one place value unit (i.e., hundred) of the leading place value of the larger number. This procedure can be applied to a wide range of problems. You will obtain more accurate results if you round *both* numbers to the place value that corresponds to the leading digit of the subtrahend.

Rounding to Leading Place Value In Each Number	*Rounding to Leading Place Value in the Minuend*

$$
\begin{array}{rcr}
1462 & \rightarrow & 1000 \\
-\,399 & \rightarrow & -\,400 \\
\hline
\text{Actual} \quad 1063 & \rightarrow & 600 \quad \text{Estimate}
\end{array}
\qquad
\begin{array}{rcr}
1462 & \rightarrow & 1500 \\
-\,399 & \rightarrow & -\,400 \\
\hline
\text{Actual} \quad 1063 & \rightarrow & 1100 \quad \text{Estimate}
\end{array}
$$

Notice the estimate of the difference in the first case (600) is within 1000 of the correct answer. In the second case (1100), the estimate is within 100 of the correct answer. A sophisticated use of rounding can help students mentally estimate the results of their computations to check the reasonableness of answers and appropriateness of operations.

Make a Systematic List

Even poor readers can practice using solution strategies by presenting problem situations using concrete materials and graphics. For example, using a pan balance and four mass pieces weighing 1, 2, 4, and 8 pounds, what are all the possible amounts of rice that can be measured? Figure 9.49 shows four results for this problem. Young children can use a balance to weigh various amounts of rice, and older children can use pictures to work the problem. Possible solution strategies that were discussed in Chapter 4 include *conduct an experiment, draw a sketch*, and *make a systematic list or table*. Try solving this problem. The answer may surprise you. A variation of this problem involves including a fifth, 16-pound weight.

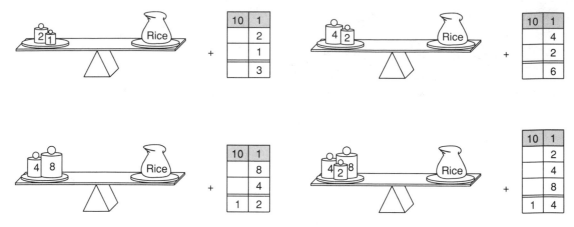

FIGURE 9.49 Balance Problems

Guess-and-Test

A well-developed sense of the magnitude of numbers is particularly useful when solving story problems and nonroutine problems using the trial-and-error, or guess-and-test, strategy. This technique involves making an educated guess, then testing to see how closely the result resembles a solution. A new estimate is then derived, based on the evaluation of each trial, until a satisfactory solution is reached.

For a story problem, suppose we need to find out how many $1.39 notebooks can be purchased for $10.00. An initial guess might be five notebooks. This solution is verified by adding $1.39 + $1.39 + $1.39 + $1.39 + $1.39 (or 5 × $1.39) and checking to see how close the result is to $10.00. The sum $6.95 is sufficiently less than $10.00 to allow the purchase of more notebooks. Increasing the estimate to seven notebooks gives a sum of $9.73, which satisfies the conditions of the problem (if we ignore sales tax). With a well-developed notion of rounding ($1.39 is about $1.50), a willingness to take risks (I think 10 notebooks are too many, so I'll try five), and the ability to adjust an estimate appropriately after an initial guess has been tested (buying five notebooks leaves about $3.00, so I'll try seven), children will be better able to tackle similar problems in the future.

Of course, alternate solution strategies can be applied to the same problem. In this case, repeatedly subtracting $1.39 from $10.00 also gives the answer 7. It is instructive to note that procedures requiring the fewest steps (like the choice to do repeated subtraction or to divide instead of guess-and-test) are often the strategy of choice only *after* several similar problems have been successfully solved.

Draw a Sketch

Often problems are easier to solve if the student initially takes time to *draw a sketch*. For example, the Diffy Game involves choosing any four numbers and positioning them on the corners of a square. Find the difference for each adjacent pair and put the answer on the midpoint between the two values. Connect adjacent differences with new line segments that form a new, rotated square inside the original. Continue the process, as shown in Figure 9.50, until all four differences equal 0. This example shows how an activity can encourage students to engage in extensive practice with basic number facts in pursuit of a higher level objective.

Ask children if they think any set of four numbers will generate a difference of 0 using this procedure? Have them try to find four numbers for the Diffy Game that require 5, 6, 7, or more cycles to arrive at a 0 difference. Observe the pattern of results when multiples of 2, 3, 4, and so on, are selected as the initial four values (e.g., 4, 8, 12, 16). What happens if the four values, 1, 2, 4, 8, are used as starting values? What happens when very large numbers are used? A calculator can be useful in solving this problem.

CALCULATOR AND COMPUTER APPLICATIONS

Calculator Activities

Replace by Zero.   This game provides practice with place value, addition, and subtraction. Pairs of children use one or two calculators. The players take turns entering values. The first player enters a number (say, 1562) and hands the calculator to the other child. The second player selects one of the digits (e.g., 5). By adding or subtracting one value, the second player tries to replace the digit with a 0 without changing any of the other digits (e.g., 1562 − 500 = 1062). If a zero fails to appear in the correct place value on the first try, the original value is reentered and the process repeated. Have players record the number of times it takes to complete the task and write the final solution for each. The problem can be adapted for many grade

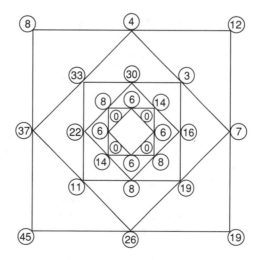

FIGURE 9.50   Diffy Game

levels by using appropriately sized starting values. The problem can be extended by changing the target digit from 0 to any other digit, or the leader can specify one or more target digits *and* associated place values (for 2357, put a 1 in the hundreds column and a 5 in the ones column—possible answer: 2357 − 232 = 2125).

Preprogrammed Calculators.   Many inexpensive electronic games are now available that are designed to provide the carefully structured rehearsal of basic facts. The better devices embed the recall drills in a game context, where the player rapidly inputs number facts to win. The user can generally select the level of difficulty. A summary of correct responses may be displayed at the end of the session. For in-school use, the teacher can provide activity cards or worksheets so the student can generate a written record of each practice session for later review. Examples of the many inexpensive electronic drill devices available from local toy stores include *The Talking Teacher* (Coleco) and *The Little Professor* (Texas Instruments).

Characteristics of Quality Drill-and-Practice Software

Microcomputer supported drill-and-practice programs are widely available. Though certainly not the most creative use of these versatile devices, programs for micro-computers are efficient in presenting graded exercises, keeping track of individual responses, and summarizing the results. Hundreds of arithmetic drill-and-practice software packages are available through local computer-user groups or commercial outlets. In selecting software packages for classroom use, the teacher should consider the following criteria:

1. Flexibility—Does the program offer a feature that allows the teacher and/or student to set the difficulty level, type of positive feedback, number of incorrect responses allowed before the correct answer is given, and the session duration?
2. Ease of Entry—Is it easy for the students to change an incorrect entry?
3. Appropriate Content—Do the screen display and the sequence of numerical entry correspond to the standard algorithm? Often programs require intermediate steps to be worked out on paper and the final solution entered left to right. This becomes particularly clumsy with multidigit problems that, except for division, are generally solved right to left.
4. Reporting Results—Does the program keep cumulative summary records on each user? Can the teacher examine individual results by referring to each student's name, and are class summaries computed?
5. Documentation—Does the software package include a complete and easy to understand instruction booklet? Does the program itself display sufficient instructions on the screen to set-up and operate the system?
6. Aesthetic Factors and Usability—Are letters, numerals, and graphics easy to recognize? Is a manageable amount of text presented at one time? Is it easy to circumvent initial instructions if the student already knows what to do? Is there a consistent way to get instructions displayed if a student forgets what to do?

It is important for teachers to learn how to load programs from disks, manage computer files, and trouble-shoot simple hardware and system level software problems. Local teacher user groups are often the best source of information and help.

Teachers also need to review new software carefully prior to introducing it to the class (see chapter 5 for suggested criteria). Look for ease of data entry, informative and motivating displays, and useful student summaries.

Once appropriate software has been identified, it is often helpful to schedule initially two children at a time on the computer. Students working in pairs are more likely to remind each other of easily overlooked computer operating procedures without disrupting the remainder of the class. For programs that require greater problem-solving or writing skills, two students can distribute responsibilities for entry and can often achieve more than they would individually in twice the allotted time.

Representative examples of mathematics drill-and-practice software include: the *Mastering Math Series* (MECC), which provides structured computation practice interspersed with games and other reward displays; *Challenge Math* (Sunburst Communications), which creates a setting where strange characters roam an old mansion and which requires accurate use of basic whole number operations; and *Survival Math* (Sunburst Communications), which embeds computational practice within simulated business activities such as the *Travel Agent Contest* and the *Hot Dog Stand*.

Concept Development Software

A wide range of software provides concept development for place value and the number operations. Examples include *Base Ten On Basic Arithmetic* (MECC) and *Math Ideas with Base Ten Blocks* (Cuisenaire Company of America, Inc.). These programs allow children to select the difficulty level for different types of exercises (counting, number comparison, addition, subtraction, multiplication, and division). The exercises are displayed using proportional base-10 displays. The user is asked a series of questions that directs the program to carry out the required actions. For example, when presented with two numbers to be added, *Math Ideas* collects all the unit blocks in the ones column at the bottom of the display and asks if the result needs to be grouped. If the user answers yes (and yes is correct), the program collects 10 units and moves 1-tens block into the tens column. The process continues until all the columns are added and regrouped (see Figure 9.51).

After each example, this program displays how many questions the students answered correctly. Some tutorial assistance is available for each operation. This type of software seems particularly useful for students who continue to experience confusion about number operation concepts in the upper grades, students who may have missed introductory lessons, or primary students who need additional follow-up activities after initial instruction using base-10 manipulatives.

Problem-Solving Software

Several good programs are available for integration into classroom problem-solving instruction.

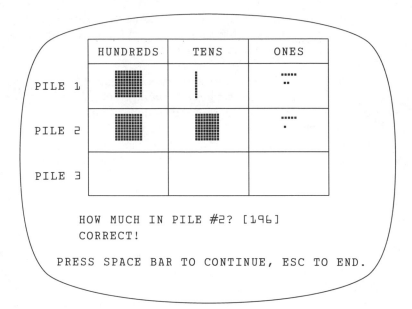

FIGURE 9.51 Math Ideas with Base-10 Blocks. *Note:* From *Math Ideas with Base Ten Blocks* [Computer Program] by Cuisenaire Company of America, 1982. Copyright 1982 by Cuisenaire Company of America, Inc. Reprinted by permission.

Arith-Magic. This software contains three useful problem-solving activities, *Diffy, Tripuz,* and *Magic Squares* (Quality Educational Designs). *Diffy* is a computer-supported version of the game described in the problem-solving section of this chapter. It allows the student to select four numbers, prompts the user to enter successive differences, and counts the number of cycles required to achieve a difference of zero. The user will discover that it is difficult to pick four whole numbers requiring more than eight cycles (see Figure 9.52). The program also allows the use of decimal values and fractions.

Tripuz offers similar computation practice using all four basic operations. Six numbers are arranged in a triangular shape. The three numbers that form the vertices of the triangle are hidden. The values shown at each midpoint are the sums (differences, products, or quotients) of each pair of vertices (see Figure 9.53).

The objective is to find the three missing values. *Guess-and-test, write a systematic list, work backwards,* and *find a rule* strategies can all be employed to solve this problem. The program checks the choices supplied by the user.

Magic Squares provides addition fact practice as the means to solve a more complex problem. The program displays a 3 × 3 array. It shows three initial values, and the students try to place the remaining digits 1–9 in the six remaining locations so that each row, column, and diagonal has the same sum (see Figure 9.54).

There are many possible solutions to this problem using all the digits 1–9. Students should look for patterns in successful solutions to help them find new

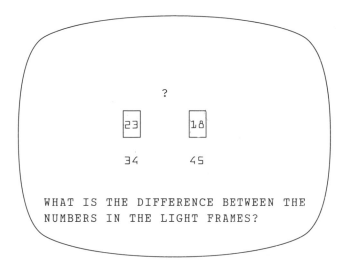

FIGURE 9.52 Diffy. *Note:* From *Arith-Magic* [Computer program] by J. Benton, 1979. Copyright 1979, 1983 by Quality Educational Designs, Inc. Reprinted by permission.

ones. Activities like these, which require students to use basic skill knowledge in the service of solving more interesting problems, can be particularly useful additions to the mathematics curriculum. See chapter 4 for additional problem-solving software.

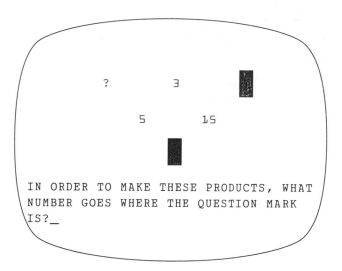

FIGURE 9.53 Tripuz. *Note:* From *Arith-Magic* [Computer program] by J. Benton, 1979. Copyright 1979, 1983 by Quality Educational Designs, Inc. Reprinted by permission.

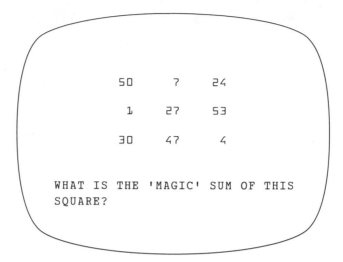

FIGURE 9.54 Magic Squares. *Note:* From *Arith-Magic* [Computer program] by J. Benton, 1979. Copyright 1979, 1983 by Quality Educational Designs, Inc. Reprinted by permission.

ADAPTING INSTRUCTION FOR CHILDREN WITH SPECIAL NEEDS

Alternative Addition Algorithms

Special-needs children who exhibit protracted difficulty learning to add and subtract may respond well to the introduction of alternative *low stress* algorithms (Hutchings, 1976). For example, column addition involving three or more values can be facilitated by using *half-space* notation. The first pair of numbers is added and the sum written slightly below the second value with the tens digit on the left and the ones digit on the right. The next number is added to the ones digit (3 + 6 = 9) and written to the right of the 6 (do not bring down the 1-ten in the previous sum). Add the next number to the ones digit above (9 + 5) and write the sum to the right and left of the 5 as shown below. The final result is found by writing the ones digit (4) below the line, counting the number of tens (2), and writing this digit in the tens column.

Standard Notation	*Half-Space Notation*
6	6
7	$_1{}^7{}_3$
6	6_9
+ 5	$+\,_1{}^5{}_4$
24	24

Values larger than 9 can also be added using this method. First, add the ones column (6 + 8 = 14; 4 + 5 = 9). Write the ones digit for the sum (9) and regroup the 1-ten (①) into the tens column. Add the pairs of values in the tens column (1 +

FIGURE 9.55 Place Value Table with Working Space

$2 = 3$; $3 + 7 = 10$; $0 + 3 = 3$), write the tens digit result in the tens column, count the number of hundreds (1), and write the hundreds digit in the hundreds column.

$$
\begin{array}{c c c}
 & {}^{①}2_3 & 6 \\
 & {}^{1}7_0 {}^{1}8_4 & \\
+ & 3_3 & 5_9 \\
\hline
1 & 3 & 9
\end{array}
$$

Another low stress algorithm involves using a *working space* to reduce the burden on the working memory when adding multidigit numbers. As shown earlier for subtraction, working space is used like a calculator memory to store values temporarily while each column is regrouped into the proper form. As shown in Figure 9.55, the working space for addition is inserted between the answer box and the line beneath the addends in the place value table to allow sums larger than nine to be temporarily stored prior to regrouping.

During computation, the sum of each column is written in the working space *without* regrouping. This memory aid allows children to focus their attention on the regrouping task. The amount to be regrouped is subsequently moved to the top of the next column, and the remainder is moved down to the answer space. The same procedure is followed for each column. The final result is the number that falls through the dotted line into the answer box after each place value is regrouped.

Another alternative algorithm requires a good understanding of place value but simplifies the regrouping process. As shown in Figure 9.56, numbers represented by the digits in each place value are added and the sums written in separate rows in a working space beneath the addends. These partial sums are subsequently added giving the final answer.

100	10	1	
2	4	7	
3	8	5	+
	1	2	⇐ (7 + 5)
1	2	0	⇐ (40 + 80)
5	0	0	⇐ (200 + 300)
6	3	2	⇐ Sum

FIGURE 9.56 Partial Sum Addition Algorithm

When using this algorithm, children must pay attention to the actual value represented by each digit when adding. Notice the columns can be added in any order using this procedure. This process is inefficient with numbers larger than 1000 because the column of partial sums gets long. Though it may use a bit more paper, this algorithm is sometimes a useful alternative for children who are unable to add consistently using the standard algorithm.

Alternative Subtraction Algorithms

Alternative algorithms can also be helpful for teaching special-needs children to subtract. First, ensure that these children are provided with ample experience working with proportional and nonproportional representations prior to the exclusive use of symbolic algorithms. The consistent use of addition and subtraction tables can also reduce the memory demands required for multidigit exercises.

Figure 9.57 shows the standard decomposition subtraction algorithm along with two alternative algorithms. The low stress alternative moves the working space between the minuend and subtrahend. This positions the regrouped minuend directly above the subtrahend to minimize the interference of the crossed-out digits. Using this algorithm, children can carry out all the regrouping steps prior to subtracting the columns. For the low stress example shown in Figure 9.57b, first regroup 1-ten into the ones column, write 12 in the working space ones box, and write 4 in the working space tens box. Next, regroup 1-hundred into 10-tens, then write 1 in front of the 4 in the working space tens box, making 14, and 2 in the working space hundreds box. Once all the regrouping is completed, the columns can be subtracted in any order.

For the equal additions algorithm, instead of regrouping a unit from the adjacent column, one unit is added to *both* the subtrahend and the minuend in different columns. Since the same amount is added to both numbers (though in different columns), the overall *difference* is unchanged. For the example shown in Figure 9.57c, we cannot subtract the ones column, so we add 10-ones to the minuend in the ones column and 1-ten to the subtrahend in the tens column. Because we have added ten to the minuend and the subtrahend, the difference stays the same. Subtracting the ones column gives 8-ones. The tens column cannot be subtracted, so we add 10-tens to the minuend of the tens column and 1-hundred to the subtrahend of the hundreds column. The tens and hundreds columns can now be subtracted and the results written in the answer box. Although this procedure does not follow the place value development of the algorithm, equal additions subtraction may be useful in

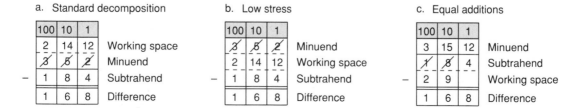

FIGURE 9.57 Alternate Subtraction Algorithms

cases where efforts to teach the decomposition or low stress algorithms have failed. This algorithm is used in parts of the United States and in other countries.

Solve the following subtraction exercises using the decomposition, low stress, and equal additions subtraction algorithms.

1. $365 - 278 = ?$

2. $232 - 89 = ?$

3. $417 - 45 = ?$

4. $208 - 29 = ?$

5. $1005 - 245 = ?$

Alternative algorithms offer upper-grade children who have experienced protracted difficulty with arithmetic a fresh opportunity to succeed. These alternative procedures minimize the requirements of keeping unwritten digits in mind throughout the computation. Exercises are broken into smaller, more manageable steps. The written record of each step can also help the teacher diagnose possible procedural errors. Alternative algorithms can also offer gifted and talented students an opportunity to explore familiar content in a new and interesting way.

SUMMARY

Addition and subtraction are fundamental concepts in elementary mathematics education. As Piaget observed, a thorough knowledge of class inclusion and reversibility contributes to a meaningful understanding of the addition and subtraction operations. Both the union of sets and measurement models can be used to introduce addition at the primary levels. Place value is introduced for sums larger than nine. The commutative, associative, and identity properties of addition can be employed to minimize the memorization requirements of the basic addition and subtraction facts. Closure, the fourth addition property, finds application in expanding the number system in order to solve new classes of problems. Addition and subtraction are inverse operations that reinforce each other's development when introduced concurrently at the primary level. Three types of subtraction can be observed: take-away, missing addend, and comparative subtraction. Each is modeled differently and has applications in real-world problems. Unifix cubes, base-10 blocks, bean sticks, chip trading materials, the abacus, place value tables, and to a lesser extent, the number line, can be used to introduce the addition and subtraction concepts and develop the symbolic algorithms. Developmental activity sequences (DAS) can be employed as a management technique to teach addition and subtraction concepts and skills to groups of children who are at various levels of development and experience. Children begin using concrete materials when introduced to a new concept or skill, recording their actions and results in place value tables. Some children may require intermediary work with nonproportional materials such as chip trading or the abacus, while others begin working exercises directly in place value tables or using the standard symbolic algorithm. An increased emphasis on mental computation promotes a deeper understanding of number properties, assists in analysis of problem situations, develops an improved number sense, encourages the design of alternative algorithms, and promotes estimation skills. Story and nonroutine problems involving addition and subtraction provide excellent opportunities for applying newly developed mathematics and language skills. Children need to possess the necessary computation skills to solve assigned problems, or calculators and/or tables should be made available. Problem-solving experiences should not be used as an excuse simply to practice number operation skills. Excellent calculator and computer software instructional materials are available to help with addition and subtraction drill and practice, concept, and problem-solving instruction. Introducing alternative low stress algorithms may be useful when working with special-needs children who have experienced protracted difficulty learning the addition and subtraction algorithms.

COURSE ACTIVITIES

1. Look through several resource books and develop an addition and subtraction basic fact practice kit. The kit should include at least five different games (slide rule, dice games, game boards, etc.) designed to provide a motivating environment for developing and practicing the basic addition and subtraction facts (0–9). Discuss your kit with your peers and, if possible, try out the activities with students in a classroom setting.

2. Write a lesson plan involving the use of proportional base-10 materials to introduce single and multidigit addition problems. Carry out the necessary manipulations yourself and record the actions and results in place value tables. Practice demonstrating the necessary manipulative actions needed to solve addition exercises. When you are confident discussing the procedures and correcting potential student errors, present the lesson to your peers and, if possible, a group of elementary students.

3. Using materials of your choice, set up several subtraction problems with zeros in the minuend. Without pointing or touching the materials, give verbal instructions to an assistant to carry out the necessary regrouping and subtracting actions. Try the same task blindfolded. Write the sequence of steps required to complete this task. Give the list to a novice, observe and record his or her efforts to complete the task. Discuss the results with your peers.

4. Write a lesson plan to introduce subtraction with regrouping across a zero using an abacus or chip trading materials. Practice modeling subtraction problems with addends of three or more digits. Record the actions and results in place value tables. Give the lesson to your peers and, if possible, to children in a classroom.

5. Write two different story problems to go with each of the following number sentence headlines:

   a. $12 + 12 = 30 - 6$
   b. $\frac{1}{2} + \frac{1}{2} + \frac{1}{2} = 1\frac{1}{2}$
   c. $(4 + 4) + 2 = ?$

6. List several ways a teacher can alter the presentation of word problems to enhance instruction. Select five story problems from a mathematics textbook and use

them to make up a series of exercises based on the techniques presented in this chapter. Solve them.

7. Design a problem that incorporates graphics, symbols, and at most, four words. The problem should require the application of addition and subtraction. Exchange the problems in class, comment on each other's designs, and publish a class book of wordless problems.

8. Find two nonroutine problems in an elementary textbook or another reference and try to solve them using the *guess-and-test, make a sketch*, or other strategy.

9. Review the Hutchings article listed in the references or other source to increase your repertoire of alternative computation algorithms. Find at least one alternative addition or subtraction algorithm that is not presented in this chapter. Discuss the similarities and differences among the various addition or subtraction procedures. If possible, introduce an alternative algorithm to a small group of children. Keep a record of the experience to discuss with your peers.

10. Include several examples of addition and subtraction activities, organized by grade level, in your lesson idea file.

11. Read one of the *Arithmetic Teacher* articles listed in the reference section. Write a brief report summarizing the main ideas of the article and describe how the recommendations for instruction might apply to your own mathematics teaching.

MICROCOMPUTER SOFTWARE

Arith-Magic Three strategy games, Diffy, Tripuz, and Magic Squares, each requiring the user to complete a number of computations to solve a problem successfully (Quality Educational Designs).

Base Ten On Basic Arithmetic Multiplication and place value practice (MECC).

Challenge Math Provides practice using basic whole number and decimal operations in an environment containing space intruders, a dinosaurlike creature, and a mysterious mansion (Sunburst Communications).

The Factory: Exploration in Problem Solving Practice in developing strategies for solving word problems (Sunburst Communications).

Mastering Math Series Computation drill and practice series interspersing exercises with games and reward displays (MECC).

Match Game Concentration game with flexible, user-defined lessons (Teaching Tools).

Math Ideas With Base Ten Blocks Addition practice using graphic representations of proportional base-10 materials (Cuisenaire Company of America).

Multiploy Arithmetic drill in a game format (Reston Publishing Company).

Place Value Place Two programs, one a representational calculator allowing exploration of regrouping during addition and subtraction and the other a strategy game employing place value concepts (Inter-Learn Inc.).

Stickybear Math: An Addition and Subtraction Program for Children 6 to 9 A series of well-designed drill and practice activities for the early primary grades (Weekly Reader Family Software).

Survival Math Computation practice for upper-grade and middle school level employing work situations (Sunburst Software).

REFERENCES AND READINGS

Baratta-Lorton, M. (1976). *Mathematics their way.* Menlo Park, CA: Addison-Wesley.

Beattie, I. (1986). Modeling operations and algorithms. *Arithmetic Teacher, 33*(6), 23–28.

Burton, G. (1984). Teaching the most basic Basic. *Arithmetic Teacher*, *32*(1), 20–25.

California State Department of Education (1985). *Mathematic framework for California public schools*. Sacramento, CA.

Case, R. (1982). General developmental influences on the acquisition of elementary concepts and algorithms in arithmetic. In T. Carpenter, J. Moser, & T. Romberg (Eds.), *Addition and subtraction: A cognitive perspective* (pp. 156–170). Hillsdale, NJ: Erlbaum.

Driscoll, M. (1980). *Research within reach: Elementary school mathematics*. St. Louis: CAMREL

Duea, J., & Ockenga, E. (1982). Classroom problem solving with calculators. *Arithmetic Teacher*, *29*(6), 50–51.

Greene, G. (1985). Math-facts memory made easy. *Arithmetic Teacher*, *33*(4), 21–25.

Greenes, C., Immerzeel, G., Ockenga, E., Schulman, J., & Spungin, R. (1980). *Teacher's commentary, techniques of problem solving, problem deck A–D*. Palo Alto, CA: Dale Seymour.

Hutchings, B. (1976). Low stress algorithms. In D. Nelson & R. Reys (Eds.), *Measurement in school mathematics: 1976 Yearbook* (pp. 218–239). Reston, VA: National Council of Teachers of Mathematics.

Pereira-Mendoza, L. (1984). Using English sentences and pictures for practice in mathematics. *Arithmetic Teacher*, *32*(1), 34–38.

Reisman, F. (1977). *Diagnostic teaching of elementary school mathematics: Methods and content*. Chicago: Rand McNally.

Reys, B. (1985). Mental computation. *Arithmetic Teacher*, *32*(6), 43–46.

Reys, B. (1986). Estimation and mental computation: It's "about" time. *Arithmetic Teacher*, *34*(1), 22–23.

Schoen, J. (Ed.). (1986). *Estimation and mental computation: 1986 Yearbook*. Reston VA: National Council of Teachers of Mathematics.

Souviney, R., Keyser, T., & Sarver, A. (1978). *Mathmatters: Developing computational skills with developmental activity sequences*. Glenview, IL: Scott, Foresman.

Sowell, E. (1987). Developmental versus practice lessons in the primary grades. *Arithmetic Teacher*, *34*(7), 6–8.

Suydam, M., & Reys, R. (Eds.). (1978). *Developing computational skills: 1978 Yearbook*. Reston, VA: National Council of Teachers of Mathematics.

Thompson, C., & Hendrickson, D. (1986). Verbal addition and subtraction problems: Some difficulties and some solutions. *Arithmetic Teacher*, *33*(7), 21–25.

Thornton, C., & Wilmot, B. (1986). Special learners. *Arithmetic Teacher*, *33*(6), 38–41.

Touger, H. (1986). Models: Help or hindrance? *Arithmetic Teacher*, *33*(7), 36–37.

Trafton, P. (1978). Estimation and mental arithmetic: Important components of computation. In M. Suydam & R. Reys (Eds.), *Developing computational skills: 1978 Yearbook* (pp. 196–213). Reston, VA: National Council of Teachers of Mathematics.

Wirtz, R. (1974). *Drill and practice at the problem solving level*. Carmel, CA: Curriculum Development Associates.

Young, J. (1984). Uncovering the algorithms. *Arithmetic Teacher*, *32* (3), 20.

10

Multiplication and Division of Whole Numbers

Probably nothing . . . could have astonished a Greek mathematician more than . . . the whole population . . . perform[ing] . . . division for the largest numbers. . . . Our modern process of easy reckoning . . . is the most miraculous result of a perfect notation.

—Alfred North Whitehead

Upon completing Chapter 10, the reader will be able to

1. Describe the developmental concept associated with the understanding of multiplication and division.
2. Describe an instructional hierarchy for introducing the whole number multiplication and division operations.
3. Describe three multiplication models and give an appropriate lesson for each.
4. Describe three division models and give an appropriate lesson for each.
5. Describe the inverse relationship between multiplication and division and give an example of an appropriate classroom lesson.
6. Describe several techniques to help children memorize the multiplication and division facts.
7. List four properties of multiplication and give examples of how each is used in the development of multiplication concepts and skills.
8. List five instructional materials that can be used to introduce the multiplication and division algorithms and give sample lessons for each.
9. Describe examples of developmental activity sequences that introduce the multidigit multiplication and division algorithms.
10. Describe how rounding off can be used to estimate partial quotients in division problems and give an appropriate lesson.
11. Describe three ways to introduce multiplication and division story problems to primary and upper-grade children.
12. Describe nonroutine multiplication and division problems and give a sample lesson.
13. Describe how calculators and computers can be used to teach multiplication and division.
14. Describe alternative algorithms for multiplication and division and discuss their application in the instruction of special-needs children.

Teaching multiplication and division of whole numbers follows an instructional hierarchy that includes early counting experiences, concept development, the memorization of basic facts, and work with algorithms. An example of one possible instructional hierarchy for multiplication and division is shown in Figure 10.1. Reference to an instructional hierarchy can help with the organization of activities into effective lesson sequences.

Children are introduced to informal multiplication and division situations in the early primary grades. Formal work generally begins in Grade 2 or 3 for multiplication and in Grade 3 for division; instructon can extend through Grade 6 for more complex exercises.

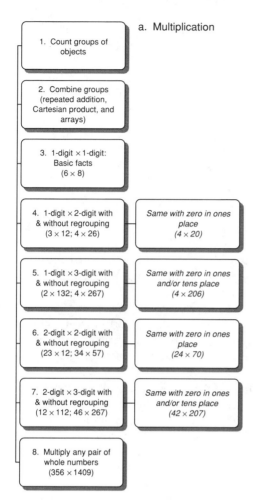

a. Multiplication

1. Count groups of objects

2. Combine groups (repeated addition, Cartesian product, and arrays)

3. 1-digit × 1-digit: Basic facts (6 × 8)

4. 1-digit × 2-digit with & without regrouping (3 × 12; 4 × 26) — Same with zero in ones place (4 × 20)

5. 1-digit × 3-digit with & without regrouping (2 × 132; 4 × 267) — Same with zero in ones and/or tens place (4 × 206)

6. 2-digit × 2-digit with & without regrouping (23 × 12; 34 × 57) — Same with zero in ones place (24 × 70)

7. 2-digit × 3-digit with & without regrouping (12 × 112; 46 × 267) — Same with zero in ones and/or tens place (42 × 207)

8. Multiply any pair of whole numbers (356 × 1409)

FIGURE 10.1 Instructional Hierarchies

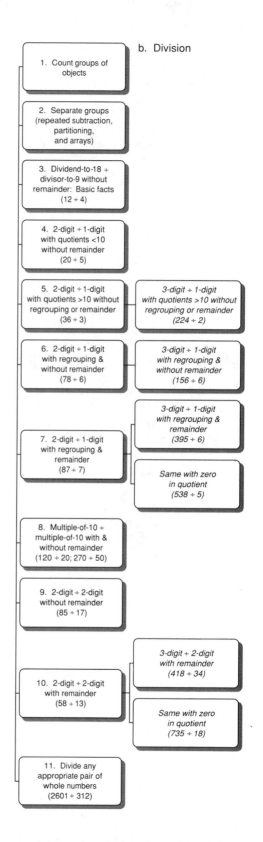

b. Division

1. Count groups of objects

2. Separate groups (repeated subtraction, partitioning, and arrays)

3. Dividend-to-18 ÷ divisor-to-9 without remainder: Basic facts (12 ÷ 4)

4. 2-digit ÷ 1-digit with quotients <10 without remainder (20 ÷ 5)

5. 2-digit ÷ 1-digit with quotients >10 without regrouping or remainder (36 ÷ 3) — 3-digit ÷ 1-digit with quotients >10 without regrouping or remainder (224 ÷ 2)

6. 2-digit ÷ 1-digit with regrouping & without remainder (78 ÷ 6) — 3-digit ÷ 1-digit with regrouping & without remainder (156 ÷ 6)

7. 2-digit ÷ 1-digit with regrouping & remainder (87 ÷ 7) — 3-digit ÷ 1-digit with regrouping & remainder (395 ÷ 6) / Same with zero in quotient (538 ÷ 5)

8. Multiple-of-10 ÷ multiple-of-10 with & without remainder (120 ÷ 20; 270 ÷ 50)

9. 2-digit ÷ 2-digit without remainder (85 ÷ 17)

10. 2-digit ÷ 2-digit with remainder (58 ÷ 13) — 3-digit ÷ 2-digit with remainder (418 ÷ 34) / Same with zero in quotient (735 ÷ 18)

11. Divide any appropriate pair of whole numbers (2601 ÷ 312)

INTRODUCING THE CONCEPT OF MULTIPLICATION

Just as subtraction *undoes* addition, division is the **inverse** operation of multiplication. Generally, it is more efficient to introduce an operation and its inverse simultaneously.

Inverse Operations—
$$\rightarrow \text{Addition} \qquad 3 + 4 = 7$$
$$\rightarrow \text{Subtraction} \qquad 7 - 4 = 3$$

Inverse Operations—
$$\rightarrow \text{Multiplication} \qquad 4 \times 5 = 20$$
$$\rightarrow \text{Division} \qquad 20 \div 5 = 4$$

The ability to mentally reverse operations enables children to use their knowledge of addition and multiplication to understand subtraction and division.

Piaget observed that the transitive relationship, or the concept of **multiequivalence**, is an important factor in the development of a meaningful understanding of multiplication and division. Transitivity involves the transfer of a relationship from one situation to another. For example, suppose a child is asked to remove all the eggs from a standard carton, place them in a bowl, refill the carton from another supply, and then put these eggs in a plate. A child who recognizes, without having to recount, that the bowl and plate now contain an equal number of eggs understands the notion of transitivity (see Figure 10.2).

Multiequivalence is an extension of the conservation of number concept. An understanding of multiequivalence requires students to coordinate the transformation of more than two equal groups at a time. This concept forms a basis for understanding multiplication and division. For example, if eighty pennies are systematically counted out into ten stacks of equal height, a child who understands transitivity can count one stack and be confident that the others contain the same amount. Also, the mental ability to reverse the process is an important factor in assuring the child that the total number of distributed objects, plus any remainder, is equal to the initial amount.

Children who have not attained transitivity and reversibility may find it helpful to use base-10 materials throughout the development of the multiplication and division concepts and algorithms. In fact, research shows that the premature introduction

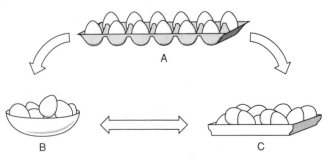

A = B and A = C, therefore B = C

FIGURE 10.2 Multiequivalence

of symbolic algorithms may be a primary cause for the development of children's arithmetic error patterns (Ashlock, 1976; Driscoll, 1980).

Types of Multiplication

There are three common models of multiplication:

☐ *repeated addition* multiplication
☐ *cartesian product* multiplication
☐ *array* multiplication

Each represents a different type of multiplication situation found in real-world situations.

Repeated Addition Multiplication

Children's early experiences with multiplication often begin with **skip counting**. For example, children learn to skip count by twos when counting shoes, socks, or mittens; by threes when counting the utensils (knives, forks, and spoons) needed for dinner or the wheels on a group of tricycles; by fours when counting the number of horseshoes in a horse race or the tires on cars in a parking lot; and by fives when counting fingers and toes in the classroom or the amount of money in a stack of nickels. Skip counting is a handy way to determine the total number of objects in several identical groups (see Figure 10.3).

Adding a column of identical values is called **repeated addition**. Once these special sums are memorized, children no longer need to skip count or calculate to determine the solution. To **multiply,** then, children need to know the number of equal groupings (e.g., three pairs of socks) and the number in each group (two in each pair). The formal terms for these values—**multiplier** (first factor), **multiplicand** (second factor), and **product** (answer)—are normally introduced in Grade 3.

2 + 2 + 2 = 6

FIGURE 10.3 Skip Counting Pairs of Socks

$$\textit{Repeated Addition} \qquad \textit{Multiplication}$$

$$2 + 2 = 4 \quad \rightarrow \quad 2 \times 2 = 4$$

$$2 + 2 + 2 = 6 \quad \rightarrow \quad 3 \times 2 = 6$$

$$2 + 2 + 2 + 2 = 8 \quad \rightarrow \quad 4 \times 2 = 8$$

Simple multiplication, then, can be thought of as the memorized sums of repeated addition exercises. There are one hundred sums involving 0–9 equal groups, each containing 0–9 objects. These repeated addition exercises generate the basic multiplication facts 0×0 through 9×9. Normally, only these one hundred basic facts involving single-digit factors are memorized in school. Products for all whole numbers larger than nine can be calculated using these basic facts.

How many products are there for all the combinations of single-digit and double-digit factors? Do you think it would be useful to memorize the multiplication facts to 99×99 ? Why?

Cartesian Product Multiplication

A second way to think about multiplication involves **cartesian products**. A cartesian product refers to the number of combinations that can be made using objects from two or more sets. For example, as shown in Table 10.1, if two children (Sarah and Rebecca) have three T-shirts—red, green and blue—how many different ways can they wear the three shirts?

If Sarah wears the red shirt, Rebecca can select either a green or blue one. Similarly, Rebecca has two choices if Sarah wears the green or blue shirt. A total of six different combinations is possible. We can write this combination problem as the multiplication exercise:

$$2 \times 3 = 6$$

How many different ways can three children wear two T-shirts? Make a table to find out. How many different ways can two children wear three shirts and four hats? Try to find a pattern relating the number of objects in two or more groups with the number of resulting combinations.

TABLE 10.1 Cartesian Product

Child	Ways to Wear T-Shirts					
Sarah	R	R	G	G	B	B
Rebecca	G	B	R	B	R	G

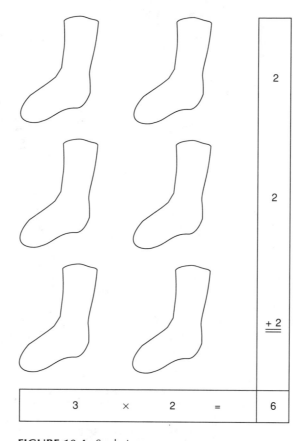

FIGURE 10.4 Sock Array

Array Multiplication

To show products of larger values efficiently, an alternate display of repeated addition can be introduced. Instead of putting each addend in a separate group, you can place objects into an **array,** a systematic arrangement of objects that has equal rows and equal columns (see Figure 10.4).

Arrays made from squared paper can be used to display multiplication problems. The factors are the length and width of a rectangle, and the product is the area (see Figure 10.5).

$3 \times 2 = 6$

FIGURE 10.5
Squared Paper Array

Using centimeter squared paper, cut out arrays representing the following multiplication exercises.

1. 4 × 6 = ?

2. 3 × 3 = ?

3. 5 × 8 = ?

4. 8 × 5 = ?

5. 7 × 1 = ?

A distinct advantage of array displays becomes apparent when double-digit factors are introduced. It is possible to take advantage of the structure of base-10 materials to separate large multiplication exercises into manageable chunks. The manipulations and display correspond to the multiplication algorithm. For example, the exercise 6 × 14 = 84 is displayed as an array in Figure 10.6a. The factors are positioned outside the vertical and horizontal axes, and the product is displayed as a rectangular array inside. Problems involving two double-digit factors can also be displayed using an array (see Figure 10.6b).

Products larger than 999 are more difficult to display using concrete materials. By the time children are asked to work with larger products, they should be proficient in using the symbolic algorithm.

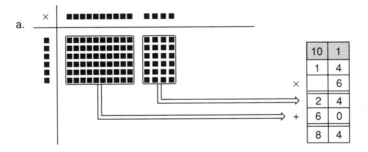

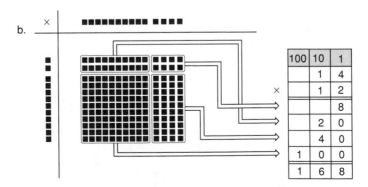

FIGURE 10.6 Base-10 Array

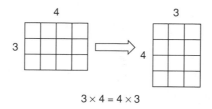

$$3 \times 4 = 4 \times 3$$

FIGURE 10.7 Commutative Arrays

The Properties of Multiplication

Several properties of multiplication are introduced at the elementary grade level. The **commutative property** ($3 \times 4 = 4 \times 3$) can be observed in the array display of multiplication. Reversing the factors reorients the display but does not alter the product (see Figure 10.7). In the section on learning the multiplication facts, the commutative property is used to reduce the number of facts to be memorized.

The **associative property** for multiplication is introduced when more than two factors are multiplied. This property is demonstrated in Figure 10.8, where the rectangular construction has a 3-cube $\times$ 3-cube base and is 4-cubes tall. The total number of cubes in this construction is $(3 \times 3) \times 4 = 36$. If a rectangular prism is constructed with a 3×4 base and a height of 3 cubes, there would still be a total of 36 cubes ($(3 \times 4) \times 3 = 36$). Have children find several other ways to construct a rectangular prism from 36 cubes and record the results as a number sentence. Regardless of the order of the factors, the product remains the same.

The associative property can also be used as a strategy for mentally calculating products of some groups of three or more values. For example, $6 \times 5 \times 5 = 6 \times 25 = 150$; or, $23 \times 2 \times 50 = 23 \times 100 = 2300$.

The term **distributive property of multiplication over addition** means that multiplying a number by the sum of any two values gives the same result as multiplying the number by each of the two values and adding the products ($5 \times (4 + 3) = (5 \times 4) + (5 \times 3)$). Children can use number facts they have already memorized to find the products of larger numbers. For example, using the distributive property, the exercise 4×12 can be restated as

$$4 \times 12 = 4 \times (6 + 6) = (4 \times 6) + (4 \times 6) = 24 + 24 = 48$$

Children can also verify the distributive property using arrays. First, they can display the exercise as shown in Figure 10.9a. The second factor can then be grouped

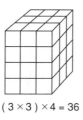

$$(3 \times 3) \times 4 = 36$$

FIGURE 10.8

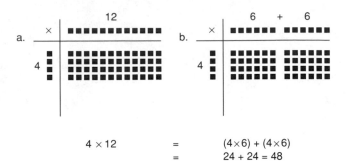

$$4 \times 12 \qquad = \qquad (4 \times 6) + (4 \times 6)$$
$$= \qquad 24 + 24 = 48$$

FIGURE 10.9 Multiplication Using Distributive Property

as shown in Figure 10.9b and the two 4×6 arrays constructed. (Other groupings, such as $(4 \times 2) + (4 \times 10)$, would also be possible.) The product is written below each array and the sum computed. Children then compare the two arrays side by side to verify that the answer remains the same.

The **identity element** for multiplication is 1 ($1 \times n = n$). Have children use arrays to show that any whole number multiplied by 1 gives a product equal to the original number. Also, arrays can be used to show that any whole number multiplied by 0 equals 0 (0×4 gives an array containing no items).

INTRODUCING THE DIVISION CONCEPT

There are three common models of division:

☐ *repeated subtraction* division
☐ *partitioning* division
☐ *array* division

Repeated Subtraction Division

Measurement, or **repeated subtraction,** division is the inverse of repeated addition multiplication. For example, suppose a family has 18 ears of corn. If each person can eat 3 ears, how many persons will the corn feed (see Figure 10.10)?

Distributing ears of corn in groups of three and counting the number of groups can be recorded using the following algorithm.

6 groups of 3

FIGURE 10.10 Division Using Repeated Subtraction

```
Number in each group →  3 )‾1‾8‾
                          − 3    1
                          15
                          − 3    1
                          12
                          − 3    1
                           9
                          − 3    1
                           6
                          − 3    1
                           3
                          − 3    1
       Remainder →         0     6  ← Number of groups
```

Partitioning Division

A second way to think about division involves **partitioning** sets into equal groups. For example, to share 18 ears of corn (3 ears each), children partition the ears into 3 equal piles (each person gets one ear from each pile) and count the number in each group to see how many people can be fed (see Figure 10.11).

The distributing and counting procedure can be recorded using the standard division algorithm.

```
                                    6  ← Number in each group
        Number of groups  →   3 )‾1‾8‾
                                − 18
                                   0  ← Remainder
```

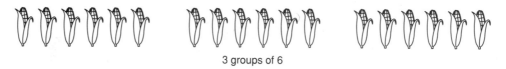

3 groups of 6

FIGURE 10.11 Division Using Partitioning

Array Division

Introducing the array representation helps children coordinate the repeated-subtraction and partitioning views of division. Using the organizing mat shown in Figure 10.12a, the division exercise is set up with the **divisor** (the number of groups or items in each group) along the vertical axis, the **dividend** (the number of objects to be distributed) inside. The **quotient** (the number of objects in each group) is subsequently displayed above the completed array. These formal terms for the three values in a division exercise are normally introduced in Grade 4. The division task is to organize the dividend of 18 into an array with 3 equal rows. Using *repeated subtraction,* we can systematically distribute objects one at a time in each row, keep-

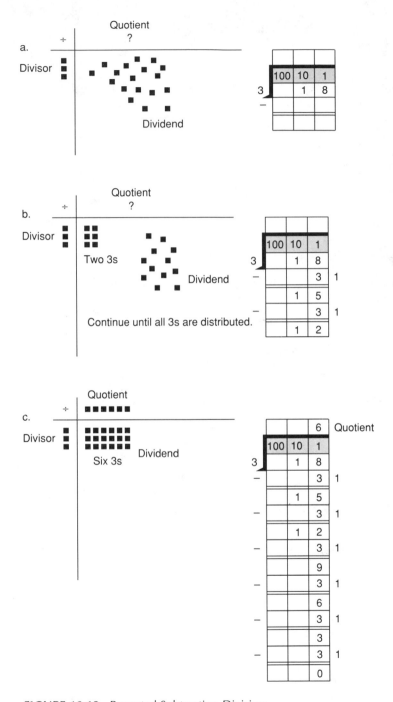

FIGURE 10.12 Repeated Subtraction Division

ing them equal in length. When an insufficient number of objects to be distributed remains, the row length gives the number of 3s contained in 18 (see Figure 10.12b,c).

Partitioning offers a more efficient procedure for constructing the division array when large dividends are involved and facilitates the development of the standard symbolic algorithm. To use partitioning as a division method, first estimate the number of items that can be distributed equally among the groups, then check to see if enough objects are available in the dividend to distribute. For example, Figure 10.13a shows that for 18 objects, an estimated 5 objects can be equally distributed among the 3 groups. The first estimate of 5 objects in each group leaves a remainder of 3. As shown in Figure 10.13b, distributing these 3 objects gives the final quotient of 6.

Children can also demonstrate the special role in division of the identity value 1. Any value divided by 1 gives a quotient equal to the original number (see Figure 10.14). Also, any number divided by itself equals 1.

Using counters, construct arrays that show the partitioning solution for each of the following division exercises.

1. $2\overline{)8}$	**4.** $8\overline{)32}$	**7.** $1\overline{)8}$
2. $3\overline{)12}$	**5.** $7\overline{)35}$	**8.** $1\overline{)12}$
3. $5\overline{)25}$	**6.** $6\overline{)36}$	**9.** $8\overline{)8}$

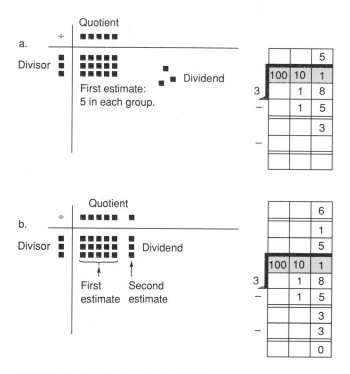

FIGURE 10.13 Division Using Partitioning

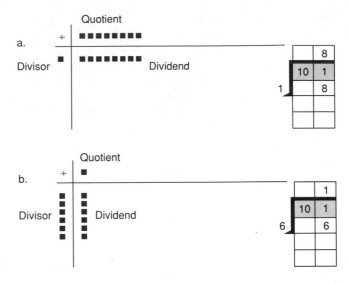

FIGURE 10.14

Language

Several English words are commonly used to specify the multiplication and division operations. Most phrases that suggest multiplication can be transcribed symbolically as they are read from left to right. For example:

☐ I'll take six *of* those cartons of eggs—($6 \times 12 = 72$).
☐ Take two *of* the six-packs—($2 \times 6 = 12$).
☐ Find six *times* eight—($6 \times 8 = 48$).
☐ *Multiply* three and four—($3 \times 4 = 12$).
☐ Find the *product* of six and seven—($6 \times 7 = 42$).

Since multiplication can be thought of as a special kind of addition, phrases that imply summing identical addends can also suggest multiplication. For example:

Three people *each have* six shells. Find the number of shells they have altogether—($3 \times 6 = 18$).

Because multiplication can be represented by an array, phrases that describe the counting of items displayed in equal rows and columns can refer to multiplication as well. For example:

Find the *area* of a rectangle four meters by five meters—($4 \times 5 = 20$).

Finally, cartesian products introduce additional phrases that signal a multiplication operation.

If a fast food restaurant offers 4 kinds of burgers and 3 drinks, *how many different* meals are possible—($4 \times 3 = 12$).

> Write a story or draw a picture of a situation representing each of the three models of multiplication: repeated addition, cartesian products, and arrays. Exchange the problems with a partner and write the type of multiplication model represented next to each problem.

When translating story problems involving division, it is important that children have experience identifying examples of the different models of division in real-world situations. For example:

Repeated subtraction: How many *students* will be able to share 40 pencils if each receives 10 pencils?

$$\frac{4 \; students}{10 \overline{)40}}$$

Partitioning: If there are 10 students, how many *pencils* will each receive if 40 pencils are equally shared?

$$\frac{4 \; pencils}{10 \overline{)40}}$$

The first example seeks the number of *students*, while the second asks for the number of *pencils*. Both examples imply that division should be used to solve the problem.

Problem situations that involve arrays of objects can also require division for their solution. For example:

Array: Using a bag of 400 seeds, how many rows of corn can be planted if each row contains 25 seeds?

$$\frac{16}{25 \overline{)400}}$$

Like subtraction, reading division exercises can be confusing at first because English provides both left-to-right and right-to-left interpretations. For example, if 3 is the divisor and 12 the dividend, the description can read

☐ *Left to right*—how many threes are contained *in* twelve: $3 \overline{)12}$
☐ *Right to left*—twelve divided *by* three: $(12 \div 3)$

Students often confuse the dividend with the divisor, particularly when the dividend is smaller than the divisor ($2 \div 4 = \frac{1}{2}$ or 0.5). When introducing division, initial confusion may be reduced by consistently using the left-to-right description. The right-to-left description can be introduced later as part of the discussion of rational numbers and ratio.

Depending on the context, some of the words and phrases generally used to specify multiplication or division can imply other operations. For example, the word *of* used previously to specify multiplication can also imply subtraction, as in, "I want five of your ten cookies." Children need to practice interpreting a variety of multiplication and division representations to improve their ability to translate English story problems into symbolic exercises.

RECALLING THE MULTIPLICATION AND DIVISION FACTS

The Multiplication Facts

Before memorizing the one hundred basic facts, children should thoroughly understand the concept of multiplication and be able to recognize the multiplication operation in various real-world situations. Memorization of the multiplication table generally begins in Grade 2 or 3. While many children have mastered the basic facts by the end of Grade 4, some may take longer. The process is a formidable task for many children, even when the teacher provides several alternative drill activities and memory aids. Despite the teachers' and students' efforts, some children may not memorize the entire table by the end of Grade 6. The introduction of important topics in mathematics, such as problem solving, geometry, graphing, statistics, and probability, should not be delayed because a student lacks perfect mastery of the basic facts. Regular use of a multiplication table or a calculator may be warranted for Grade 6–8 children who cannot quickly recall the basic facts. Such tools can help special-needs children participate more effectively in the full range of mathematics instruction during the middle school years and beyond.

Two-Entry Multiplication Table

In general, it is an inefficient use of time for students to approach the task of learning the multiplication facts by memorizing the entire **two-entry multiplication table**. Children can employ general rules and patterns to minimize the long-term memory requirements (Brownell & Carper, 1943; Williams, 1971). For example, the *commutative property,* introduced as an aid for memorizing the addition facts in Chapter 9, is a very useful strategy for learning many of the multiplication facts as well.

Since zero times any value equals zero, the nineteen **zero facts** ($0 \times n$ or $n \times 0$) do not have to be memorized. Children should work with 0 factors until they can readily apply the multiplication by zero rule:

☐ The product of 0 and any number n is 0 (i.e., $0 \times n = 0$).

Eliminating the zero facts leaves eighty-one multiplication facts to memorize. The commutative property considerably reduces the number of these facts that must be committed to memory. Knowing the product of 6×8 obviates the need to also memorize 8×6. Of the eighty-one basic facts displayed in Figure 10.15, nine have **double factors** (e.g., 4×4). Of the remaining seventy-two facts, only thirty-six must be memorized, as both members can be recalled by knowing only one of each commuted pair.

×	1	2	3	4	5	6	7	8	9
1	1	2	3	4	5	6	7	8	9
2	2	4	6	8	10	12	14	16	18
3	3	6	9	12	15	18	21	24	27
4	4	8	12	16	20	24	28	32	36
5	5	10	15	20	25	30	35	40	45
6	6	12	18	24	30	36	42	48	54
7	7	14	21	28	35	42	49	56	63
8	8	16	24	32	40	48	56	64	72
9	9	18	27	36	45	54	63	72	81

FIGURE 10.15 Multiplication Table

The list of forty-five facts (the thirty-six unshaded, commuted pairs plus nine doubles in Figure 10.15) can be further reduced with the application of the identity rule:

☐ The product of 1 and any number n is n (i.e., $1 \times n = n$).

Children can apply this rule rather than memorize the nine remaining one facts. This leaves only thirty-six basic facts to commit to memory. Of these remaining facts, the two-facts, five-facts, and double-facts are easiest for children to recall: the two facts because it is possible to quickly use skip counting or repeated addition to find the product ($2 \times 8 = 8 + 8 = 16$); the five facts because it is easy to skip count by fives (the product always ends in a 0 or 5 digit); and the double facts because the identical factors place less demand on the working memory.

Once students demonstrate the ability to recall, or quickly compute, the two, five, and double facts, only fifteen multiplication facts remain. These must be memorized by using a combination of mnemonics, rhymes, patterns, and drills.

Children can develop their own rhymes (e.g., Six times eight is forty-eight) to help them remember elusive facts. Patterns can be used to help confirm a product. For example, several patterns can be found in Table 10.2. An interesting story about one pattern involves a fellow named Gus who always had trouble with his nine facts. He knew $1 \times 9 = 9$ but could not remember the rest. In desperation on a test one day, he carefully counted the number of exercises he was sure he would miss. Starting at the bottom, he wrote the digits 1 through 8 as he counted up to his one right

TABLE 10.2 Nine Facts

$1 \times 9 =$	9
$2 \times 9 =$	18
$3 \times 9 =$	27
$4 \times 9 =$	36
$5 \times 9 =$	45
$6 \times 9 =$	54
$7 \times 9 =$	63
$8 \times 9 =$	72
$9 \times 9 =$	81

answer, $1 \times 9 = 9$. Still frustrated, he counted the wrong items again by writing the digits 1 through 8 from the top down. He was greatly surprised when he scored 100 percent!

Another nine fact pattern involves the sum of the digits of the products. Notice that adding the digits in the products always gives the answer 9 ($54 \rightarrow 5 + 4 = 9$). This pattern can be employed as a way to check the nine facts. For example, children often confuse the multiplication facts associated with the products 27 and 28, 32 and 36, 42 and 45, 54 and 56, and 63 and 64. When confirming the product of 6×9, an initial guess of 56 can be immediately ruled out since the sum of the digits is not 9. The pattern also helps with the selection of the correct product when the initial guess is in the correct decade (6×9 must be in the 50s since 6×10 is 60; therefore, the answer is 54 due to the nines pattern).

The six and nine tables can also be quickly found if the five and ten tables are already known. For example, 6×7 is 7 more than 5×7, or $35 + 7 = 42$. Similarly, 6×9 is 6 less than 6×10, or $60 - 6 = 54$.

In the upper grades, children are frequently diagnosed as not knowing their multiplication tables. However, children can generally recall many of the one hundred multiplication facts: they may have difficulty remembering ten or so facts that do not follow one of the simplifying rules or patterns that enhance recall. Teachers can help children identify the precise set of facts which they have difficulty recalling. Children can then concentrate their practice on personalized lists of facts both in school and at home. Oral, mental, and written drills, flash cards, calculators, games, rhymes, mnemonics, patterns, and mental arithmetic can be used by children to help them quickly recall the basic facts. Success at this task is important for social and academic reasons, because further mathematics study often depends on the quick recall of the basic facts.

> Write the fifteen hard multiplication facts and write a mnemonic, rhyme, or pattern as an aid for memorizing each.

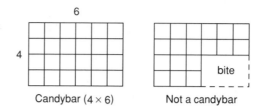

FIGURE 10.16   Candybar Multiplication

Candybar Multiplication

An interesting way to develop the two-entry multiplication table involves the use of arrays called **candybars** made of centimeter squared paper. A candybar is a rectangular array with a maximum of nine squares on an edge. There can be no *bites* removed from the candybars (see Figure 10.16).

If we assume that a 2×3 and 3×2 candybar are the same because only their orientation has changed, there are forty-five candybars in all. Children can work in teams to construct a complete set. They can determine the *name* of each candybar by counting the number of its pieces (centimeter squares). The result, of course, is the *product* of the length and width of the candybar.

Finally, children attach each candybar to a piece of tagboard (12cm $\times$ 12cm), write the candybar's name on the back as shown in Figure 10.17, and use the set as flash cards to practice the multiplication facts.

The set of the forty-five candybars (without the tagboard backing) can also be used to construct the multiplication table. Have students work in pairs, using a blank table made from centimeter squared paper as shown in Figure 10.18. Place a candybar so that one corner touches the upper left corner of the table and the edges align with the vertical and horizontal axes. Carefully lift the lower right corner of the candybar and write its name in the square directly under the corner bite. Remove the candybar, rotate it 90°, and repeat the process in its *commuted* position. Similar placement of all forty-five candybars will result in the familiar two-entry multiplication table.

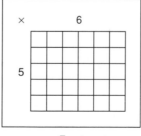
Front

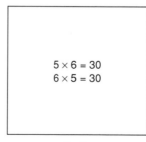
Back

FIGURE 10.17   Candybar Flashcard

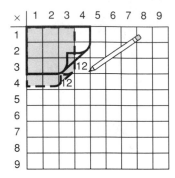

FIGURE 10.18 Constructing the Multiplication Table Using Squared Paper "Candybars"

Digit Multiplication

The multiplication facts involving the numbers 6–9 can be computed quickly by using the fingers on both hands. To solve the problem 7 × 8, for example, hold both fists in front with the thumbs on top as shown in Figure 10.19. Store the first factor 7 on the left hand by counting on from 5, extending the thumb and fingers one at a time. Similarly, count up to the second factor 8 on the right hand.

There are two extended, or **wiggly,** fingers on the left hand and three on the right. Assigning the value 10 to each wiggly gives 50. There are three folded fingers on the left hand and two on the right. Multiply these two values (3 × 2 = 6) and add the result to the wigglies, giving 50 + 6 = 56. Try other examples to see if the procedure works with all product pairs 6–9.

> Can you can figure out why this procedure works? Referring to patterns in the multiplication table may help with this investigation.

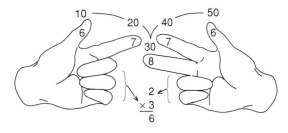

7 × 8 = 50 + 6 = 56

FIGURE 10.19 Digit Multiplication

TABLE 10.3 Family of Number Sentences

Multiplication	Division
$7 \times 8 = 56$	$56 \div 8 = 7$
$8 \times 7 = 56$	$56 \div 7 = 8$

Recalling the Division Facts

Because division is the inverse operation of multiplication, students who know the multiplication tables should find it easy to recall the single-digit quotients for divisors less than 10 and dividends less than 100. Exercises of this type are called the **division facts.** To recall the quotient for $56 \div 7$, first think of the related multiplication fact.

$$7 \times ? = 56 \qquad \text{Missing-factor division}$$
$$7 \times 8 = 56 \qquad \text{Related multiplication fact}$$

Just as the difference in a subtraction exercise can be thought of as a missing addend, a quotient can be considered a **missing factor.** Children who have memorized the multiplication table can practice recalling the second factor, or **quotient,** given the first factor and the product (e.g., $7 \times ? = 56$). Since multiplication and division are inverse operations, a **family of number sentences** can be introduced. Knowing any one member of the family shown in Table 10.3 facilitates the recall of the other three number sentences. When this process can be carried out for all number families 1–9, students will have memorized the division facts.

> Write the family of multiplication and division number sentences for the values 7, 9, and 63.

Division by Zero

It is important to point out that attempting to divide by 0 causes problems. Since the two division models make no sense when zero is used as a divisor, mathematicians say division by zero is **undefined.** For the example $8 \div 0$, neither division model makes sense.

☐ *Repeated subtraction*—How many times can **zero** be subtracted from eight? (Lots!)

☐ *Partitioning*—How many would each get if eight apples are divided equally among **zero** people? (The situation makes no sense.)

Quotient

÷	1	2	3	4	5	6	7	8	9
1	1	2	3	4	5	6	7	8	9
2	2	4	6	8	10	12	14	16	18
3	3	6	9	12	15	18	21	24	27
4	4	8	12	16	20	24	28	32	36
5	5	10	15	20	25	30	35	40	45
6	6	12	18	24	30	36	42	48	54
7	7	14	21	28	35	42	49	56	63
8	8	16	24	32	40	48	56	64	72
9	9	18	27	36	45	54	63	72	81

Divisor

Dividend

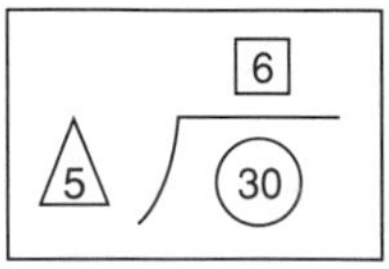

FIGURE 10.20 Division Table

Two-Entry Division Table

Initially, it is useful for students to refer to a table of division facts when introduced to more complex division exercises. The multiplication table can serve as such a reference: the numbers along the vertical axis become the divisors, the numbers interior to the table are the dividends, and the numbers along the top of the table represent the missing factors, or quotients (see Figure 10.20).

Students should be encouraged to practice using this table until they can confidently find quotients for the division facts. Notice the process *undoes* the related multiplication exercise.

Flashcards

A set of semicircular flashcards like the six tables example shown in Figure 10.21 can be used by pairs of students to practice the multiplication and division facts. One child shows the other either the multiplication or division side of the card and places a pencil in one of the holes to indicate the basic fact the second child is to recall. The partner states the answer and checks the result on a calculator if available (the correct answer is displayed on the leader's side of the card). The children then reverse roles.

One flash card is constructed for each of the factors 0–9. Use a compass to make circles (about fifteen centimeters in diameter) from tagboard and cut the circles in half along the diameter. Punch ten holes equally spaced around the arc of the

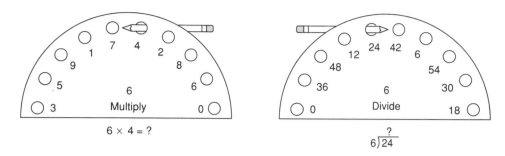

FIGURE 10.21 Multiplication and Division Flashcards

semicircle. Write the first factor for each card (0, 1, 2, . . . , or 9) on both sides near the center. Write the second factors (0 through 9) in random order beneath each hole on the multiplication side and the product under the corresponding hole on the division side.

Board Games

To practice the recall of multiplication and division facts, have each child write the facts with which they have difficulty on red cards with the answers on the back and the well-known facts on identical white cards (multiplication candybars discussed earlier can be used as well, circling the difficult facts with a red marker).

Make a game board similar to the one pictured in Figure 10.22. Students play in groups of two to four. The object is to recall multiplication facts correctly in order to move each rocket along its trajectory. A red card moves the rocket three spaces, the

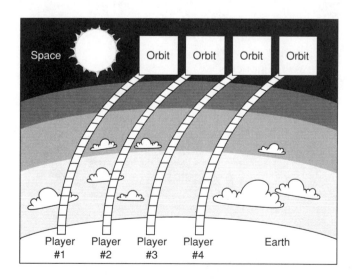

FIGURE 10.22 Space Shuttle Orbit Game

white card one space. Players take turns drawing from their own red or white piles. For each correct answer, the player moves ahead accordingly. For wrong answers, players move back the appropriate number of spaces. If the rocket moves back to earth after it takes off, it crashes. The last move must put the player's rocket exactly into orbit by landing on the *orbit insertion window*. Players who overshoot and become lost in space can continue playing but are unable to maneuver their rockets.

Encourage students to invent other board games to help them practice the basic arithmetic facts.

INTRODUCING THE MULTIPLICATION ALGORITHM

Multiplying by 10 and 100

Multiplication by 10 and 100 (and later, multiples of 10 and 100) provides a transition between the basic facts and multidigit exercises. As with addition and subtraction, proportional base-10 materials (bean sticks or base-10 blocks) can be used to display multiplication exercises as arrays. Concrete actions and results can be simultaneously recorded in place value tables to introduce the multiplication algorithm.

In Figure 10.23, the first factor (5) is placed along the vertical axis and the second factor (10) along the horizontal axis. To find the product, an array of five rows and ten columns is constructed. By using 10-rods instead of rows of ten separate units, the array can be quickly constructed.

Point out that multiplying a number by 10 has the effect of multiplying the value by 1 and moving the result one place value to the left. A 0 is entered into the ones column as a place holder as shown in Figure 10.23b.

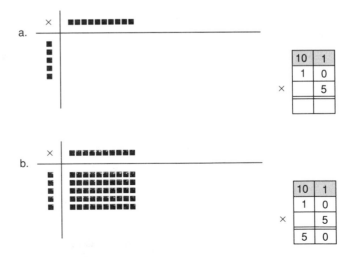

FIGURE 10.23 Multiplication Array Using Base-10 Materials

Using a multiplication mat and proportional base-10 materials, carry out the necessary actions to complete the following exercises. Record the actions and results in place value tables.

1. 3 × 10 = ?

2. 5 × 10 = ?

3. 7 × 10 = ?

4. 1 × 10 = ?

5. 15 × 10 = ?

The array procedure for multiplying by 100 is unwieldy because of the large quantity of base-10 materials involved. After they have had sufficient practice making arrays that show products with 10 as a factor, children find it easier in exercises involving the factors 100 and 1000 to extend the pattern of moving the digits of the first factor to the left in the place value table.

The array procedure can also display exercises involving a single-digit value times a multiple of 10 (see Figure 10.24). When products reach 100, the results are regrouped into the hundreds column.

After extensive experience using base-10 materials to multiply multiples of 10 by single-digit numbers, children can be introduced to multiplication exercises involving multiples of 100 and 1000, using symbols and place value tables only. Several practice worksheets with examples such as those shown in Figure 10.25 should be prepared.

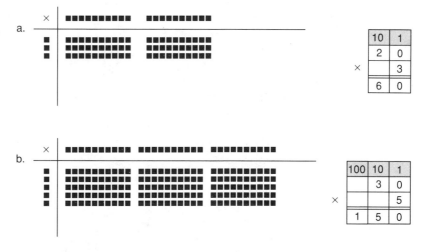

FIGURE 10.24

a.

1000	100	10	1
	3	0	0
×			9
2	7	0	0

b.

1000	100	10	1
	4	0	0
×			5
2	0	0	0

c.

1000	100	10	1
2	0	0	0
×			4
8	0	0	0

FIGURE 10.25

Complete the following multiplication exercises and record the results in place value tables.

1. 3 × 20 = ?

2. 5 × 20 = ?

3. 4 × 40 = ?

4. 3 × 200 = ?

5. 4 × 400 = ?

Single-Digit Times Double-Digit Values

Exercises involving single- and double-digit factors can be displayed using the multiplication mat and base-10 materials. Children construct an appropriate sized array using the largest possible base-10 blocks to simplify the process. The first factor is placed along the vertical axis and the second factor along the top. For example, 4 × 12 is displayed as in Figure 10.26. To find the product, fill in the first row with 1-ten and 2-ones. The rest of the array can then be filled in and the results recorded in a place value table. The two subproducts 4 × 2 and 4 × 10 are displayed by the two parts of the array. The total product is the sum of the two values: 8 + 40 = 48. This is a useful application of the distributive property of multiplication over addition.

Figure 10.27 shows how exercises with products larger than 99 can be displayed using the same procedure.

The **display algorithm** presented in a place value table appears somewhat different from the standard multiplication algorithm. Each subproduct is displayed separately in the display algorithm, whereas the standard procedure is to combine sets of subproducts on one line. For much of the work in elementary school, the display algorithm works as well as the standard algorithm. As the size of the factors increases, however, the number of subproducts grows quickly.

Double-Digit Times Double-Digit Values

The product of two double-digit factors can be displayed using procedures similar to those employed in the previous section. When constructing the product array, it is convenient to use the largest possible base-10 blocks to speed the process. For

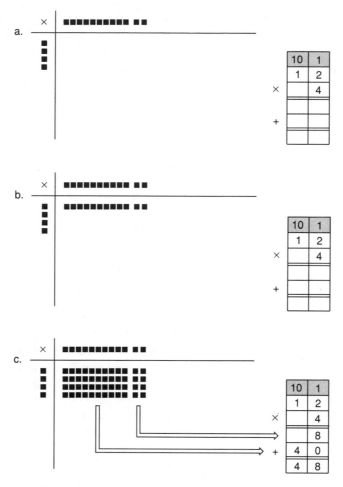

FIGURE 10.26

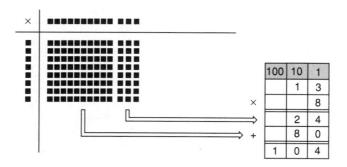

FIGURE 10.27

example, using one 100-flat is easier than constructing a 10×10 array from unit blocks or 10-rods.

Figure 10.28 shows the steps to compute the product of 12×13. The factors are first positioned on the multiplication mat as shown in Figure 10.28a. After the first two rows of 13 are positioned, it is much more efficient to include one 100-flat (10×10) and complete each of these 10 rows with three additional unit blocks (see Figure 10.28c). Positioning the required 30-ones is tedious, however, and as shown in Figure 10.28d, these 30-ones can be replaced by three 10-rods arranged vertically. This arrangement provides the additional 3-ones needed for each of the 10 rows in the 100-flat. The 12×13 product array is now complete.

The value of each subproduct can be recorded in a place value table. In this case, there are four subproducts, each represented by a part of the array. The sum of these four values gives the product. Note that the subproducts are recorded in the same order as with the standard algorithm.

1. Ones × ones
2. Ones × tens
3. Tens × ones
4. Tens × tens

Since addition is commutative, the order of subproducts makes no difference when using the display algorithm. To ensure a smooth transition to the standard algorithm, however, it is useful to follow the standard sequence.

When the number of subproducts becomes large (i.e., for triple-digit or larger factors), it is more efficient to introduce the standard multiplication algorithm. The relationship between the display and standard algorithms can be made explicit, as shown in Figure 10.29. The groups of subproducts recorded separately in the display algorithm are mentally combined when using the standard algorithm.

Figure 10.30 shows exercises that require regrouping between the first and second subproducts.

The standard algorithm can be further abbreviated by omitting the zero-place holders in each subproduct. Over the course of instruction, children begin with concrete displays of product arrays and record the results using the display algorithm. Once the display algorithm is mastered for double-digit factors, with and without concrete arrays, the standard algorithm is introduced. Finally, the abbreviated standard algorithm can be introduced to save time when writing subproducts:

	Display Algorithm		*Standard Algorithm*		*Abbreviated Algorithm*
	24		24		24
Concrete	36		36		36
Product	$\rightarrow$ 24	$\rightarrow$	144	$\rightarrow$	144
Arrays	120		720		72
	120		864		864
	600				
	864				

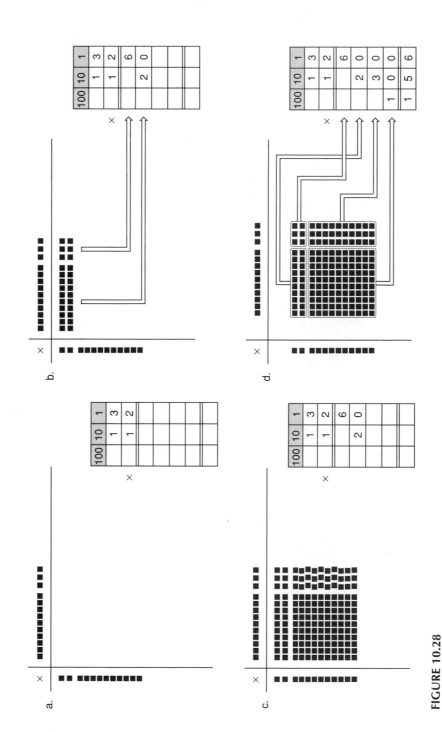

FIGURE 10.28

373

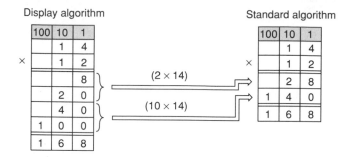

FIGURE 10.29

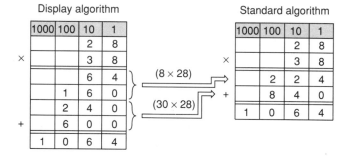

FIGURE 10.30

Multiplication of Larger Values

When factors exceed roughly 20, or products 999, the use of arrays becomes unwieldy. By the time such problems are introduced, children should have mastered working smaller exercises, using concrete materials and recording the results in place value tables. Exercises involving larger numbers can be introduced using squared paper place value tables and standard symbols (Figure 10.31).

Teacher-prepared worksheets presenting exercises in place value tables are appropriate at this stage of development. Textbooks are a useful source of exercises that are organized according to the level of difficulty. As factors get large, the number of subproducts increases to the extent that the display algorithm may become tedious, encouraging the shift to the standard algorithm.

Using proportional base-10 materials, carry out the necessary actions to complete the following multiplication exercises. Record the actions and results in place value tables.

1. $4 \times 12 = ?$

2. $8 \times 15 = ?$

3. $12 \times 13 = ?$

4. $11 \times 316 = ?$

5. $12 \times 223 = ?$

Display algorithm

1000	100	10	1
	3	2	4
×			5
		2	0
	1	0	0
+ 1	5	0	0
1	6	2	0

Standard algorithm

1000	100	10	1
		2	7	1
×			2	4
	1	0	8	4
+ 5	4	2	0
6	5	0	4

FIGURE 10.31

INTRODUCING THE DIVISION ALGORITHM

After they can readily recall the division facts, children can be introduced to selected exercises that have multiples of 10 and 100 for quotients. Figure 10.32 shows how to work the problem 30 ÷ 3 using base-10 materials.

The task of constructing an appropriate array from the dividend 30 is simplified by exploiting the structure of the base-10 materials. The 3-tens could be traded for 30-ones and distributed in groups of three by arranging them into three rows (i.e.,

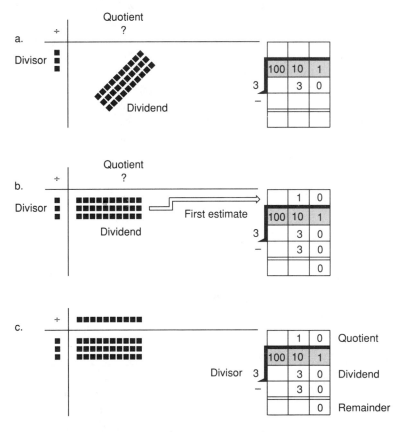

FIGURE 10.32 Division Using Base-10 Materials

repeated subtraction). A quicker method is to *estimate* if the dividend is large enough so that each of the three rows can contain at least a 10-rod and *check* to see if there are enough tens to construct an appropriate array (see Figure 10.32b).

To show the quotient, a spare 10-rod is placed above the horizontal axis to match the length of the array (see Figure 10.32c). As there are no blocks left undistributed, the **remainder** (the amount left over) is zero. In this case, the first estimate was also the best estimate, because the dividend was completely distributed. This is not always the case, of course. The process of dividing involves making an initial quotient estimate, checking to see if it is too large, and if not, continuing to distribute the remaining blocks until no more can be distributed equally.

The symbolic record of the division process shown in Figure 10.32c is the *display algorithm*. This algorithm encourages students to focus on the underlying place value concept when dividing and facilitates the estimation of partial quotients.

If there are not enough blocks to complete an equal distribution, the initial quotient estimate is too large. A smaller estimate must then be selected. The sequence of actions involved in making a second estimate when the first was too small is shown in Figure 10.33.

The first estimate of 10 left 4-tens (40) undistributed. The second estimate of one additional 10-rod left a remainder of zero. Since all the dividend was distributed, the quotient is the *sum* of the two estimates $10 + 10 = 20$, and the remainder is 0. Of course, as children learn to make better initial estimates, the number of steps needed to complete a division exercise will decrease. In this case, a perceptive student might have made an initial guess of 2-tens (20) and completed the exercise in one step.

For quotients larger than 40, it becomes tedious to use arrays. Once they are adept at using arrays to solve exercises with small multiple-of-10 quotients, students can tackle exercises involving larger multiples of 10 and 100 using the display algorithm in place value tables. Place value tables help children write digits in the proper column. Figure 10.34 shows exercises solved in place value tables using *multiple* and *best estimates* for quotients.

Front-End Estimation of Quotients

Mastery of exercises that have multiples of 10 as quotients can help children make more accurate initial quotient estimates. For single-digit divisors, **front-end estimation** can be used to help make these estimates. Children round down the dividend to the leading digit, as shown in Figure 10.35. The dividend (368) is thought of as 30-tens. An initial quotient estimate would be $368 \div 4 \approx 300 \div 4 = 30\text{-tens} \div 4 \approx 7\text{-tens}$, or 70.

The best quotient estimate shown in Figure 10.35 will be close to this front-end estimate. More accurate quotient estimates can be found if dividends are rounded down to the second largest place value. For example, $368 \div 4 \approx 360 \div 4 = 36\text{-tens} \div 4 = 9\text{-tens}$, or 90. Encourage children to use a calculator to check the accuracy of initial quotient estimates. Estimating quotients using the front-end estimation method is generally introduced early in the development of the division algorithm.

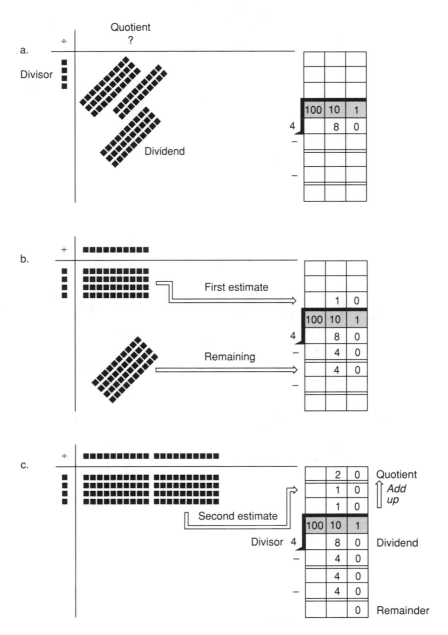

FIGURE 10.33

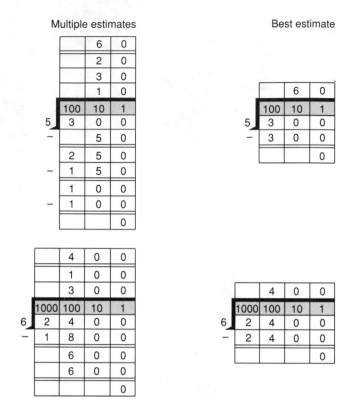

FIGURE 10.34 Division Using Place Value Tables

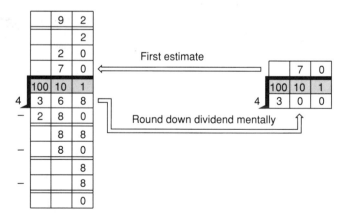

FIGURE 10.35 Rounding Down to Estimate Quotients

> Use front-end estimation to mentally compute an initial quotient for each of the following exercises. Use base-10 materials to show the concrete actions involved.
>
> **1.** $3 \overline{)85}$
>
> **2.** $4 \overline{)109}$
>
> **3.** $6 \overline{)222}$
>
> **4.** $5 \overline{)179}$
>
> **5.** $8 \overline{)566}$

Single-Digit Divisors

Division involves estimating quotients by exploiting the inherent structure of base-10 materials. The first step is to estimate how many of the dividend can be *equally* distributed to each unit in the divisor. Children can take advantage of the ones, tens, and hundreds blocks in the base-10 materials and the front-end estimation method to select an initial quotient. For example, the front-end quotient estimate for $84 \div 6$ is $80 \div 6 = 8\text{-tens} \div 6 \approx 1\text{-ten}$, or 10 (see Figure 10.36).

Since there are not enough 10-rods left to distribute among the six rows, the 2-tens and 4-ones remaining must be regrouped into 24-ones for the next distribution. The second distribution can be calculated directly from the basic facts ($24 \div 6 = 4$). Smaller initial quotient values could be used, though additional steps would be required (i.e., first distribute three units to each row, then distribute one more).

Figure 10.37 shows how larger examples can be worked if the quotients are reasonably small. For $345 \div 8$, the initial quotient estimate can be found using front-end estimation ($345 \div 8 \approx 300 \div 8 = 30\text{-tens} \div 8 \approx 3\text{-tens}$, or 30). A better initial quotient estimate can be found by rounding down the dividend to the next to the largest place value ($345 \div 8 \approx 340 \div 8 = 34\text{-tens} \div 8 \approx 4\text{-tens}$ or 40).

> Using proportional base-10 materials, carry out the necessary actions to solve each division exercise below. Use front-end estimation and rounding down to the second largest place value to estimate the initial quotient for each exercise. Use place value tables to record the actions and results.
>
> **1.** $5 \overline{)75}$
>
> **2.** $6 \overline{)108}$
>
> **3.** $8 \overline{)128}$
>
> **4.** $5 \overline{)195}$
>
> **5.** $9 \overline{)216}$

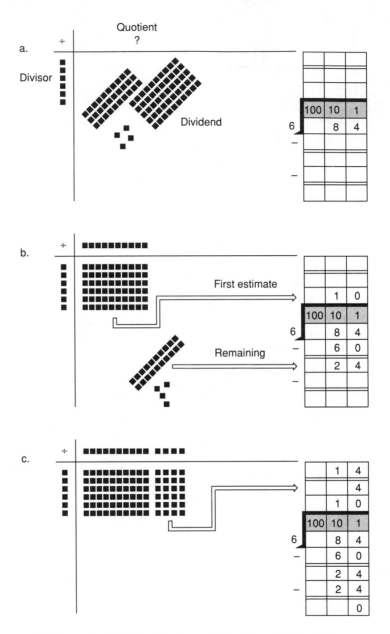

FIGURE 10.36 Division Mat Setup Using Base-10 Materials

Double-Digit Divisors

Determining initial quotient estimates for exercises involving double digit divisors is somewhat more complicated. For these exercises, it is generally useful to round off the divisor and the dividend to the leading place value of each. Figure 10.38 shows the

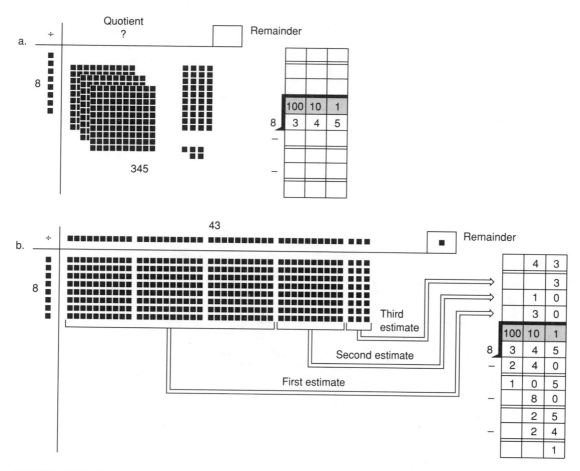

FIGURE 10.37 Division with Remainders

steps for computing 382 ÷ 18. First, have children round off the divisor and dividend to their respective leading digits and try to estimate the quotient mentally (382 ÷ 18 ≈ 400 ÷ 20 = 20). Using the initial quotient estimate of 20, part of the dividend can be arranged into an array as shown in Figure 10.38b. Finally, Figure 10.38c shows how 18 of the remaining undistributed dividend of 22 can be regrouped and positioned in the array. The remainder is 4.

If this method gives a quotient estimate that is too large, as in the example 282 ÷ 34 ≈ 300 ÷ 30 = 10, the rounding-off method makes it easier to identify an accurate second estimate (if 10 is too large, the quotient must be 8 or 9).

Exercises involving divisors and/or quotients less than 40 can be readily displayed using array representation. Figure 10.39 shows examples of worked exercises. Skill in making accurate initial quotient estimates can be enhanced with sufficient practice using arrays and the rounding-off method.

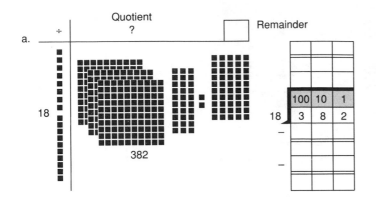

For first quotient estimate, round off divisor and dividend and mentally calculate quotient.

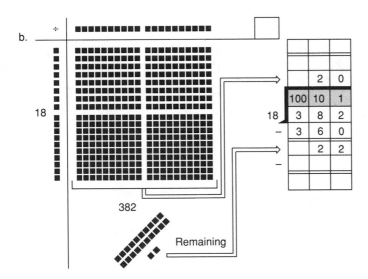

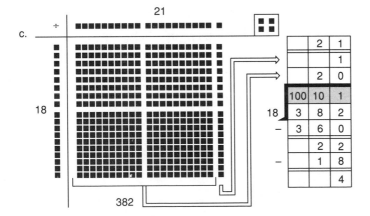

FIGURE 10.38 Rounding Off to Estimate Quotients

Round off the divisors and dividends in each exercise below to each leading digit and estimate the initial quotient estimates. Use base-10 materials to demonstrate the steps.

1. 13)175

2. 14)224

3. 17)361

4. 25)445

5. 18)272

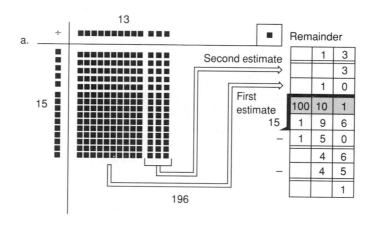

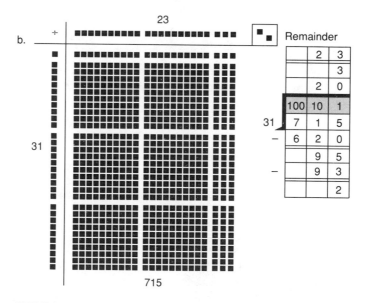

FIGURE 10.39

Using place value tables, carry out the necessary manipulations to solve each division exercise below. Compare the display algorithm to the standard algorithm in each case.

1. $12 \overline{)75}$
2. $11 \overline{)100}$
3. $13 \overline{)125}$
4. $15 \overline{)194}$
5. $21 \overline{)205}$

Division Involving Larger Values

Exercises involving divisors and/or quotients larger than 40 are more readily solved using place value tables. Once children have mastered division exercises with quotients smaller than 40, the standard division algorithm can be introduced by comparison with the steps in the display algorithm. If the initial quotient estimate is the largest possible value, it is the *best* estimate. When the best estimate is found for each place value, the display algorithm parallels the standard algorithm (see Figure 10.40).

With the standard division algorithm, children are expected to make accurate best estimates for each partial quotient. The value for each partial quotient is also represented by one digit in its proper column, as opposed to the entire number as in the display algorithm. In Figure 10.40, the best estimates are 2-tens and 1-one. The remainder is the amount left when no further whole number divisions are possible. This value is commonly displayed along with a capital R (for remainder) next to the quotient.

Solve each division exercise below using the display algorithm in place value tables. Relate the display algorithm to the standard algorithm in each case.

1. $22 \overline{)175}$
2. $63 \overline{)2105}$
3. $35 \overline{)1125}$
4. $83 \overline{)4394}$
5. $121 \overline{)5205}$

The use of proportional base-10 materials and the display algorithm facilitates conceptual understanding of division and provides a flexible algorithm for computing. Skill in making initial quotient estimates improves as children learn to apply their concrete manipulations to symbolic exercises. As with previous number operations, the instructional procedures are intended to serve as a dynamic system of support

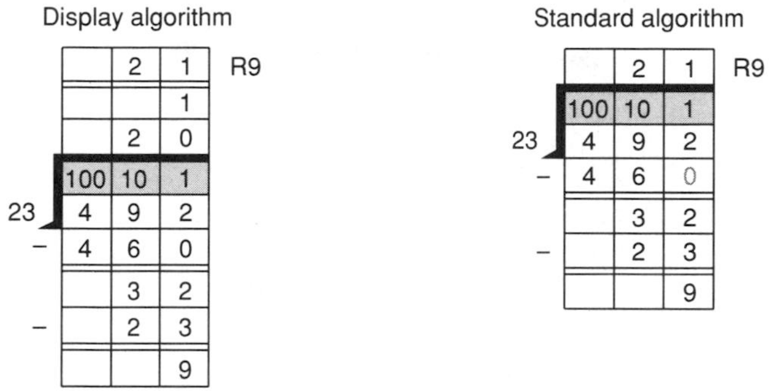

Display algorithm

		2	1	R9
			1	
		2	0	
	100	10	1	
23)	4	9	2	
−	4	6	0	
		3	2	
−		2	3	
			9	

Standard algorithm

		2	1	R9
	100	10	1	
23)	4	9	2	
−	4	6	0	
		3	2	
−		2	3	
			9	

FIGURE 10.40

for children learning to divide. As their ability increases, children take on greater responsibility for coordinating elements within the task.

The recent availability of inexpensive calculators has reduced the importance of long division computations involving large numbers. However, a thorough understanding of the types of division situations, skill in mentally estimating quotients, and an ability to carry out calculations involving small numbers are needed to facilitate effective problem solving.

CLASSROOM ACTIVITIES

Developmental Activity Sequence

A sequence of activities providing different levels of perceptual support can help children with a range of abilities develop the concept of multiplication and its associated algorithm. A direct relationship can be established between the concrete actions involved in constructing a product array for two single- or double-digit factors and the symbolic manipulations required by the algorithm. Figure 10.41 shows the step-by-step relationship that exists between the same multiplication exercise worked using an array and using the display algorithm. Each subproduct calculated in the sequence (4×3, 2×20, 10×3, and 10×20) is displayed as a part of the array and is recorded in the place value table.

Many children experience difficulty with the transition from the concrete representation of multiplication exercises to the use of an algorithm. As an intermediate level between concrete base-10 materials and the symbolic algorithm, exercises can be worked using the graphic representation of **dots, lines,** and **squares.** In this scheme, dots represent ones, lines represent tens, and squares represent hundreds (see Figure 10.42).

This graphic representation of base-10 materials allows children to work more efficiently with larger factors. Worksheets can be prepared displaying blank array mats and place value tables such as the example in Figure 10.42. Graded exercises can be

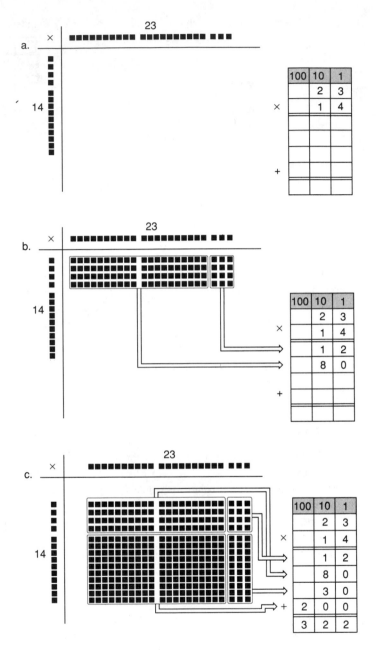

FIGURE 10.41

assigned from a textbook or other source, transferred to the array mats using dots, lines, and squares, and recorded symbolically in place value tables. The need for graphic displays will diminish with student experience. They remain a useful tool, however, for analyzing and correcting errors.

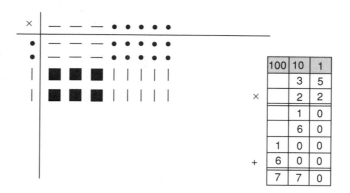

FIGURE 10.42 Multiplication Using Representational Notation

Like the other operations, division is also introduced using proportional materials. The use of arrays enables students to practice making initial quotient estimates and checking the accuracy of these estimates. Students record the results of their concrete actions using the display algorithm in place value tables.

Later, exercises can be presented symbolically. Incorrect answers can be repaired by checking calculations using arrays. If necessary, students experiencing difficulty with symbolic division can solve exercises using a transitional representation like the dots, lines, and squares shown in Figure 10.43. This representation offers children a simple way to record their work with arrays if they have difficulty using place value tables. Older students may prefer dots, lines, and squares as a substitute for base-10 materials when working division exercises.

Napier's Bones

John Napier is credited with the invention of a simple calculator that computes products. Upper-grade and gifted children may find constructing and using this device an interesting multiplication extension activity. The tool was originally constructed of nine ivory bars, or **bones,** each respectively inscribed with the first nine multiples of the digits 1–9. A simple set of bones can be constructed from tag board or tongue depressors, as shown in Figure 10.44.

To multiply 6×18, select the 1-bone and the 8-bone and place them next to each other as in Figure 10.45a. Next, look at row 6. Add the values along the diagonals, regrouping to the next column to the left as necessary (see Figure 10.45b). This calculation gives the product.

Products of a 1-digit × 3-digit or larger values can be computed by selecting bones for each digit in the largest factor. When both factors contain more than one digit, the subproducts are computed separately and combined for the final result. Point out that the subproduct must be adjusted one or more place values to the left

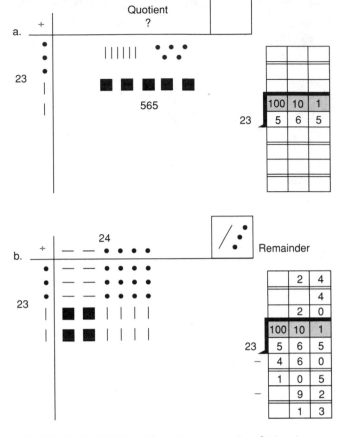

FIGURE 10.43 Division Using Representational Notation

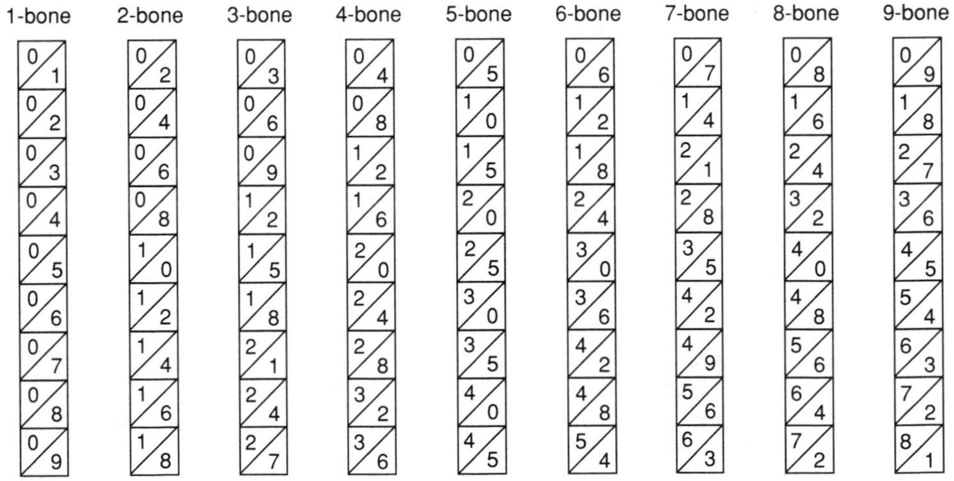

FIGURE 10.44 Napier's Bones

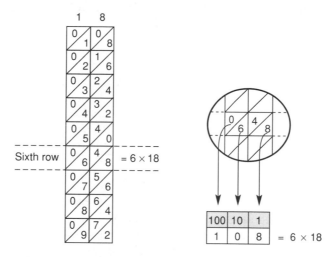

a. Multiplying using Napier's Bones   b. Close-up of 6 × 18

FIGURE 10.45

and zero-place holder(s) appended when the factor is a multiple of 10 as in Figure 10.46.

What do the digits on the diagonals stand for? Why do we get the product of two numbers when these digits are summed? Have your students try to figure out how these bones multiply.

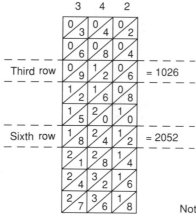

Note: Answer for sixth row
must be multiplied
by 10, so product will
be 60 × 342 = 20520.

63 × 342 = (3 × 342) + (60 × 342) = 1026 + 20520
= 21,546

FIGURE 10.46 Multiplying Large Values Using Napier's Bones

PROBLEM-SOLVING EXPERIENCES INVOLVING MULTIPLICATION AND DIVISION

Many problems require the application of repeated addition, finding area, making arrays, or finding combinations. Multiplication and division are generally involved in solving such problems. Experience with concrete problem situations embodying fundamental concepts of multiplication and division helps children develop skill in selecting appropriate strategies or operations. At first, only problems involving one operation should be presented. Later, situations involving more than one operation (e.g., multiplying, then adding) can be introduced.

> Work through each of the problems presented in the following section. Try to extend those you find of interest by inventing additional problems based on the same situation. Discuss the solution process and results with other members of your class.

Making Tables and Finding Patterns

How many eating utensils are needed for a birthday party? If each place setting includes a knife, fork, and spoon, how many will be required for twenty-two guests? Grade 4–6 children may recognize that this is a simple multiplication exercise and calculate the product: $3 \times 22 = 66$.

Younger children may need to *conduct an experiment* and have help *organizing information into a table* to solve the problem. Give each child twenty-two counters to serve as guests. On the board, draw a two-column table like Table 10.4 with the number of people in the first column and the number of eating utensils in the second. Beginning with the smallest party (one person), have each child draw three utensils,

TABLE 10.4   Table Settings

Guests	Utensils
1	3
2	6
3	9
4	12
5	15
6	18
7	21
8	24
9	27
10	30
11	?

TABLE 10.5 Repeated Addition

$$3 = 3$$
$$3+3 = 6$$
$$3 + 3 + 3 = 9$$
$$3 + 3 + 3 + 3 = 12$$
$$3 + 3 + 3 + 3 + 3 = 15$$

circle them, and place one counter on top of them. Record the number of guests (1) and the number of utensils (3) in the table. Continue drawing sets of utensils and have the children skip count to find the total. Record the results in the table until reaching ten guests.

It will take a long time to continue the table all the way to twenty-two guests, so have the children *look for a pattern* to help determine the next value without having to count all the utensils. Looking down the second column, notice that the total number of utensils can be found by adding 3 to each current value. For example, to determine the number of utensils required for eleven guests, we add 3 to the number of utensils required for ten guests. The pattern *add 3* can be extended to include any number of guests. Table 10.5 shows this application of repeated addition.

The number of utensils needed for twenty-two guests could be found with some effort using this pattern. Suppose one hundred guests showed up, or one thousand guests. The repeated addition procedure would be tiring indeed. In order to extend patterns to larger values, it is helpful to *find a rule* that will predict the number of utensils needed based exclusively on the number of guests.

To find such a rule requires paying careful attention to the relationship between the entries in first and second columns. The rule in this case for the pairs of values 1 and 3, 2 and 6, 3 and 9, 4 and 12, and 5 and 15 can be summarized as: *multiply the number of guests by 3*. A rule makes it possible to compute the number of utensils required for any number of guests without knowing the previous step. For example, if you know the function shown in Table 10.6, it is not necessary to know the number

TABLE 10.6 Table-Setting Function
(Guests $\times$ 3 = Utensils)

Guests	Utensils
1	3
2	6
3	9
4	12
5	15
.	.
.	.
.	.
87	?

of utensils needed for eighty-six guests in order to determine the number needed for eighty-seven guests.

Formula Gardening

Measuring area often involves multiplication or division. Suppose sixteen radish plants can be grown in each one square meter of land. How many plants can be grown in a rectangular garden twenty-five meters on an edge? Grade 5–6 children might first try to *draw a sketch* of the problem. Since the garden is large, the sketch will have to be scaled quite small, making it difficult to count all the plants. It may be better to sketch a smaller square garden, say two meters on an edge, and count the number of plants. Then *construct a table* that organizes the important information about a whole series of square gardens (edge length, area, number of plants) and try to *find a pattern or rule* to make the counting easier.

After constructing Table 10.7, either individually or as a whole class for less experienced problem solvers, help the children observe that the garden area can be calculated by multiplying edge length by itself (e.g., for a garden six meters on an edge, the area is: $6 \text{ m} \times 6 \text{ m} = 36 \text{ m}^2$). The number of plants can then be found by multiplying the area by 16.

A formula relating edge length (L) to the number of plants (N) can be written:

$$16 \times (L \times L) = 16 \times L^2 = N$$

Children can use this formula to find the number of plants for any size garden:

$$16 \times (50)^2 = 40,000 \text{ radish plants}$$
$$16 \times (100)^2 = 160,000 \text{ radish plants}$$
$$16 \times (273)^2 = 1,192,464 \text{ radish plants}$$

Constructing tables, analyzing patterns, and finding formulas are powerful problem-solving strategies in elementary mathematics. Not only do these strategies help

TABLE 10.7   Planting Gardens

Edge length (L)	Area (L²)	Plants (N)
1	1	16
2	4	64
3	9	144
4	16	256
5	25	400
6	36	576
.	.	.
.	.	.
.	.	.
25	?	?

children solve the immediate problem, but the procedures may be useful when applied to a whole class of similar problems encountered in the future.

> For the radish planting problem, how would the formula $16 \times L^2 = N$ change for rectangular gardens?

Missing and Superfluous Information

Everyday problems are generally embedded in a complex of simultaneous events and facts, many of which may be unrelated to the solution. For example, when comparison shopping, the ingredients, packaging, date, and brand may be determining factors in choosing a product. However, to identify which of two different sized cans of tuna is the best buy, only the weight and price are used to compute the unit price (price per gram or ounce).

Students need experience separating the key elements needed for the solution from superfluous information included in the problem description. Sometimes a problem is impossible to solve because important information is missing.

At the outset, students simply identify the values in a problem needed for the solution. Problem statements with superfluous information and those with missing values can also be introduced. It is more important for students to closely examine problem descriptions than attempt to carry out the solution operations before they are ready.

Give children examples such as those below and ask them to identify the missing and superfluous information in the problem statement.

Example 1.   MISSING INFORMATION
Darron and Rae wanted to build a ladder for their tree house. They figured it would cost about 50 cents per foot to build the ladder. If they earned $1.00 per hour mowing lawns, how long would it take for them to earn enough to buy the materials for their ladder?

(What missing information is needed to solve this problem?)

Example 2.   SUPERFLUOUS INFORMATION
Paint is sold in 1 quart and 1 gallon containers at $3.59 and $13.98 respectively. One quart will cover 100 square feet. The quart containers come in 14 colors and the gallons in 11 colors. If Mitch is planning to paint a wall 10 feet by 30 feet, which size container would be the best buy?

(Circle values needed to solve this problem.)

Have children find problems in their textbook and rewrite them to include missing and/or superfluous information. The class can exchange problems to identify the key information needed to solve each problem.

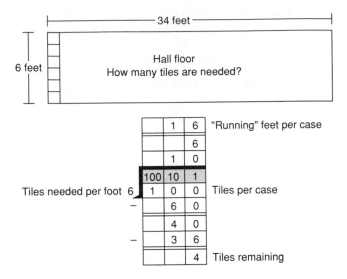

FIGURE 10.47   Floor Tiles

Job Estimates

Suppose a thirty-four foot hall is six feet wide, and we need to buy enough one-square-foot floor tiles to cover the hall. How many one-hundred-tile cases are needed for the job? Children should be encouraged to first *draw a sketch* of the problem. The sketch in Figure 10.47 shows that six tiles are needed to fit across the hall. Distributing a case of one hundred tiles in six equal rows looks like array division. One case would make rows sixteen tiles long with four tiles left over. Two cases would cover thirty-three *running* feet of hall with two left over. By completing the sketch of the array, or using division, we find that two cases would not quite finish the job.

Measuring Money

How tall is a stack of one million dollar bills? Carrying out the actual experiment could be very expensive and time-consuming. The problem can be cut down to size by first solving a simpler case of the same problem.

First, have children count the number of bills needed to make a stack *one centimeter* high. Substitute play money or pieces of paper if you're a bit short on funds. About eighty bills make a stack one centimeter high. To compute the height of a stack of one thousand bills, children need to divide, using a calculator, or paper and pencil,

$$
\begin{array}{r}
12 \leftarrow \text{centimeters} \\
80\overline{)1000} \leftarrow \text{bills} \\
\underline{800} \\
200 \\
\underline{160} \\
40
\end{array}
$$

The $1000 stack is about twelve centimeters high. Substitute one million for one thousand and follow the same procedure to find the height.

$$\begin{array}{r} 12{,}500 \quad \leftarrow \text{centimeters} \\ 80\,)\overline{1{,}000{,}000} \quad \leftarrow \text{bills} \end{array}$$

The stack would be 12,500 centimeters, or 125 meters, high—taller than a forty-story building! Children should find *reducing problems to a simpler case* to be a useful strategy for solving a wide range of problems.

> A criminal in a recent movie was carrying a briefcase that supposedly contained two million dollars in $100 bills. Does this seem possible?

CALCULATOR AND COMPUTER APPLICATIONS

Calculator Games and Activities

Twenty-one. Calculator strategy games can offer an excellent environment for examining patterns. A good example is the game *Twenty-one*. Pairs of children share one calculator for this activity. Starting with zero, players alternate adding the value 1 or 2 to the current sum. The goal is to be the first player to reach *exactly* 21. This game can also be played orally or using pencil and paper. A typical game might proceed as shown in Table 10.8.

TABLE 10.8 Game of Twenty-one

Move	Player 1	Calculator Display	Player 2	
START		0		
1	+1 →	1		
2		3	← +2	
3	+2 →	5		
4		7	← +2	
5	+1 →	8		
6		9	← +1	
7	+2 →	11		
8		13	← +2	
9	+1 →	14		
10		15	← +1	
11	+2 →	17		
12		18	← +1	
13	+1 →	19		
14		21	← +2	Player 2 wins

Several interesting patterns occur with this game. Have children play it with a partner and keep track of any patterns that may provide clues to a winning strategy. Possible patterns include the odd and even numbers, multiples of 3, and recurrence of key numbers such as 18 and 2. Have children try to devise a strategy that *guarantees* Player 2 will always win.

Play the game Twenty-one with a partner. If you discover a winning strategy, change the rules so that 21 is poison (if you get to 21 first, you lose); or allow adding 1, 2, or 3; or change the winning number to 25.

Calculator Estimation. The development of estimation skills deserves more attention in the elementary curriculum because of its importance in solving problems. The ability to mentally calculate an approximate result can be an invaluable aid in determining whether a solution strategy is reasonable.

Calculators can be used to help students practice estimating quotients when dividing (or for any other operation on the calculator). Working in pairs, have children estimate the quotients for division exercises on the same worksheet. After the estimates are completed, have each pair find the quotients using a calculator and calculate the difference between each result and its estimate. The student who makes the best estimate (i.e., the difference between the estimate and the computed answer is smallest) circles the estimated quotient. The children with the most circled estimates wins.

For younger students, select problems that give whole number quotients. Older students should round off computed quotients to the nearest tenth.

	Grades 3–4		*Grades 5–6*
1.	$6 \overline{)24}$	1.	$18 \overline{)48}$
	- - - - - - = = = = = Estimate Calculation		- - - - - - = = = = = Estimate Calculation
2.	$9 \overline{)36}$	2.	$31 \overline{)256}$
	- - - - - - = = = = = Estimate Calculation		- - - - - - = = = = = Estimate Calculation
3.	$5 \overline{)45}$	3.	$3.25 \overline{)12.3}$
	- - - - - - = = = = = Estimate Calculation		- - - - - - = = = = = Estimate Calculation

Computer Software

Several computer software products are available to help teach multiplication and division. Concept tutorials, estimation activities, and practice drills provide a range of computer experiences in the classroom or in a laboratory setting.

Power Drill: A Set of Estimation Games.   This set of software activities provides practice for Grade 4–6 children in estimating sums, differences, products, and quotients (Educational Development Center). Front-end estimation and more sophisticated rounding techniques are explored in this simple to use yet rich practice environment. For example, the program presents the user with an exercise such as

$$65, 287 \div ? = 264$$

Using front-end estimation, the child simplifies the exercise ($60,000 \div ? = 200$), mentally calculates the answer, and enters 300. The program displays the result along with the original exercise.

$$65,278 \div ? = 264$$
$$65,278 \div 200 = 326\,R\,78$$

Since the quotient is too big ($326 > 264$), the estimate must be adjusted to compensate. By refining quotient estimates, children practice using rounding and mental arithmetic skills to solve division problems. The program gives advice if the initial estimate is way off and displays the number of attempts to find the missing number. There are options for each of the four operations and three levels of difficulty.

Heath Mathematics Software: Addition, Subtraction, Multiplication, Division, and Tournament.   This series of five disks provides concept tutorials, practice exercises, and graded and timed drills appropriate for Grades 1–4 (DCH Educational Software). The tutorials introduce each operation with graphic displays of concrete materials. Multiplication and division are shown using loops filled with objects. Exercises are first presented using the concrete model and then without. The tutorial and practice activities are followed by thirty to forty-five randomly generated target practice exercises. Children who reach a preset mastery level are moved ahead, and those who do not are given additional practice at their current level. The programs maintain records of student performance. The final disk supports up to six simultaneous, timed number operation tournaments of up to fifty children each. The tournaments cover addition, subtraction, multiplication, and division and create a tournament *ladder* on which children can challenge a higher ranked player.

Number Fact Sheets.   This teacher utility prints randomly generated number operation worksheets and tests for duplication (Gamco Industries). The teacher can select the difficulty level, type of operation, optional answer key, and number of problems.

ADAPTING INSTRUCTION FOR CHILDREN WITH SPECIAL NEEDS

Low Stress Multiplication and Division Algorithms

Lattice Multiplication.   The introduction of an alternative algorithm may pique the interest of special-needs children who have failed to learn the standard algorithm.

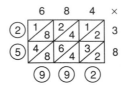

$$38 \times 684 = 25{,}992$$

FIGURE 10.48 Lattice Multiplication

Using squared paper, have children draw a rectangle with the length equal to the number of *digits* in the multiplier and the width equal to the number of *digits* in multiplicand. As shown in Figure 10.48, subproducts are computed for all pairs of digits and written with the tens above the diagonal and ones below the diagonal in the appropriate squares.

The subproducts are summed along the diagonals, regrouping to the left as necessary. The product (circled values) is read from left to right around the rectangle.

> Review the section on multiplication using Napier's Bones. What similarities do you see between multiplication with a lattice and Napier's Bones?

Division Display Algorithm. The display algorithm presented in this chapter is a useful transition algorithm for students having difficulty making accurate quotient estimates. In Figure 10.49, two alternate forms of the **display algorithm** are shown. The first is the algorithm developed previously. The second follows the same proce-

Display algorithm 1

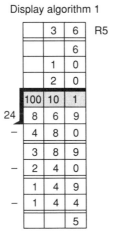

Display algorithm 2

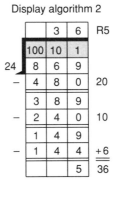

FIGURE 10.49 Alternative Division Algorithms

dure, only the quotient estimates are displayed on the right, next to each partial product. To help children who are not confident of their recall of the basic multiplication facts, have them first make a list of multiples of the divisor.

$$0 \times 24 = 0$$
$$1 \times 24 = 24$$
$$2 \times 24 = 48$$
$$3 \times 24 = 72$$
$$4 \times 24 = 96$$
$$5 \times 24 = 120$$
$$6 \times 24 = 144$$
$$7 \times 24 = 168$$
$$8 \times 24 = 192$$
$$9 \times 24 = 216$$
$$10 \times 24 = 240$$

The list can be quickly constructed by noting that the first multiple is always 0, the second is the divisor, the third is double the divisor, the fourth is the sum of the second and third, and so on. A calculator can also be used to quickly construct the multiple table. Using this list and an understanding of multiplication by 10, 100, and 1000, children can estimate initial quotients more accurately. For the example 869 ÷ 24, looking at the 24-multiple table, the child can quickly see that 20 is too small and 40 is too large for an initial quotient estimate (20 × 24 = 480, 40 × 24 = 960). Therefore, 30 might be about right. Similarly, the table gives the best quotient estimate for 149 ÷ 24 (6 × 24). Children can make a divisor multiple table for each problem they encounter.

Each of these algorithms allows students to underestimate the initial quotient and adjust subsequent estimates accordingly without having to start the exercise over. The procedure is often more satisfying because it encourages students to proceed to the next step without having to erase their work.

Children who exhibit particular difficulty learning to multiply and divide may need more help with the transition between concept building activities involving base-10 arrays and the symbolic algorithms. To assist with the transition, graphic representations (e.g., dots, lines, and squares) can also be emphasized to a greater extent than normal. In such cases, it may be appropriate to allow the continued use of the display algorithm instead of requiring a transition to the standard algorithm.

At some point in each child's education, a judgment must be made regarding the benefit of continuing practice of the long division algorithm. For those students who have experienced minimal success by the end of Grade 6, it becomes increasingly difficult to justify the time committed to reteaching long division when inexpensive calculating devices now available carry out such computations with ease and accuracy.

How often are long division exercises hand computed by others outside the role of student or teacher? Recent research indicates that nearly all calculations actually carried out by individuals in everyday life are approximations based on some rounding procedure (Rogoff & Lave, 1983). Few mathematicians or engineers, who constantly work with numbers, would trust themselves to carry out a vital calculation using only pencil and paper. Virtually everyone who cares about the results of their computations uses some form of calculating device. It is important not to lose sight of the *purpose* of mathematics education in our attempt to satisfy performance objectives along the way.

Peasant Multiplication.   Another multiplication algorithm, still used in rural areas of Eastern Europe and Russia, can be introduced as an extension activity for gifted and talented students. The **peasant multiplication algorithm** was originally developed by the Hindus and used by Arabs, Greeks, and Romans hundreds of years ago, before place value systems were invented. Imagine the difficulty a merchant might have had multiplying large numbers using Roman numerals. The multiplication technique involves doubling and halving, an easy operation even for the Egyptian and Roman numeration systems.

To find the product of 35×186, systematically halve the first factor (ignoring any remainders) until reaching the quotient 1, as shown in Table 10.9. Next, double the second factor the same number of times. A calculator can be used to halve and double each factor, ignoring the decimal fraction parts of the quotients. In the first column, draw a line through each even number (in this case 8, 4, and 2) and the corresponding values in the second column (744, 1488, and 2976). The sum of the remaining values in the second column is the product $35 \times 186 = 6510$.

Explaining why this procedure works is an interesting problem. One way to explore the procedure is to look at the problem backward. In Table 10.10, notice that a third column has been added that keeps track of the number of times 186 is added to itself to make each value in the second column.

When the *even* lines are extended through column 3, the sum of the remaining values in the third column $(1 + 2 + 32)$ equals the multiplier (35). The sum of the remaining values in the second column, then, must equal 186 added to itself 35 times, or 35×186.

TABLE 10.9   Peasant Multiplication

Halve First Factor	Double Second Factor
35	186
17	372
~~8~~	~~744~~
~~4~~	~~1488~~
~~2~~	~~2976~~
1	5952
Total	$6510 = 35 \times 186$

TABLE 10.10 Peasant Multiplication

Halve First Factor	Double Second Factor	Multiple of Second Factor
35	186	1
17	372	2
~~8~~	~~744~~	~~4~~
~~4~~	~~1488~~	~~8~~
~~2~~	~~2976~~	~~16~~
1	5952	32
Total	6510 (Product)	35 (First Factor)

Multiply other values using peasant multiplication and look for patterns to help explain the underlying process.

SUMMARY

Multiplication and division are fundamental concepts in elementary mathematics education. As Piaget observed, a thorough knowledge of transitivity contributes to a meaningful understanding of these operations. Repeated addition, Cartesian products, and arrays are three ways children can think about multiplication. Division can be modeled using repeated subtraction, partitioning, or arrays. Base-10 blocks, bean sticks, chip trading materials, dots, lines, and squares, and place value tables can be effectively used to introduce the multiplication and division concepts and develop the symbolic algorithms. Multiplication and division are inverse operations that reinforce each other's development when introduced concurrently. The commutative property, associative property, and distributive property of multiplication over addition are used by children to help develop the multiplication algorithm and to learn the multiplication facts. The special roles of 0 and 1 in multiplication and division are also important concepts to be introduced at the elementary level. Developmental activity sequences (DAS) can be used to teach multiplication and division concepts and algorithms to groups of children who are at various levels of development and experience. Initially, the class is introduced to a new concept or skill using concrete materials, recording their actions and results in place value tables. Some children may require intermediary work with nonproportional materials such as dots, lines, and squares, while others begin working exercises directly in place value tables or using the standard algorithm. Story and nonroutine problems involving multiplication and division offer opportunities to apply newly developed mathematics and language skills. Each multiplication and division model has applications in real-world problem situations. An understanding of the different models helps children identify appropriate operations when they are encountered. Introducing alternative low stress algorithms may be useful when working with special-needs children who have experienced protracted difficulty learning the multiplication or division algorithm. Exploring alternative algorithms is also an interesting extension activity for gifted and talented students.

COURSE ACTIVITIES

1. Working in small groups, compile a list of real-world situations that embody the repeated addition, array, or Cartesian product notions of multiplication. Compile a class list of the situations described and design story or picture problems based on each situation. Write each problem on a separate card, shuffle the pack, and have individuals try to identify which of the three multiplication types best describes each situation.

2. Survey your friends and members of your class and make a list of the rhymes and mnemonic devices they used to memorize the multiplication facts. Organize these techniques by category (songs, rhymes, etc.) and compile a class list.

3. Construct a set of eighty-one multiplication *candybars* using centimeter squared paper. Make flash cards by attaching each candybar to the front of a file card and writing the corresponding multiplication fact on the back. Write a lesson plan to introduce the multiplication facts to a group of Grade 3 children employing these materials. If possible, implement the lesson and discuss the results.

4. Construct a board game designed to give students practice recalling the multiplication facts. Try it out with your peers or, if possible, a small group of children. Revise the game based on your observations and student responses. Exchange games with other members of your class and compile a set of at least five games for future classroom use.

5. Using a set of proportional base-10 materials, practice setting up multiplication prob-

lems, carrying out the necessary manipulations, and recording the actions and results in place value tables. With your peers, practice demonstrating multiplication exercises with products less than 1000 until you can confidently verbalize the procedures and evaluate procedural errors. Write a lesson plan to introduce one-digit × two-digit multiplication to a small group of Grade 3–4 students and record the interaction on audio or video tape. Review the session with a peer and look for points where your instructions were confusing or where you misinterpreted a student response.

6. Write story or picture problems that are examples of the repeated subtraction, partitioning, and array models of division. Working with a group of peers, mix the cards and try to choose those problems that are examples of each division model. Justify your selections.

7. Using a set of proportional base-10 materials, set up and solve division problems with divisors and quotients less than 40. Practice until you can consistently make accurate initial quotient estimates. Record the concrete actions and results in place value tables. Write a lesson plan to introduce division using arrays to a small group of peers and have an observer watch you implement the lesson. Discuss the experience with the observer and other members in your class.

8. Select several word problems from an upper-grade textbook or other resource and rewrite them to include superfluous information, or remove one or more of

the required values. Write each problem on a separate card. Exchange the cards with a partner and separate problems with missing values from those that have superfluous information.

9. Review a problem-solving resource book or elementary textbook to find problems that can be solved by *organizing data in tables, looking for patterns and functions,* and *working backward.* Solve them and discuss the procedures with your peers. If possible, introduce a problem to a group of Grade 4–6 students. Work through the problem with them. Give them a similar problem to solve without your assistance. Discuss the results with your peers.

10. Make a set of Napier's Bones and practice multiplying numbers of various sizes. Write directions that would explain to a novice how the bones can be used to multiply numbers. If possible, introduce the activity to a group of Grade 5–6 students. Arrange for an observer to watch the interaction. Review the session immedi-

ately afterward with the observer and take note of points of confusion, successful and unsuccessful management techniques, and student responses to the presentation.

11. Discuss the alternative multiplication algorithms described in this chapter and others used by your peers. Solve the problem $27 \times 36 = ?$ on separate index cards using each algorithm and include them in your idea file.

12. Solve $692 \div 24 = ?$ using the two display algorithms shown in this chapter. Coordinate the results of each with the standard algorithm.

13. Include examples of several multiplication and division lessons in your idea file. Organize them according to appropriate grade level.

14. Read one of the *Arithmetic Teacher* articles listed in the reference section. Write a brief report summarizing the main ideas of the article and describe how the recommendations for instruction might apply to your own mathematics teaching.

MICROCOMPUTER SOFTWARE

Base Ten on Basic Arithmetic Practice with basic multiplication facts and multiples of ten (MECC).

Heath Mathematics Software: Addition, Subtraction, Multiplication, Division, Tournament Number operation tutorials, practice exercises, and timed tournament drills (DCH Educational Software).

Read and Solve Math Problems Tutorial for solving word problems (Educational Activities Inc.).

Math Ideas with Base-10 Blocks Multiplication practice using graphic representations of proportional base-10 materials (Cuisenaire Company of America).

Math Mastery Series Set of programs providing practice in number skills and the four operations (MECC).

Number Fact Sheets Prints graded worksheets and tests with answer key for duplication by teacher (Gamco´Industries).

Power Drill: A Set of Estimation Games Practice using front-end and other rounding techniques to solve addition, subtraction, multiplication, and division exercises (Educational Development Center).

REFERENCES AND READINGS

Ashlock, R. (1976). *Error patterns in computation: A semi-programmed approach.* Columbus, OH: Merrill.

Baratta-Lorton, M. (1976). *Mathematics their way.* Menlo Park, CA: Addison-Wesley.

Bates, T., & Rousseau, L. (1986). Will the real division algorithm please stand up? *Arithmetic Teacher, 33*(7), 42–46.

Broadbent, F. (1987). Lattice multiplication and division. *Arithmetic Teacher, 34*(5), 28–31.

Brownell, W., & Carper, D. (1943). *Learning the multiplication combinations*. Durham, NC: Duke University Press.

Burns, M. (1978). *The book of think,* New York: Little, Brown.

Cheek, H., & Olson, M. (1986). A den of thieves investigates division. *Arithmetic Teacher, 33*(9), 34, 35.

Downie, D., Slesnick, T., & Stenmark, J. (1981). *Math for girls and other problem solvers*. Berkeley: Regents of the University of California.

Driscoll, M. (1980). *Research within reach: Elementary school mathematics.* St. Louis: CAMREL.

Greenes, C., Immerzeel, G., Ockenga, L., Schulman, J., & Spungin, R. (1980). *Techniques of problem solving*. Palo Alto, CA: Dale Seymour.

Greenes, C., Spungin, R., & Dombrowski, J. (1977). *Problem-mathematics: Mathematical challenge problems with solution strategies*. Palo Alto, CA: Creative Publications.

Haag, V., Kaufman, B., Martin, E., & Rising, G. (1986). *Challenge: A program for the mathematically talented*. Menlo Park, CA: Addison-Wesley.

Hall, W. (1983). Division with base-ten blocks. *Arithmetic Teacher, 31*(3), 21–23.

Hamic, E. (1986). Students' creative computations: My way or your way. *Arithmetic Teacher, 34*(1), 39–41.

Immerzeel, G., & Ockenga, E. (1977). *Calculator activities: Book 1 and 2*. Palo Alto, CA: Creative Publications.

Lessen, E., & Cumblad, C. (1984). Alternatives for teaching multiplication facts. *Arithmetic Teacher, 31*(5), 46–48.

Pearson, E. (1986). Summing it all up: Pre-1900 algorithms. *Arithmetic Teacher, 33*(7), 38–41.

Quintero, A. (1985). Conceptual understanding of multiplication: Problems involving combination. *Arithmetic Teacher, 33*(3), 36–39.

Robold, A. (1983). Grid arrays for multiplication. *Arithmetic Teacher, 30*(5), 14–17.

Rogoff, B., & Lave, J. (1983). *Everyday cognition: Its development in social context*. Cambridge: Harvard University Press.

Sharron, S. (Ed.). (1979). *Applications in school mathematics: 1979 Yearbook*. Reston, VA: National Council of Teachers of Mathematics.

Silvey, L., & Smart, J. (Eds.). (1982). *Mathematics in the middle grades: 1982 Yearbook*. Reston, VA: National Council of Teachers of Mathematics.

Souviney, R., Keyser, T., & Sarver, A. (1978). *Mathmatters: Developing computational skills with developmental activity sequences*. Glenview, IL: Scott, Foresman.

Suydam, M. (Ed.). (1978). *Developing computational skills: 1978 Yearbook*. Reston, VA: National Council of Teachers of Mathematics.

Williams, J. (1971). *Teaching techniques in primary mathematics*. Slough, UK: Nuffield Foundation for Educational Research.

11

Patterns and Functions

Beauty in mathematics is seeing the truth without effort.

—George Polya

Upon completing Chapter 11, the reader will be able to

1. Describe the concept of number pattern and give several examples of appropriate classroom activities.
2. Construct a two-column table to record patterns and give examples of appropriate introductory activities.
3. Describe the concept of a function and its relationship to patterns.
4. Graph the (x,y) coordinates of a function expressed in a two-column table and give examples of appropriate classroom activities.
5. Describe the relationship between a function and its coordinate graph.
6. Give several examples of patterns that can be expressed as functions and describe appropriate classroom activities.
7. Describe how children can use the finite-difference method to uncover functions.
8. Describe geometric and other nonroutine problems involving patterns and functions and give sample lessons.
9. Describe how patterns can be used to instruct special-needs children.

Pattern refers to any systematic configuration of geometric figures, sounds, symbols, or actions. Much of what we learn stems from successfully unraveling the patterns embedded in the events and ideas we encounter in life. Language develops as a consequence of imitating systematic recurrences of speech. In Chapter 8, we explored patterns based on attributes such as color, texture, size, and number that are used to classify objects into manageable categories. Early number and counting development is closely associated with the ability to classify objects according to measurement and geometric attributes and arrange them into systematic patterns.

NUMBER PATTERNS

Number patterns are the outcome of counting and classification activity. In mathematics, the term **pattern** takes on a precise meaning when applied to arithmetic and geometric sequences. The **terms**, or values, in these sequences are related to each other according to a consistent rule. For example, the terms of the odd number sequence are related by the rule, *add 2 to each succeeding term*. Similarly for the even numbers.

☐ Odd number sequence—1, 3, 5, 7, 9, . . .
　　Odd rule—(1 + 2 = 3; 3 + 2 = 5; etc.)
☐ Even number sequence—0, 2, 4, 6, 8, . . .
　　Even rule—(0 + 2 = 2; 2 + 2 = 4; etc.)

Give children a supply of small blocks and have them construct the odd and even numbers. Each value in the sequence should be represented by a separate stack. Ask them to describe any patterns they see in their block sequence and draw a sketch of the results.

Arrow diagrams such as those shown in Figure 11.1 can be used to introduce number patterns. Children fill in the circles by carrying out the indicated operation. This activity can be adapted to several grade levels by varying the values and operations involved. Children can make up arrow diagram problems for each other as well.

Other rules relate more than two terms in a sequence. An interesting example was described over seven hundred years ago by the Italian mathematician Leonardo Fibonacci. Starting with 0 and 1, subsequent terms in Fibonacci's sequence are computed by adding each previous pair of terms.

Fibonacci sequence—0, 1, 1, 2, 3, 5, 8, 13, 21, . . .

This sequence appears in a surprising number of contexts, from biology to music. For example, Fibonacci noted that the pattern of reproduction of a pair of rabbits follows this sequence. Rabbits must mature two months before reproducing. Subsequently, each parent pair produces one new pair of offspring each month. If no rabbits die, the rabbit population (not including the original parents) will multiply as as shown in the two-column Table 11.1.

Other interesting patterns can be generated based on the Fibonacci sequence. For example, each term in the sequence can be squared and these values summed pairwise as in the original sequence:

Fibonacci sequence—0, 1, 1, 2, 3, 5, 8, 13, . . .

Square each term—0, 1, 1, 4, 9, 25, 64, 169, . . .

Add pairs—1, 2, 5, 13, 34, 89, 233, . . .

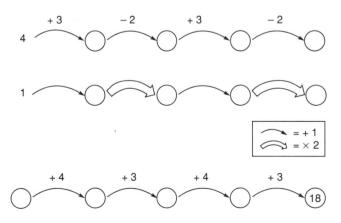

FIGURE 11.1 Arrow Diagrams

TABLE 11.1 Rabbit Reproduction

Month	Pairs of Rabbits
0	0
1	1
2	1
3	2
4	3
5	5
6	8
.	.
.	.
.	.

> Compare the squared Fibonacci sequence to the original Fibonacci sequence. Notice that the sum of consecutive pairs of squares seems to give alternating terms of the original Fibonacci sequence. Carry out the process a few more terms to see if the pattern continues. What is the property of the Fibonacci sequence that causes this pattern?

POLYGON AND POLYHEDRON NUMBERS

Historically, number patterns based on the geometric arrangement of objects have taken on special, even mystical significance. The sequence of square numbers (1, 4, 9, 16, and so on) is a familiar example. Other number sequences are also suggested by geometric shapes.

Triangle Number Sequence

The triangle numbers are constructed by arranging identical counters into a stair-step pattern as shown in Figure 11.2. Beginning with one cube, children continue building

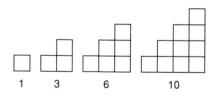

FIGURE 11.2 Triangle Numbers

TABLE 11.2 Triangle Number
Pattern

Steps	Triangle Number
1	1
2	3
3	6
4	10
.	.
.	.
.	.

larger configurations in the triangle pattern. The triangle numbers grow according to the number of objects in each step. Have children build the triangle numbers using blocks. Ask them to predict the number of blocks in the tenth stack.

The pattern of values can be organized into a table according to the number of steps in each triangular stack of cubes (see Table 11.2).

Several interesting patterns can be found in the triangle number sequence. By examining the figure associated with each triangle number, notice that each subsequent term can be found by adding a new row to the bottom of the preceding value (see Figure 11.3).

To compute the next triangle number, simply add the current term and the number equal to the **position value** of the new term. For example, to find the sixth term we add six to the value of the fifth term:

$$1, \ 3, \ 6, \ 10, \ 15, \ 21, \ \dots$$
$$\bigvee$$
$$+6$$

Compute fifteen terms of the triangle numbers. Record your results in a two-column table.

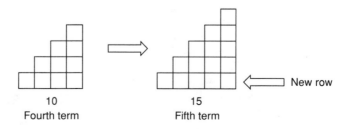

10
Fourth term

15
Fifth term

New row

FIGURE 11.3 Triangle Numbers

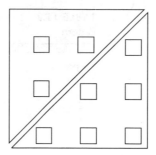

FIGURE 11.4 Relationship Between Triangle and Square Numbers

Square Number Sequence

Children can combine adjacent pairs of triangle numbers to show a new sequence called the **square numbers** (see Figure 11.4).

The square numbers derive their name from square arrays that can be constructed using 1, 4, 9, 16, 25, and so on, counters. Several patterns can be explored based on the square number sequence. Using centimeter squared paper, have children draw the sequence of square numbers with edge lengths of 1–10 centimeters. Have them color or cut out the largest possible square array within each square, leaving an L-shaped figure as shown in Figure 11.5. Using blocks, demonstrate the process on the overhead projector while the children work at their seats. The **L** has equal arms (not including the corner square) that grow one unit for each step in the sequence. Ask if the arms will always be equal regardless of the size of the original square. If so, then the number of blocks in both arms must be even (two times any whole number (**n**) is even). When the corner object is included, the result must *always* be odd:

$$2 \times n = \text{even number}$$

$$(2 \times n) + 1 = \text{even number} + 1 = \text{odd number}$$

The L-shaped figure is the result of subtracting adjacent pairs of square numbers. The **difference sequence** is the odd numbers:

Square numbers— 1, 4, 9, 16, 25, 36, . . .

Difference— 3, 5, 7, 9, 11, . . .

Using blocks, construct the first ten elements of the square number sequence. Record your results in a table. Subtract adjacent elements in the sequence and record the results in a table. Why must the difference sequence include all the odd numbers larger than 3?

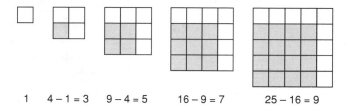

FIGURE 11.5 Difference Between Adjacent Square Numbers

Tetrahedral Numbers

Have children construct piles of marbles with equilateral triangle bases as shown in Figure 11.6. Use a heavy shag carpet scrap to keep the marbles from rolling. These structures are three-dimensional representations of the **tetrahedral number** sequence. The sequence is named for the regular geometric space-figure constructed from four equilateral triangles, the *tetrahedron*.

Cannon balls were commonly stacked according to this pattern. Each stack comprises a triangular arrangement of balls. The largest triangular configuration forms the base. Each subsequent layer is the next smaller term in the triangle number sequence. A tetrahedral number, then, is the sum of all triangle numbers from 1 to the value of its base. Children can use this pattern to find the number of cannon balls in a tetrahedral stack with ten balls along each edge of its base (see Figure 11.7).

Compute the first ten tetrahedral numbers. A calculator may be useful for this exercise. Record your results in a table.

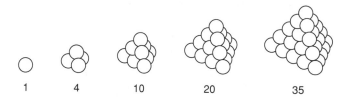

FIGURE 11.6 Tetrahedral Numbers

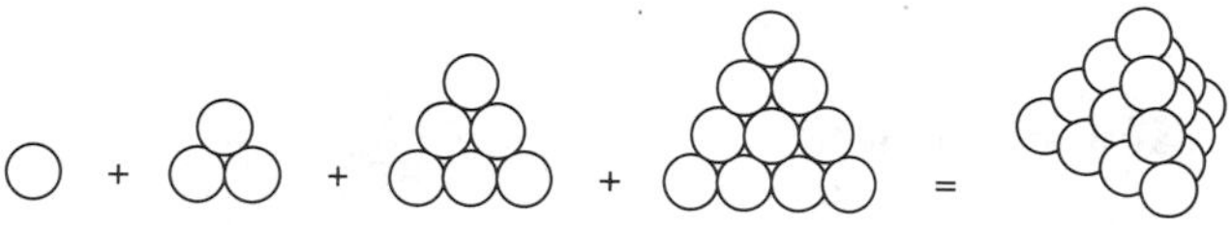

$$1 + 3 + 6 + 10 = 20$$

FIGURE 11.7 Sum of Triangle Numbers = Tetrahedral Number

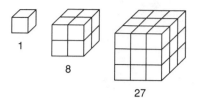

FIGURE 11.8 Cubic Numbers

Cubic Numbers

Children can construct a model of the **cubic numbers** by stacking blocks to form square cubes (see Figure 11.8).

The results can be organized into a table where the first column indicates the position of the box in the sequence, and the second column lists the corresponding terms of the sequence (see Table 11.3).

Children should notice that this sequence grows very rapidly compared to previous sequences. Each new cubic term increases the value of the previous term by the total number of the cubes added to all three faces of previous cubes. This increases the value of subsequent terms more quickly than in the triangle, square, and tetrahedral number sequences.

> Compute ten terms of cubic number sequence. Record your results in a table.

TABLE 11.3 Cubic Numbers

Position	Term
1	1
2	8
3	27
4	64
5	125
.	.
.	.
.	.

USING NUMBER PATTERNS

Making Noodles

Noodles are made by hand in some Chinese restaurants. Dough is prepared and stretched using a doubling process. The first doubling gives two very thick noodle

TABLE 11.4 Noodle Numbers

Doublings	Noodles
1	2
2	4
3	8
4	16
5	32
6	64
.	.
.	.
.	.

strands. The second gives four noodles. Continuing the doubling and stretching process soon produces a large number of very thin, delicious noodles. Children can use a long piece of string to demonstrate this doubling process. Then have them try to find a pattern that will allow them to determine how many doublings are required to produce 1024 noodles by continuing the sequence started in Table 11.4.

Compute ten terms of noodle number sequence. Record your results in a table.

Prime and Composite Numbers

A whole number that has exactly two whole number factors, 1 and itself, is called a **prime number**. All other whole numbers except 1 are called **composite numbers**. The number 1 is special because it is a factor of all values and, therefore, is considered neither prime nor composite.

Arrays can be used to show the special characteristics of prime numbers. Using squared paper, have children cut rectangular arrays representing the whole numbers 1–10 (3 → 1 by 3; 4 → 1 by 4 and 2 by 2). Make a classroom chart to keep track of how many different arrays there are for each value (count commutes as the same array). Primes have *only* one array representation, and composites have more than one (see Figure 11.9).

Larger prime numbers can be identified using the **Sieve of Eratosthenes**. Using a hundred chart, children first circle 1, the identity element that is neither prime nor composite. The next value is 2, so cross out all the multiples of 2 (except 2, of course). The next uncrossed value is 3, so cross out all the multiples of 3 (except 3). Skip 4, because it is already crossed out. Continue this process until all possible multiples are crossed out. The values remaining are primes, and the ones crossed out are composite numbers. Ask the children why the Sieve leaves only the prime values (see Figure 11.10). How many multiples need to be checked before no more

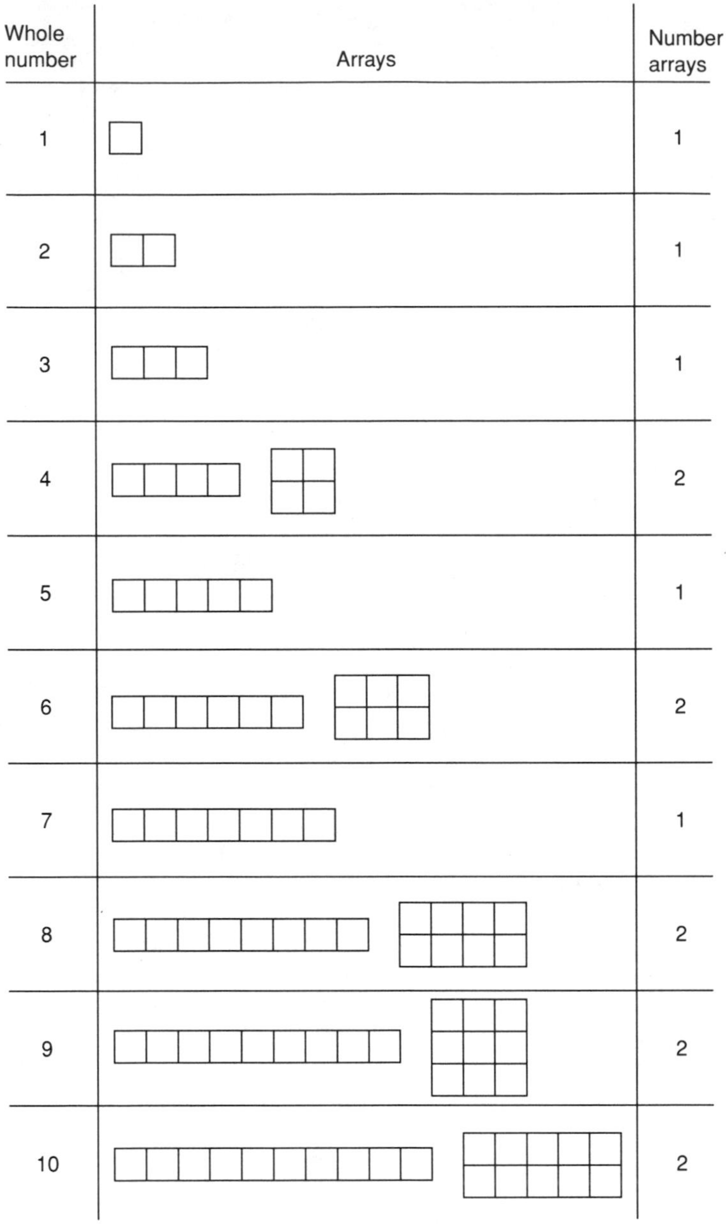

Whole number	Arrays	Number of arrays
1		1
2		1
3		1
4		2
5		1
6		2
7		1
8		2
9		2
10		2

FIGURE 11.9 Prime and Composite Number Arrays

values will be crossed out? Why? How could this process be extended to identify even larger prime values?

Primes listed from small to large form the sequence of prime numbers. Children with a special interest in this topic can look for **twin primes** (e.g., 3 and 5; 5 and 7;

①	2	3	⊠	5	⊠	7	⊠	⊠	⊠
11	⊠	13	⊠	⊠	⊠	17	⊠	19	⊠
⊠	⊠	23	⊠	⊠	⊠	⊠	⊠	29	⊠
31	⊠	⊠	⊠	⊠	⊠	37	⊠	⊠	⊠
41	⊠	43	⊠	⊠	⊠	⊠	⊠	⊠	⊠
⊠	⊠	53	⊠	⊠	⊠	⊠	⊠	59	⊠
61	⊠	⊠	⊠	⊠	⊠	67	⊠	⊠	⊠
71	⊠	73	⊠	⊠	⊠	⊠	⊠	79	⊠
⊠	⊠	83	⊠	⊠	⊠	⊠	⊠	89	⊠
⊠	⊠	⊠	⊠	⊠	⊠	97	⊠	⊠	⊠

FIGURE 11.10 Sieve of Eratosthenes

41 and 43) in the sequence of prime numbers:

$$2, 3, 5, 7, 11, 13, 17, 19, 23, 29, 31, 37, 41, 43, 47, \ldots$$

A way to test a value to see if it is prime is to divide it by all smaller whole numbers and check for zero remainders. For large numbers, the procedure becomes tedious even when using a calculator. The amount of work can be reduced by noticing that whole numbers larger than one-half of the value being tested can never give a whole number quotient (except the value itself, which gives a quotient of 1). For example, to test if 51 is prime, check each divisor 2 through 25. Except for 51 itself, no divisor 26 through 50 can give a whole number quotient.

The problem can be further simplified by checking only *prime* divisors less than or equal to one-half of the number being tested. To test 51, divide by 2, 3, 5, 7, 11, 13, 17, 19, and 23. Since both 3 and 17 give zero remainders, 51 is *not* prime. The same test shows that 53 is prime. Though this procedure will determine if a given number is prime, no one has yet discovered a pattern that will generate, say, the 100th or the 167th value in the prime number sequence.

Look for patterns in the sequence of prime numbers. Show that it is only necessary to check for factors up to the square root of any value to determine if it is prime.

Casting-Out Nines

Interesting number patterns emerge from the process called **casting-out nines**. This procedure is based on the observation that for any multiplication fact involving the value 9, the sum of the product's digits is equal to 9 ($3 \times 9 = 18 \rightarrow 1 + 8 = 9$).

The process of adding the digits that comprise any whole number exposes interesting patterns. By adding a value's digits and repeating the process for each result, the sum eventually becomes a single digit which is called the **digital root**.

Original Number		First Sum		Digital Root
237	$\rightarrow$	$2 + 3 + 7 = 12$	$\rightarrow$	$1 + 2 = 3$

Have children find the digital roots of several two- and three-digit numbers. Have them look for other patterns for combining digits. Have them try adding the digit in the ones column to the remaining digits read as a separate number. The process is repeated until arriving at a single digit ($371 \rightarrow 37 + 1 = 38 \rightarrow 3 + 8 = 11 \rightarrow 1 + 1 = 2$). Does this process seem to give the same value as the digital root? What if other groupings of digits were allowed? Would the result be the same as the digital root? For example

$$35,482 \rightarrow 35 + 482 = 517 \rightarrow 5 + 17 = 22 \rightarrow 2 + 2 = 4 \rightarrow \text{digital root?}$$

Children can also verify that the digital root of any whole number is equal to the remainder resulting from dividing the value by 9. If the digital root of a number is 9, then the remainder resulting from dividing that number by 9 will be 0 (see Table 11.5).

Grade 4–6 children often find it challenging to verify that large numbers with the digital root 9 are really divisible by 9. This activity can motivate students to practice long division and calculator skills.

Casting-out nines can also be used as an alternative way to check the accuracy of arithmetic exercises. To check an addition problem, find the digital root of each of the addends and the sum. If the answer is correct, adding the digital roots of the addends will give a digital root equal to that of the sum.

TABLE 11.5 Casting-Out Nines

Number	Digital Root	Division by Nine
27	9	3 R0 9)27
111,111,111	9	12,345,679 R0 9)111,111,111

$$
\begin{array}{rcr}
254 & \to & 2 \\
+\,187 & \to & +7 \\
\hline
441 & \to & 9 \quad \text{Digital Root}
\end{array}
$$

If the answer is off by exactly a multiple of 9, this checking procedure will, of course, *incorrectly* validate that the answer is right. For example, if 432 was thought to be the sum for 254 + 187, the digital roots would incorrectly indicate that the answer is right. Caution should be exercised when children apply digital roots as a checking method. In particular, two digits can be easily reversed when copying columns of numbers. This error always gives a result that is off by *exactly* a multiple of nine. Can you show why is this true?

More experienced students can use similar checks for multiplication, subtraction, and division. Like addition, multiplication can be checked by carrying out the same operation on the digital roots as on the original values in the exercise.

$$
\begin{array}{rcr}
46 & \to & 1 \\
\times\,23 & \to & \times\,5 \\
\hline
1058 & \to & 5 \quad \text{Digital Root}
\end{array}
$$

Note that subtraction exercises sometimes give a *negative* difference between the digital roots (55 − 29 gives the digital roots $1 - 2 = {}^{-}1$). To avoid work with negative values, use the procedure of adding the difference and the subtrahend to check if it equals the minuend.

$$
\begin{array}{rcll}
251 & \to & \underline{8} & \text{Digital Root} \\
-\,67 & \to & +4 & \\
\hline
184 & \to & 4 &
\end{array}
$$

$$
\text{or} \quad
\begin{array}{rcll}
213 & \to & \underline{15} & \to 6 \quad \text{Digital Root} \\
-\,89 & \to & +8 & \\
\hline
124 & \to & 7 &
\end{array}
$$

For division, it is more convenient to check by using the inverse operation, multiplication.

$$
\begin{array}{l}
\overset{\textstyle 14}{26\,\overline{)364}} \to 26 \times 14 = 364 \to 4 \\
\qquad\quad\; \downarrow \quad\;\; \downarrow \qquad\qquad\quad \updownarrow \\
\qquad\quad\; 8 \times 5 = 40 \to 4 \quad \text{Digital Root}
\end{array}
$$

The casting-out nines procedure is a direct result of the base-10 structure of our numeration system. It can be explained by using techniques of modular arithmetic where 0 and 9 are considered equivalent (as 0 and 12 are considered equivalent on the clock). Mathematically mature Grade 5–6 students can use this idea to understand why it is possible to add 9 to negative digital root differences found in some subtraction exercises and continue the checking process (for 213 − 89 = 124, the digital root difference is $6 - 8 = {}^{-}2 \to {}^{-}2 + 9 = 7$, which equals the digital root for 124).

Using casting-out nines procedures, verify that the following computations are correct.

1. 245 + 816 = 1061

2. 57,234 − 45,785 = 11,449

3. 22,111 − 7289 = 14,822

4. 143 × 217 = 31,031

5. 16,875 ÷ 456 = 37 R3

Divisibility Tests

It is sometimes helpful for children to be able to quickly determine whether a value is evenly divisible by some small number. **Factoring**, renaming fractions in simplest terms, and work with ratio and proportion can be facilitated by the use of **divisibility rules**. For example, when renaming fractions in simplest terms, it is necessary to find common factors for the denominators. For the fraction $\frac{21}{51}$, a child can find the digital roots for 21(3) and 51(6) and quickly note that both are divisible by 3 (both numbers are therefore divisible by 3 but not 9). Therefore, $\frac{21}{51}$ can be renamed in simplest terms as $\frac{7}{17}$.

Rules have been invented that, in many cases, allow children to predict whether a number is evenly divisible by the divisors 1 through 9 without actually carrying out the division. If readily available, a calculator can serve the same purpose. However, verifying why these rules work can be an interesting mathematical exploration for mathematically able students.

Some of the divisibility rules are easy to verify. For example, *all* whole numbers are evenly divisible by 1. All *even* numbers are divisible by 2. To help children understand the divisibility rule for 4, list all the multiples of 4 to 100. Notice that every multiple of 4 can be evenly divided by 2 at least two times (36 ÷ 2 = 18 and 18 ÷ 2 = 9). Look at larger multiples of 4 such as 112 or 448. To test if 112 is divisible by 4, we first note that its ones digit is even, so it is divisible by 2 (112 ÷ 2 = 56). Since 56 is even, it must also be divisible by 2. To be divisible by 4, then, we are only concerned that the last digit of the quotient be even each of the two times we divide by 2. Regardless of the digits in the larger place-values, if the number represented by the ones and tens digits is divisible by 4 (i.e., 2, twice), then the original number is divisible by 4. Using the same procedure, children can verify that we only need to check the value represented by the final three digits of a number to test for divisibility by 8.

In the previous section, we noted that if a digital root is 9, the number is divisible by 9. A number divisible by 9 is also divisible by 3. Further, because the digital root of a number is equal to the remainder after dividing by 9, numbers with digital roots 3, 6, or 9 have remainders divisible by 3 and, therefore, are themselves divisible by 3. Have children test several values using the digital root method and check the results on a calculator. Then ask them to explain why the divisibility rule for 6 works.

The following table lists divisibility rules for divisors 1 through 9. Similar tests for larger divisors can been invented as well. Some children may find it challenging to find divisibility tests for divisors 10 through 15 (11 is particularly interesting). When introducing greatest common factor and least common multiple, encourage children to employ these rules to help them identify factors (see Table 11.6).

Try to explain why each divisibility rule works. Use divisibility tests to verify the following statements. Check the results with a calculator.

1. The number 111,002,436 is divisible by 3.

2. The number 111,111,112 is divisible by 2.

3. The number 123,456,789 is divisible by 9.

4. The number 6034 is divisible by 7.

5. The number 999,888,064 is divisible by 8.

TABLE 11.6 Whole Number Divisibility Rules

Divisor	Test
1	All whole numbers.
2	Units digit is even.
3	Digital root is a multiple of 3.
4	Value represented by last two digits is a multiple of 4.
5	Units digit is 0 or 5.
6	Tests for divisors 2 and 3 apply.
7	Doubling units digit and subtracting result from remaining digits read as separate number is a multiple of 7 (may need to repeat process).
8	Value represented by last three digits is a multiple of 8.
9	Digital root is 9.

Example 1.   Is 3472 divisible by 4?

Test: $\begin{array}{r} 18 \\ 4\overline{)72} \end{array}$ → Verdict—Yes

Example 2.   Is 62,482 divisible by 6?

Test: last digit is *even*
but digital root = 4   → Verdict—No

Example 3.   Is 5761 divisible by 7?

Test: $\begin{array}{r} 576 \\ -\ 2 \\ \hline 574 \end{array} \rightarrow \begin{array}{r} 57 \\ -8 \\ \hline 49 \end{array}$ → A multiple of 7, so verdict—Yes

TABLE 11.7 Greatest Common Factor

Value	Factors	GCF
18	1, 2, 3, 6, 9, 18	
24	1, 2, 3, 4, 6, 8, 12, 24	6

Greatest Common Factor

The ability to identify the *largest* factor common to two or more numbers is necessary for work with fractions, ratios, proportions, and probability. One method for finding the **greatest common factor** (GCF) is to list all the factors of each number and check to see where the two sets of factors intersect (i.e., the values the two sets have in common). The largest of these common factors is called the GCF. For example, give children pairs of values and have them list all the factors for each, as shown for the numbers 18 and 24 in Table 11.7. Have them draw a loop around the pairs of common factors (in bold) and circle the largest common factor, the GCF (6).

For large numbers with many factors, a second method for finding the GCF is helpful. Using the previous example, list all the *prime* factors for each number (see Table 11.8). Note that the prime factorization may include repeated values (i.e., 2, 3, 3).

The values 18 and 24 have one 2 and one 3 in common. The product of these common prime factors equals the GCF (2 × 3 = 6). This technique is often convenient since only prime factors must be found for each value, making the divisibility checks less tedious.

Factor trees can be used to help children find the prime factorization of a number. To make a factor tree, have children write the value to be factored at the top of a piece of paper. Using divisibility rules, have the class write one pair of factors beneath the value and draw a line to each as shown in Figure 11.11. This process is continued until all the factors are prime numbers. The prime factors of a number are the set of values displayed at the end of each branch of the tree. Have children select different sets of initial factors for a number, complete a factor tree, and compare results. Does everyone eventually get the same set of prime factors? The important observation that every number has a *unique* prime factorization is called the **fundamental theorem of arithmetic**. Though elementary children may not be able to prove this theorem, the class can verify that any whole number has only one set of prime factors (ignoring order, of course).

TABLE 11.8 Prime Factors

Value	Prime Factors	GCF
18	2, 3, 3	
24	2, 2, 2, 3	2×3 = 6

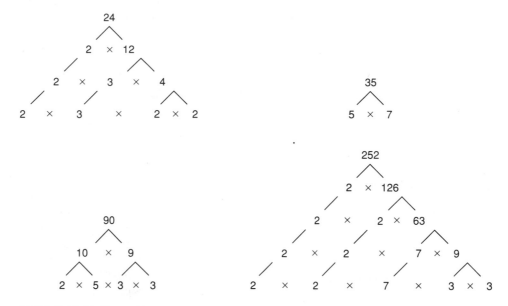

FIGURE 11.11 Factor Trees

Once the prime factors for two values are determined, the GCF can be found by multiplying all the prime factors both numbers have in common (e.g., 2, 3, and 3 for 90 and 252). Have children write the prime factors as in Table 11.9, loop the common factors (in bold), and multiply the set of common factors to find the GCF.

Use factor trees to find the set of prime factors for each pair of numbers and compute the GCF for each. Use a different pair of initial factors for one exercise and see if you get the same sets of prime factors and GCF.

1. (12, 15)
2. (18, 24)
3. (12, 30)
4. (9, 14)
5. (105, 252)

TABLE 11.9 Greatest Common Factor

Value	Prime Factors	GCF
90	2, 5, 3, 3	
252	2, 2, 7, 3, 3	$2 \times 3 \times 3 = 18$

TABLE 11.10 Least Common Multiple

Value	Multiples	LCM
6	6, 12, 18, 24, 30, 36, 42, 48, ...	
8	8, 16, 24, 32, 40, 48, 56, ...	24
12	12, 24, 36, 48, 60, ...	

Least Common Multiple

When adding fractions, it is sometimes necessary to find a common denominator. This number must be a *common multiple* of all of the denominators. You can always find a common multiple of two or more numbers by simply computing their product. For example, a common multiple of 6 and 8 is 48. However, 24, which is also a multiple of both 6 and 8, is smaller and may therefore be easier to work with. The number 24 is the **least common multiple** (LCM) of 6 and 8 because it is the *smallest* common multiple of the two numbers.

One method for finding the LCM for two or more values is to list several multiples of each number, determine where the sets of multiples intersect, and select the smallest common multiple. For example, have children list the multiples of 6, 8, and 12 as shown in Table 11.10 and loop the common multiples (in bold).

Both 24 and 48 are common multiples for the 6, 8, and 12. The smallest common multiple, 24, is the LCM.

Children can also use prime factorization to identify the LCM of a set of numbers. After finding the prime factors for each number using divisibility rules or factor trees, select the *largest* set of *each* factor represented. As shown in Table 11.11, for the factors 2 and 3, the 2 appears a maximum of three times (as factors of 8) and the 3 appears at most one time (as a factor of 6 and 12). Their product gives the LCM = $2 \times 2 \times 2 \times 3 = 24$.

Work with the GCF and LCM is often introduced as an application of the divisibility tests and prime numbers. GCF and LCM reappear later in work with fractions and ratios.

TABLE 11.11 Prime Factors

Value	Prime Factors	LCM
6	**2, 3**	
8	**2, 2, 2**	$2 \times 2 \times 2 \times 3 = 24$
12	2, 2, 3	

USING PATTERNS TO DEVELOP FUNCTIONS

One of the powerful tools available in elementary mathematics is the notion of **function**. A function is a special kind of mathematical relation where each object in one set is related to *only* one object in a second set. For example, to count sheep the hard way, count the number of feet and divide by 4. This functional relationship can be recorded as shown in Table 11.12.

An inequality is one example of a mathematical relation that is *not* a function. For example, if everyone in a class has a pencil longer than three inches, the relationship of children to pencil lengths is not a function, because we cannot predict the exact length of a pencil for any one child.

Functional relations are often useful for solving problems. For example, if every truck has eighteen tires, the number of tires depends on, or *is a function of*, the number of trucks. Many functions can be written as a number sentence, or *equation*. In the case of feet and sheep, a simple equation that gives the number of sheep for any number of feet is

number of feet ÷ 4 = number of sheep

> If the average human head has 50,000 strands of hair, write an equation for counting people the hard way.

The functional relationship in many arithmetic or geometric sequences can also be written as an equation. An equation relates each term in a number sequence to its position in the sequence. For example, to find the tenth even number, have children list the position numbers in the first column and the corresponding terms of the even number sequence in the second column (see Table 11.13).

Children can easily extend the pattern to the tenth term. However, to find the one hundredth, thousandth, or billionth even number, the process would become tedious

TABLE 11.12 Feet and Sheep Function

Feet	Sheep
4	1
8	2
12	3
16	4
.	.
.	.
.	.

TABLE 11.13

Position Numbers	Even Numbers
1	2
2	4
3	6
4	8
5	10
6	12
7	?
8	?
9	?
10	?

indeed. Instead, have students work out a systematic rule relating each position number with its corresponding term in the sequence. Rather than computing each subsequent term in the second column, invent a rule relating each position value to its corresponding even number term. In Table 11.14, note that doubling the position number *always* gives the corresponding even number term. Introduce the place holder $\triangle$ to represent any position number and $\square$ to represent the corresponding term in the number sequence. Later, letters like x and y can be substituted for these variable place holders.

To compute the hundredth even term, the ninety-ninth even number is no longer required. Applying the equation **2 × position = even number** (or $2 \times \triangle = \square$) gives the hundredth even number term.

$$2 \times 100 = 200$$

Guess My Rule

Pairs of children can play the game *Guess My Rule* where each child makes up a secret rule (e.g., $\triangle + 5 = \square$) and creates a two-column table showing the first six pairs of

TABLE 11.14 Even Numbers

Position $\triangle$	Term $\square$	
1	2	
2	4	
3	6	$2 \times \triangle = \square$
4	8	
.	.	
.	.	
.	.	

values based on this rule. They exchange tables and try to discover the rule that was used to create the table. This game can also be used with the whole class by writing a table on the overhead or board and having the group work on the rule. At first, limit the rules to simple, linear equations involving small whole number constants and coefficients.

Function Machines

The concept of function can be introduced using function machines. Any number put in the machine will be changed according to some rule. A record of the inputs and corresponding outputs can be maintained in a two-column table. Figure 11.12 shows the function machine × **3**, where every number entered will be multiplied by 3. Two or more machines can be combined so that the output from the first enters the second, making a new number. Worksheets of simple or combined function machines can be prepared to provide practice of basic skills. Given several inputs and their corresponding output values, children can work out the rule for one or more function machines. The problem can be made more challenging by leaving out the interim results.

> Verify that the equations written next to Tables 11.15–19 relate each x-value to the corresponding y-value. To do this, substitute the position numbers for each x-value and check to see if the results correspond to the associated y-value.

Coordinate Graphs

A graphic display of functional relationships can be created using a **coordinate system** of graphing. *Ordinate* refers to a position on a number line. A *coordinate*

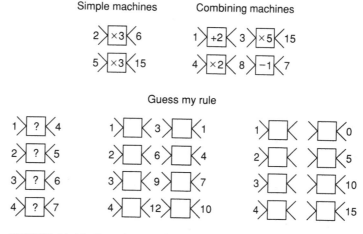

FIGURE 11.12 Function Machines

TABLE 11.15 Odd numbers

x	y
1	1
2	3
3	5
4	$7 \rightarrow 2x - 1 = y$
5	9
6	11
.	.
.	.
.	.

TABLE 11.16 Multiples of Five

x	y
1	5
2	10
3	15
4	$20 \rightarrow 5x = y$
5	25
6	30
.	.
.	.
.	.

TABLE 11.17 Square Numbers

x	y
1	1
2	4
3	$9 \rightarrow x^2 = y$
4	16
5	25
6	36
.	.
.	.
.	.

TABLE 11.18 Triangle Numbers

x	y
1	1
2	3
3	$6 \rightarrow \frac{1}{2}x^2 + \frac{1}{2}x = y$
4	10
5	15
6	21
.	.
.	.
.	.

TABLE 11.19 Tetrahedral Numbers

x	y
1	1
2	4
3	$10 \rightarrow \frac{1}{6}x^3 + \frac{1}{2}x^2 + \frac{1}{3}x = y$
4	20
5	35
6	56
.	.
.	.
.	.

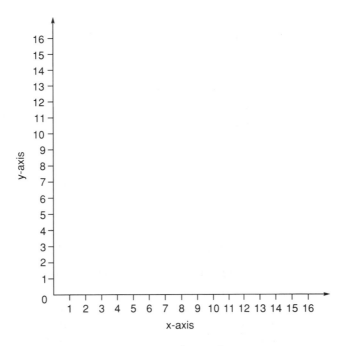

FIGURE 11.13 Cartesian Coordinate Plane

refers to a pair of such position values on a two-dimensional surface. The **Cartesian coordinate plane** is formed by positioning two number lines perpendicular to each other so that they cross at zero (see Figure 11.13).

The horizontal number line is called the **x-axis**, and the vertical line, the **y-axis**. Associated values (coordinate pairs) in a two-column function table can be displayed, or graphed, by first counting on the x-axis the number of spaces indicated by the x value, followed by the number of spaces indicated by the y value in the direction parallel to the y-axis. For example, have children graph the coordinate pairs (1, 1), (2, 3), and so on, in the odd number table on the coordinate graph as shown in Figure 11.14.

Note that the graph of this set of coordinate pairs forms a pattern of points on a straight line. This type of graph is called a **linear graph**. Extending the line makes it possible to predict terms further along in the sequence. The process of extending the graph of a functional relation to predict further (x, y) values is called **extrapolation** (see Figure 11.15).

Many functions have linear graphs. Verify that the even number and multiples-of-5 sequences generate linear graphs. What characteristics of these functions generate a linear graph? Observing the graph of the square numbers shown in Figure 11.16 may help solve this problem.

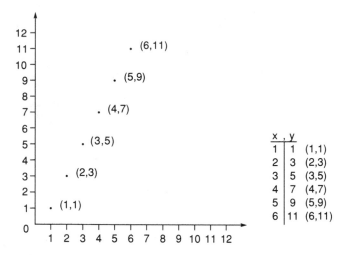

FIGURE 11.14 Graphing Odd Numbers

The graph of the square numbers forms a curved line, called a **parabola**. The graph of the triangle numbers is also a curved line (see Figure 11.17).

Equations like $2x = y$, $2x - 1 = y$ and $5x = y$ generate linear graphs. For each of these functions, the x term has an exponent (power) equal to 1 (if there is no exponent shown, it is assumed to be 1). The equations $x^2 = y$ and $\frac{1}{2}x^2 + \frac{1}{2}x = y$,

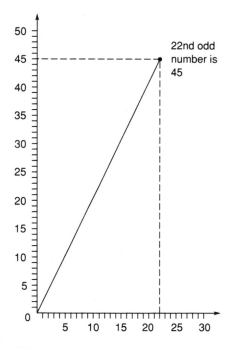

FIGURE 11.15 Extrapolation of Odd Numbers

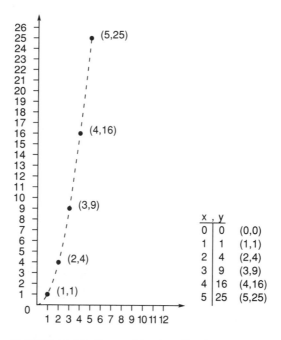

FIGURE 11.16 Square Number Graph

called **quadratic equations**, contain an x term with the largest exponent equal to 2. Both functions generate a curved, parabolic graph. Such functions may also contain an x term with an exponent equal to 1, but the largest power of x must be 2.

Graphs of functions with exponents larger than 2 generate curved line graphs as well. However, these curves are not parabolas. Graphing the third degree tetrahedral numbers (the largest power of x is 3) gives a curved graph that looks similar to the graph of the square numbers. However, extending the coordinate plane to include negative values shows the graphs to be distinctly different (see Figure 11.18).

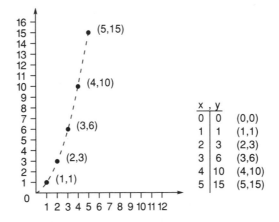

FIGURE 11.17 Triangle Number Graph

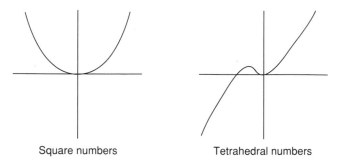

Square numbers   Tetrahedral numbers

FIGURE 11.18

Finite Differences

It is relatively easy for children to generate a table of (x, y) values if the functional relationship is already known. For example, given the equation $5x + 1 = y$, the following table can be generated by substituting values for x and computing the corresponding y values (see Table 11.20).

A table of (x, y) values may be generated by collecting data from an experiment. For example, in Figure 11.19, how many blocks will be needed for a building ten stories high? The task is to determine the function that relates the number of stories (x) with the number of blocks (y).

A procedure called **finite differences** offers a systematic way of finding a function from a table of (x, y) values. This procedure can assist more experienced Grade 5–6 students to write an equation for a function table. Linear functions can be uncovered quite readily with this procedure, though higher order functions are somewhat more complex.

As shown in Table 11.21, children first compute the difference between adjacent pairs of y values (notice this is the same function as shown in Table 11.20.)

TABLE 11.20 Function Table

x	y
0	1
1	6
2	$11 \rightarrow 5x + 1 = y$
3	16
4	21
5	26
.	.
.	.
.	.

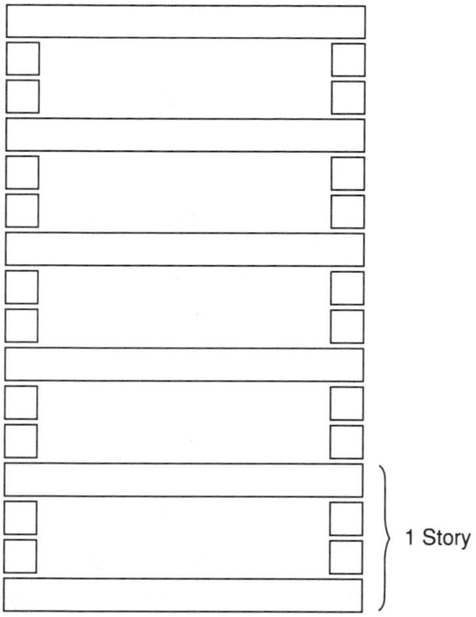

Stories	Blocks
0	1
1	6
2	11
3	16
4	21
5	26
6	?
7	?
8	?
9	?
10	?

FIGURE 11.19 How Many Blocks?

Notice the difference between adjacent y values is always 5. As we already know the function, let's see if this information gives a clue to the components of the function $5x + 1 = y$. The common difference 5 is equal to the coefficient (the number multiplied by a variable) of x. The results for the functions $3x + 3 = y$, $4x + 3 = y$ and $5x + 3 = y$ are shown in Table 11.22.

The common difference in each case corresponds to the coefficient of the x term. Also, the y value corresponding to x value 0 matches the constant number added to the x term in each function.

TABLE 11.21 Finite Differences

x	y	Difference
0	1	
1	6	5
2	11	5
3	16	5
4	21	5
5	26	5
.	.	
.	.	
.	.	

TABLE 11.22 Pattern of Finite Differences

3x+3=y			4x+3=y			5x+3=y		
x	y	Dif.	x	y	Dif.	x	y	Dif.
0	3		0	3		0	3	
		3			4			5
1	6		1	7		1	8	
		3			4			5
2	9		2	11		2	13	
		3			4			5
3	12		3	15		3	18	
		3			4			5
4	15		4	19		4	23	
		3			4			5
5	18		5	23		5	28	
.	.		.	.		.	.	
.	.		.	.		.	.	
.	.		.	.		.	.	

Make tables for $2x + 4 = y$ and $2x + 5 = y$ and verify that the finite difference procedure can be used to find the coefficient and constant for each equation.

Functions involving x terms with powers of 2 (quadratic functions) can also be uncovered from tables of values using finite difference methods. Notice that, in Table 11.23, the square numbers require two differences before a constant difference arises.

Patterns for quadratic functions emerge that can be verified by working examples. Again, the y value corresponding to the x value 0 gives the constant added to the x^2 term in the equation. The equation for the example in Table 11.23 can be written $x^2 + 0 = y$. In Table 11.24, because the y value corresponding to the x value 0 equals 2, the equation is written $x^2 + 2 = y$.

The coefficient of the x^2 term can be found by observing the common *second* difference. Both of the preceding examples have a common difference of 2 and an

TABLE 11.23 Square Numbers

x	y	First	Second
0	0		
		1	
1	1		2
		3	
2	4		2
		5	
3	9		2 → $x^2 = y$
		7	
4	16		2
		9	
5	25		
.	.		
.	.		
.	.		

TABLE 11.24 Finite Differences for Quadratic Functions

x	y	First	Second
0	2		
1	3	1	2
2	6	3	2
3	11	5	2 → $x^2 + 2 = y$
4	18	7	2
5	27	9	
.	.		
.	.		
.	.		

TABLE 11.25 Predicting Coefficients for Quadratic Functions

x	y	First	Second
0	1		
1	3	2	4
2	9	6	4
3	19	10	4 → $2x^2 + 1 = y$
4	33	14	4
5	51	18	
.	.		
.	.		

x^2 coefficient of 1 (i.e., $1 \times x^2 = y$ and $1 \times x^2 + 2 = y$). Table 11.25 shows that, when the x^2 term has a coefficient of 2, the common difference is 4.

> Test several examples to see if the x^2 term coefficient equals one-half the second common difference as shown in Table 11.26.

The following three rules will help mature Grade 5 and 6 children uncover the functions using two-column tables.

1. The number of finite differences required to arrive at a common value indicates the *largest* power for the x term (i.e., 2 differences means it is a x^2 function).
2. The y value corresponding to the x value *0* is the constant added to the x term(s).

TABLE 11.26 Coefficient of x^2 Term

x	y	First	Second
0	$\frac{1}{2}$		
1	$3\frac{1}{2}$	3	$6 \times \frac{1}{2}$
2	$12\frac{1}{2}$	9	6
3	$27\frac{1}{2}$	15	6
4	$48\frac{1}{2}$	21	6
5	$75\frac{1}{2}$	27	

$$3x^2 + \tfrac{1}{2} = y$$

3. The common difference is related to the coefficient of the x term with the largest exponent. The common difference *equals* the coefficient for linear functions and is *twice* the coefficient for quadratic functions (other rules govern higher order functions).

Functions may also have additional x terms. For example, look at the function tables (Tables 11.18 and 11.19) for the triangle numbers ($\frac{1}{2}x^2 + \frac{1}{2}x = y$) and tetrahedral numbers $\frac{1}{6}x^3 + \frac{1}{2}x^2 + \frac{1}{3}x = y$. Try to extend the finite difference rules to uncover these more complex equations.

CLASSROOM ACTIVITIES

Geometric Patterns and Functions

Geometry offers a rich source for the investigation of patterns and functions. For example, how many triangles pointing upward can you count in the diagram shown in Figure 11.20?

There are three small triangles and one large triangle. The pattern of triangles-in-a-triangle is summarized in Table 11.27, where x is the number of small triangles on an edge and y is the total number of triangles pointing up. Notice that the sequence of y values is identical to that of the tetrahedral numbers.

Other geometric patterns are shown in Figures 11.21–11.23. These problems can be solved by looking for patterns and working out functional relationships. Many similar problem solving situations can be found in the references listed at the end of this chapter.

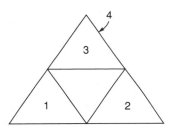

FIGURE 11.20   Triangles in a Triangle

Milk Carton Computers

A convenient way for students to generate guess-my-rule number sequences to explore is with a simple milk carton computer. The device is not really a computer but merely flips over previously prepared cards to give the illusion of an input/output operation. The *x* values are on the front, and the corresponding *y* values are on the back of each card (see Figure 11.24).

The *x* values (say, 0 to 9) are written in red on one side of a 2 cm × 4 cm card. On the back of each card, corresponding *y* values are written in blue. Each card is inserted as **input** into the top of the computer with the *x* value showing. An internal channel turns the card over and presents it at the slot in the bottom of the computer as **output**. The input *x* and output *y* can then be organized into a table. Functional relationships can be worked out by carefully observing the patterns in the table or by applying the finite difference method.

Alphanumerics

An interesting activity for Grade 4–6 children involves looking for words that equal 1,000,000 (Bain, 1987). Each letter in the alphabet is assigned a number 1–26. The product of the letters gives the value of the word (*lumps* → 12 × 21 × 13 × 16 ×

TABLE 11.27   Triangles in a Triangle

x	y
0	0
1	1
2	4
3	10 → $\frac{1}{3}x^3 + \frac{1}{2}x^2 + \frac{1}{6}x = y$
4	20
5	35
.	.
.	.
.	.

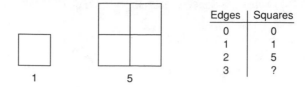

Edges	Squares
0	0
1	1
2	5
3	?

1   5

FIGURE 11.21 How Many Squares in a Square?

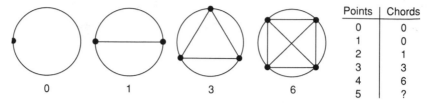

Points	Chords
0	0
1	0
2	1
3	3
4	6
5	?

0   1   3   6

FIGURE 11.22 Chords in a Circle

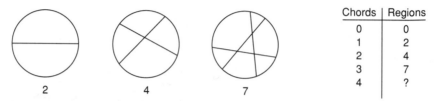

Chords	Regions
0	0
1	2
2	4
3	7
4	?

2   4   7

FIGURE 11.23 Regions in a Circle

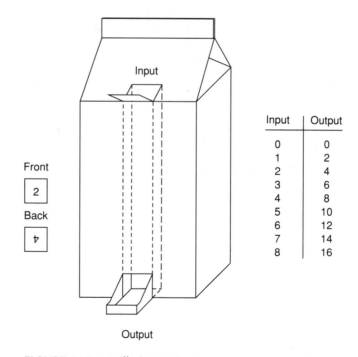

Input	Output
0	0
1	2
2	4
3	6
4	8
5	10
6	12
7	14
8	16

FIGURE 11.24 Milk Carton Computer

19 = 995,904). Use only English words and no proper nouns. Children can use a *guess-and-test strategy* or may use *prime* factors to identify special letters that are factors of one million (i.e., 2, 2, 2, 2, 2, 2, 5, 5, 5, 5, 5, 5—prime factors of one million). This is a good homework problem that can involve the entire family in finding a solution.

DEVELOPING FUNCTIONS TO SOLVE PROBLEMS

The finite difference method can be used as a tool to uncover functions associated with problem situations. Once the general form of a function is known, an x value can be used to derive the corresponding y value.

For example, a good long distance runner can run a mile every six minutes. Table 11.28 summarizes the distance d covered in time t.

As only one step is required to generate a common difference, the x term exponent is 1. The coefficient equals the first difference 6, and the y value corresponding to the x value 0 gives the constant 0. Therefore, the function is $6x^1 + 0 = y$, or $6x = y$. Using this equation, we can quickly estimate the time required to run a marathon of 26.3 miles.

$$6 \times 26.3 = y$$
$$157.8 = y$$
$$= 2 \text{ hours } 37.8 \text{ minutes}$$

The result can also be displayed on a graph as shown in Figure 11.25.

Earlier, we discussed the extrapolation process for extending a graph beyond those values initially given in a table. This technique is useful for computing larger values of known functional relationships. Values can also be computed for points between existing solutions. For example, it is easy to plot the distance corresponding to a fifteen-minute run even though the time values are given in six-minute intervals. This process, called **interpolation**, is shown in Figure 11.26.

TABLE 11.28 Runner's Record

Miles d	Minutes t	First Difference
0	0	
1	6	6
2	12	6
3	18	6
4	24	6
5	30	6
6	36	6
.	.	
.	.	
.	.	

$\rightarrow 6d = t$

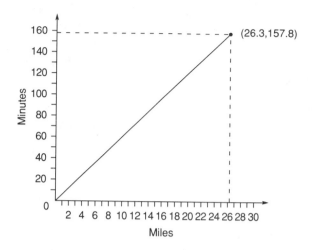

FIGURE 11.25 Running Graph Extrapolation

Table 11.29 gives the distance (d) an object falls in quarter-second time intervals (t). Find the function associated with this table and graph of the results using Figure 11.27. Determine how far an object will fall in 10 seconds (40 quarter seconds); in 1 minute; in $1\frac{7}{8}$ seconds.

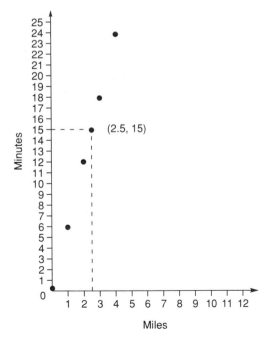

FIGURE 11.26 Running Graph Interpolation

TABLE 11.29 Falling Objects

t ($\frac{1}{4}$ sec)	d (ft)
0	0
1	1
2	4
3	9
4	16
5	25
6	36
.	.
.	.
.	.

COMPUTER APPLICATIONS

Problem-Solving Software

King's Rule. This program offers practice discovering number patterns. A sequence of numbers is presented that represents an underlying mathematical pattern. Users try to guess the pattern and test hypotheses by entering sets of values. For example, Figure 11.28 shows the sequence of trials entered by a student to discover the relationship represented by the sequence 8, 9, 73.

Once users think they know the rule, they can request a quiz. If the questions presented are answered correctly, a graphic reward is displayed. The program offers six levels of difficulty and requires only elementary arithmetic skills. *King's Rule* can

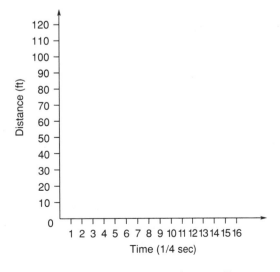

FIGURE 11.27 How Fast Do Objects Fall?

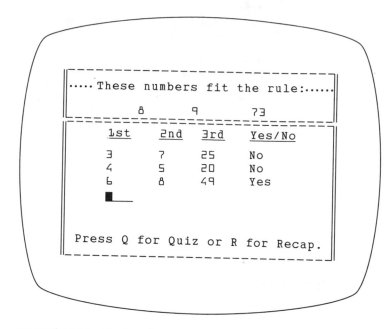

FIGURE 11.28 King's Rule. *Note:* From "King's Rule" [Computer program] by Sunburst Communications, 1985. Copyright 1985 by Sunburst Communications. Reprinted by permission.

be used effectively by individuals or groups of students to explore number patterns using the *guess-and-test* and *make a table* problem-solving strategies.

ADAPTING INSTRUCTION FOR CHILDREN WITH SPECIAL NEEDS

Factor and Number Pattern Activities

Checking Factor Candidates. To help special-needs children be more systematic when computing the factors of numbers, have them write the number above a two-column table and list **candidate factors** (1, 2, . . . , and so on) in the left column (Dearing & Holtan, 1987). Using a calculator, children divide each of the candidate factors into the number. If the decimal fraction part of the quotient has digits other than zero, cross out the divisor as a factor candidate. Continue the process until all factor candidates have been tested. Initially, the teacher can prepare worksheets with factor candidates listed up to at least the square root of the number being tested. Due to the commutative property, factors larger than the square root will have been previously found. When completing factor trees, children only need to find one set of factors for any given number since they can repeat the process on each of the factors.

$$
\begin{array}{c}
87 \\
\hline
\begin{array}{c|c}
1 & 87 \\
\cancel{2} & \\
3 & 29 \\
\cancel{4} & \\
\cancel{5} & \\
\cancel{6} & \\
\cancel{7} & \\
\cancel{8} & \\
\cancel{9} & \\
\cancel{10} & \\
\end{array}
\end{array}
$$

In practice, only the small prime divisors 2, 3, 5, and 7 need to be tested as factors for values < 121 (i.e., 11 × 11). Once one pair of factors is found, the process is repeated on the nonprime factor (if any) until the prime factor tree is completed. The number 66 can be factored in two steps using this method (the prime factors are circled).

To test values 121–169 for factors, include 11 in the list of factor candidates. By including the prime numbers 17, 19, and 23 in the candidate list, values to 841 (29 × 29) can be tested for factors.

Patterns and Basic Skills. For children who find it particularly difficulty to memorize sequences of symbols like the multiplication facts, it is often helpful to use number patterns as memory cues. The following activities provide experience working with patterns involving the addition table, multiplication table, and the hundred chart. Children should be encouraged to explain the underlying structure that causes the number patterns to appear. These pattern exploration activities may also generate useful memory cues for number facts.

Addition Table Patterns. An activity that many children find interesting involves completing a partially filled in table of addition facts by exploiting various number patterns. Notice the patterns of even numbers, whole numbers, and constant numbers. Look for other patterns as well (see Figure 11.29).

Multiplication Table Patterns. The table of multiplication facts also contains many number patterns. Children can find patterns in the completed multiplication table

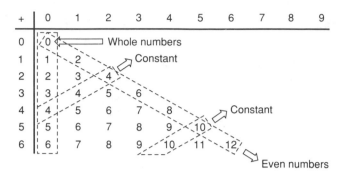

FIGURE 11.29 Addition Table Pattern

and color all the terms in a sequence with a marker. By using different colors, more than one pattern can be displayed on each table. Have children describe each pattern using a number sentence or written description below the table (see Figure 11.30).

Hundred Chart Patterns. Figure 11.31 shows number patterns on the hundred chart. Children can use crayons or felt pens to color the even and odd numbers, multiples-of-5, multiples-of-10, the prime numbers, and other patterns.

Gifted and talented children often benefit from extended explorations with patterns and functions. Such activities help children develop a better understanding of functions and their applications to practical problems. Work with two-column tables

×	0	1	2	3	4	5	6	7	8	9
0	0	0	0	0	0	0	0	0	0	0
1	0	1	2	3	4	5	6	7	8	9
2	0	2	4	6	8	10	12	14	16	18
3	0	3	6	9	12	15	18	21	24	27
4	0	4	8	12	16	20	24	28	32	36
5	0	5	10	15	20	25	30	35	40	45
6	0	6	12	18	24	30	36	42	48	54
7	0	7	14	21	28	35	42	49	56	63
8	0	8	16	24	32	40	48	56	64	72
9	0	9	18	27	36	45	54	63	72	81

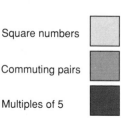

Square numbers

Commuting pairs

Multiples of 5

FIGURE 11.30 Multiplication Table Patterns

1	2	3	4	5	6	7	8	9	10
11	12	13	14	15	16	17	18	19	20
21	22	23	24	25	26	27	28	29	30
31	32	33	34	35	36	37	38	39	40
41	42	43	44	45	46	47	48	49	50
51	52	53	54	55	56	57	58	59	60
61	62	63	64	65	66	67	68	69	70
71	72	73	74	75	76	77	78	79	80
81	82	83	84	85	86	87	88	89	90
91	92	93	94	95	96	97	98	99	100

FIGURE 11.31 Can You Find Hundred Chart Patterns?

and finite differences can motivate the need for integer arithmetic and lead to work with coordinate graphs, slopes, solving equations, and later, derivatives in calculus. A good source of classroom activities is *Finite Differences: A Problem Solving Technique* (Seymour and Shedd, 1973).

SUMMARY

Patterns are systematic occurrences of a relationship within a set of objects or ideas. Number patterns and functions are important topics in elementary mathematics. Pattern activities are an important component of early number and counting development. Older children work with number patterns occurring in arithmetic, geometry, and the other strands of mathematics. Common number patterns include the even, odd, square, triangle, and Fibonacci numbers. Patterns are also involved in finding prime factors, greatest common denominators, and least common factors. Some patterns express functional relationships that can be displayed in a two-column table and graphed on the coordinate plane. Whereas patterns allow the observer to predict the next element in a sequence, a function relates each *x* value (or position number) with a corresponding *y* value (or result), enabling the observer to compute any *y* value given from a given *x* value. These functional relationships can often be written as number sentences, or equations. This powerful feature makes functions very useful in solving some types of problems. A systematic procedure called finite differences makes it possible to uncover linear, quadratic, and more complex functional relationships expressed in many problems that can be represented in two-column tables. Identifying patterns in the factor, addition, and multiplication tables can help special-needs children memorize these basic facts. Exploring number patterns and functions can also provide interesting extension activities for gifted and talented children.

COURSE ACTIVITIES

1. Write a lesson plan to introduce a number pattern to primary level children. Use physical materials to display the pattern and have children predict the next value in the sequence. Use a two-column table to record the results. If possible, implement the lesson with a group of children and discuss with your peers the implications of including pattern work in mathematics instruction.

2. Think about how to introduce the idea of *function* to intermediate grade students. What key features of this powerful concept would you include in your lesson? Make a list of examples that could be used to motivate these key features and help students understand the importance of uncovering functions when solving problems.

3. Look for patterns in **Pascal's Triangle** shown in Table 11.30. Note that each value in the interior is the sum of the two adjacent numbers directly above. List and label as many familiar sequences as possible (e.g., the Triangle Numbers).

4. The period (one back-and-forth swing) of a pendulum remains constant regardless of the weight of the bob or length of the swing. Only the length of the pendulum governs its period (disregarding air friction). Using string and a washer, make a pendulum that has a period of one second. Measure its length and record it in Table 11.31.

TABLE 11.31   Pendulum Period

Period Duration (sec)	Pendulum Length (cm)
1	?
2	?
3	?
4	?
5	?

TABLE 11.30   Pascal's Triangle

```
                        1
                    1       1
                1       2       1→Triangle numbers
            1       3       3       1
        1       4       6       4       1
    1       5      10      10       5       1
1       6      15      20      15       6       1
1   7      21      35      35      21       7       1
1   8      28      56      70      56      28       8       1
```

Next, construct a two-second pendulum. Is it twice as long as the first? Record its length in the table. Construct a three-second pendulum and record its length. Can you predict the length of a four-second pendulum from the table? A five-second pendulum? Graph the results of the pendulum experiment using Figure 11.32. What is the shape of its graph?

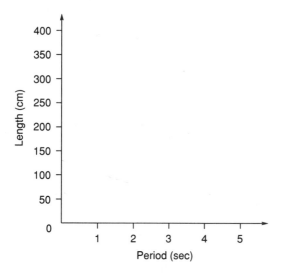

FIGURE 11.32 How Fast Does a Pendulum Swing?

5. Over two hundred years ago, the Russian mathematician Christian Goldbach found that all even numbers larger than two seemed equal to the sum of exactly two prime numbers ($12 = 7 + 5$; $4 = 2 + 2$; $48 = 31 + 17$). Using a calculator or computer, show that all the even numbers less than one hundred are the sum of two prime numbers. Find pairs of primes whose sums are 1000, 100,002, and 888,888. Try to find an even number that is not the sum of two primes. Does **Goldbach's conjecture** seem to be true? To date, no one has proved whether it is *always* possible to subdivide any even number into exactly two primes nor has anyone found an example to contradict this conjecture.

6. Review the references at the end of this chapter and add several pattern and function lessons to your classroom activity card file. If possible, implement one or more activities with a small group of children and discuss the experience with your peers.

7. Read one of the *Arithmetic Teacher* articles listed in the reference section. Write a brief report summarizing the main ideas of the article and describe how the recommendations for instruction might apply to your own mathematics teaching.

MICROCOMPUTER SOFTWARE

Bumble Plot   Coordinates graphing activities in a game format (The Learning Company).

Creative Play: Problem Solving Activities with the Computer   A collection of twenty-five programs that focus on interdisciplinary problem solving (Math and Computer Education Project).

Graphing Equations   Finding equations for given graphs and the interactive graphing game, Green Globs (Conduit Software).

King's Rule   Provides practice uncovering number patterns (Sunburst Communications).

REFERENCES AND READINGS

Bain, D. (1987). The world of alphanumerics. *Arithmetic Teacher, 35*(1), 26.

Beattie, I. (1986). Building understanding with blocks. *Arithmetic Teacher, 34*(2), 5–11.

Burns, M. (1977). *The good time math event book*. Palo Alto, CA: Creative Publications.

Burns, M. (1978). *The book of think*. New York: Little Brown.

Burns, M. (1982). *Math for smarty pants*. New York: Little Brown.

Burton, G. (1980). Definitions for prime numbers. *Arithmetic Teacher, 27*(6), 44–47.

Dearing, S., & Holtan, B. (1987). Factors and primes with a T square. *Arithmetic Teacher, 34*(8), 34.

Downie, D. , Slesnick, T. , & Stenmark, J. (1981). *Math for girls and other problem solvers*. Berkeley: University of California, Lawrence Hall of Science.

Edwards, F. (1987). Geometric figures make the LCM obvious. *Arithmetic Teacher, 34*(7), 17–18.

Gardella, F. (1984). Divisibility—another route. *Arithmetic Teacher, 31*(7), 55–56.

Greenes, C., Spungin, R., & Dombrowski, J. (1977). *Problem-mathics: Mathematical challenge problems with solution strategies*. Palo Alto, CA: Creative Publications.

Greenes, C., Willcutt, R., & Spikell, M. (1972). *Problem solving in the mathematics laboratory: How to do it*. Boston, MA: Prindle, Weber & Schmidt.

Huff, S. (1979). Odds and evens. *Arithmetic Teacher, 27*(5), 48–52.

Immerzeel, G., & Ockenga, E. (1977). *Calculator activities for the classroom*. Palo Alto, CA: Creative Publications.

Jacobs, H. (1970). *Mathematics: A human endeavor*. San Francisco: W.H. Freeman.

Judd, W. (1975). *Patterns to play on a hundred chart*. Palo Alto, CA: Creative Publications.

Krulik, S. & Reys, R. (Eds.). (1980). *Problem solving in school mathematics: 1980 yearbook*. Reston, VA: National Council of Teachers of Mathematics.

Lamb, C., & Hutcherson, L. (1984). Greatest common factor and least common multiple. *Arithmetic Teacher, 31*(8), 43–44.

Lappan, G., & Winter, M. (1980). Prime factorization. *Arithmetic Teacher, 27*(7), 24–27.

Litwiller, B., & Duncan, D. (1983). Areas of polygons on isometric dot paper: Pick's formula revised. *Arithmetic Teacher, 30*(8), 38–40.

Parkerson, E. (1978). Patterns in divisibility. *Arithmetic Teacher, 25*(4), 58.

Polya, G. (1957). *How to solve it*. Princeton: Princeton University Press.

Robold, A. (1982). Patterns in multiples. *Arithmetic Teacher, 29*(8), 21–23.

Seymour, D., & Shedd, M. (1973). *Finite differences: A problem solving technique*. Palo Alto, CA: Dale Seymour Publications.

Souviney, R. (1981). *Solving problems kids care about*. Glenview, IL: Scott, Foresman.

Tinnappel, H. (1963). On divisibility rules. In J. Hlazaty (Ed.). *Enrichment Mathematics for the Grades: 1963 yearbook*. Reston, VA: NCTM, 227–233.

Van de Walle, J., & Holbrook, H. (1987). Patterns, thinking, and problem solving. *Arithmetic Teacher, 34*(8), 6–12.

Whitin, D. (1986). More patterns with square numbers. *Arithmetic Teacher, 33*(5), 40–42.

12

Fraction Operations, Ratios, and Proportions

If you two are going to share this sandwich, one of you cut it in half, and the other one choose first.

—Mom

Upon completing Chapter 12, the reader will be able to

1. Describe the developmental concepts associated with a meaningful understanding of fractional part.

2. Describe an instructional hierarchy for introducing rational number and operation concepts.

3. Describe the concept of fractional part and list appropriate instructional materials and classroom lessons.

4. Describe rational and irrational numbers and give examples of activities to introduce these ideas to children.

5. Describe how renaming fractions is important for the development of fraction algorithms and give appropriate classroom activities.

6. Describe three alternative algorithms for adding and subtracting common fractions with unlike denominators.

7. Describe appropriate techniques for introducing multiplication and division of common fractions.

8. Describe procedures for renaming fractions in simplest terms and give appropriate classroom activities.

9. Describe proportional reasoning and its importance in mathematical problem solving.

10. Describe activities involving common fractions that offer extension experiences for gifted and talented children.

11. Describe common fraction computation errors and how to adapt fraction instruction for special-needs children.

Children begin using fractions the first time they have to share a cookie or a set of blocks with a friend. If both are satisfied with the distribution, they have been introduced to the mathematical notion of *one-half*. These early, informal experiences provide the basis for the systematic development of concepts and operations associated with common fractions and decimals, or **rational numbers**.

Many educators predict that knowledge of operations involving common fractions is likely to decrease in importance over the next decade. The ready availability of calculators and microcomputers, the increased application of metric units of measure, and the nearly universal use of decimal based money and numeration systems will make standard operations with fractions less attractive in commerce and daily life. Fraction concepts are not likely to disappear, however, due to their use in the English vernacular and their applications in more advanced mathematics topics like ratio, proportion, probability, and algebra.

The prudent choice for elementary teachers is to introduce fraction addition, subtraction, multiplication, and, to a lesser extent, division, while avoiding the use of unwieldy *junior high fractions* (fractions such as $\frac{9}{51}$, which you are unlikely to encounter anywhere except junior high school). Focusing attention on halves, thirds, fourths, fifths, eighths, tenths, and twelfths, for example, gives ample opportunity to introduce the basic fraction algorithms. Everyday problems involving fractions with less friendly denominators are likely to be solved using the decimal representation on a calculator or microcomputer. The skill of renaming fractions as decimals, therefore, is likely to be of increasing importance.

INTRODUCING RATIONAL NUMBER OPERATIONS

Decimal notation is introduced in chapter 13 following the discussion of standard fraction operations in this chapter. It is possible, however, to introduce decimal concepts concurrent with, or even prior to, the concept of standard fractions. Though there are several possible sequences that can be followed when introducing fraction and decimal concepts, some concepts depend on the prior understanding of others. For example, the ability to rename fractions in simplest terms (once referred to as reducing fractions) is facilitated by a knowledge of fraction multiplication.

Figure 12.1 shows one possible instructional hierarchy for introducing rational number concepts and operations. It may be useful to refer to the steps shown in this hierarchy when organizing activities into lesson sequences.

UNDERSTANDING FRACTIONAL PART

The concept of **fractional part** involves making *equal-sized* partitions of an object or set. For example, to specify three-fourths ($\frac{3}{4}$) of a square, the figure is first partitioned into four equal-sized areas; then, three of these *fourths* are selected. Initially, the equal-sized partitions should be **congruent** (same size and shape) as shown in Figure 12.2. Later, you can introduce noncongruent, equal-sized partitions.

Similarly, two-fifths ($\frac{2}{5}$) of a set of peanuts can be shown by partitioning the set into five equal groups and selecting two of these groups (see Figure 12.3).

As Piaget observed, the meaningful use of the equal-partitioning fraction model is related to **conservation**. A thorough understanding of conservation of area, number, volume, and mass can help children model the partitioning operation using concrete measurement experiences. Fortunately, by the time rational number operations are formally introduced in Grade 4 or 5, most children have a good understanding of these conservation concepts. Teachers can therefore be reasonably assured that most children are capable of using a range of instructional materials when learning about rational number concepts and skills.

The fractional whole–part relationship is often misinterpreted by children to include *unequal* partitionings as well as equal-sized divisions. Any of the partitionings shown in Figure 12.4 might be misinterpreted as one-half by some children.

When children are first introduced to the notion of fractional part, activities should offer physical evidence of equal partitions. For example, a balance can be used to verify that a container of sand has, indeed, been divided into two equivalent amounts (see Figure 12.5).

Symmetry can also be used to verify the equal partitioning of geometric figures. Cutting along the dotted lines in Figure 12.6 gives identical subdivisions representing fractional parts of the whole. Partitions can be superimposed to show congruence.

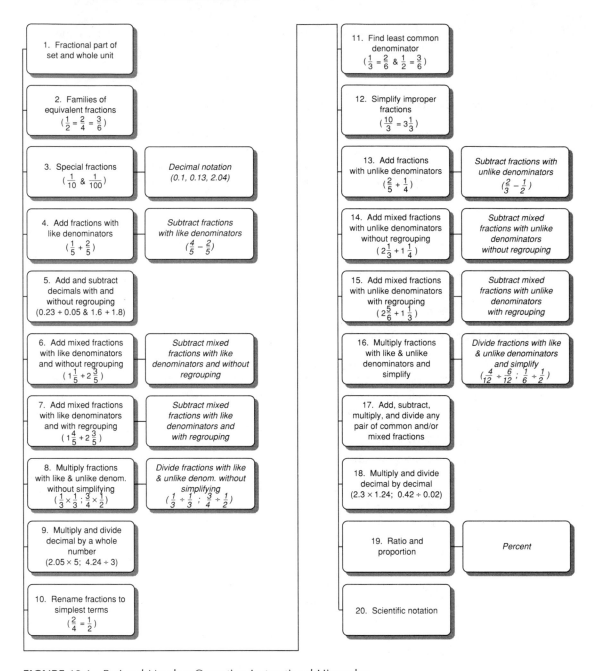

FIGURE 12.1 Rational Number Operation Instructional Hierarchy

Children can also construct **fractional parts of sets**. Peanuts, lima beans, blocks, or other uniform counters can be arranged in piles or arrays to show fractional parts of a set. To show halves, two equal stacks or rows must be formed from the given set of objects. To show thirds, three piles are constructed (see Figure 12.7).

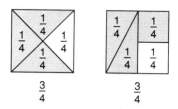

FIGURE 12.2 Squares with Three-Fourths Shaded

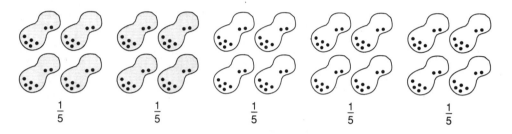

FIGURE 12.3 Two-Fifths Shaded

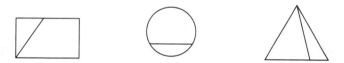

FIGURE 12.4 Two-part Partitionings Not Representing One-Half

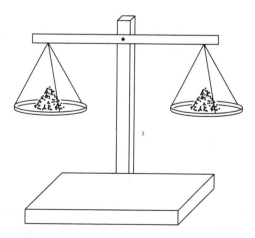

FIGURE 12.5 Balance with Two Equivalent Piles of Sand

Halves   Thirds   Fourths

FIGURE 12.6

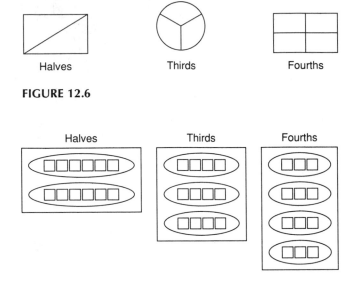

FIGURE 12.7

Division Meaning of Fractions

The division operation can be interpreted as a fraction. For example, the exercise $2 \div 3 = ?$ can be interpreted to mean *divide two into three equal parts*. The answer for this division exercise is $\frac{2}{3}$ (i.e., $\frac{2}{3} + \frac{2}{3} + \frac{2}{3} = \frac{6}{3} = 2$). The **indicated division** meaning of fractions is used to rename fractions as decimals.

Ratios

Problems in everyday life often require working out a ratio of one set of objects to another. For example, if two pencils cost nine cents, we could say a **ratio** of 2:9 (read "two to nine") exists between pencils and pennies. The yellow and green M&M candies appear in a ratio of about 3:5. Ratios do not indicate the total number of objects involved. Rather, a ratio specifies that, for each occurrence of the first value (two pencils), one should expect the occurrence of the second (nine pennies). For example, suppose it were possible to purchase two bus tickets for seventy-five cents. The ratio of tickets to pennies would be 2:75. If four tickets were required, the cost would increase *proportionally* to \$1.50, giving the ratio 4:150. Unlike fractions, however, ratios cannot be added, subtracted, multiplied, or divided (2:1 + 4:1 ≠ 6:1).

Many elementary children have difficulty using proportional reasoning. By Grades 5–6, however, most children can benefit from experiences involving ratio and proportion such as scale drawing, map reading, and percent.

FRACTION NOTATION

The symbol we use to represent fractional parts was invented by Hindu mathematicians and later adapted by the Arabs. The convention was uncommon in Europe until the Middle Ages. In some cultures today, different fraction notations are used. For example, in Japan fractions are written with the total number of partitions on top and the number of partitions being considered on the bottom—opposite to the convention used in the United States.

A **rational** number is any value that can be written in the form $\frac{a}{b}$, where a, the **numerator**, is an integer (positive or negative whole number), and b, the **denominator**, is a nonzero integer. The denominator indicates the total number of equal partitions, and the numerator specifies the number of those partitions being considered (e.g., 3 out of 5 equal partitions $= \frac{3}{5}$).

Work at the elementary level is generally limited to fractions involving positive integers. Many familiar numbers are rational. For example, all the counting numbers can be written as fractions. The negative integers, some decimals, and the value zero are also rational numbers.

$$1 = \tfrac{1}{1} \qquad {}^{-}3 = \tfrac{-3}{1} \qquad 0.5 = \tfrac{5}{10}$$

$$2 = \tfrac{2}{1} \qquad 0 = \tfrac{0}{4} \qquad 98.6 = \tfrac{986}{10}$$

Families of Equivalent Fractions

The two ratios 1:2 and 2:4 are considered equivalent because they describe a similar relationship. Likewise, the fractions $\frac{1}{2}$ and $\frac{2}{4}$ are equivalent because they represent the same total fractional part of a set or whole (see Figure 12.8).

Note that a **common unit** is required in order to compare fractional parts of objects. For example, $\frac{1}{2}$ of an orange is not equal to $\frac{2}{4}$ of an apple.

A convenient way to generate an entire **family of equivalent fractions** is to exploit the special property of multiplication-by-one, the **multiplicative identity**. Any number n multiplied by 1 gives the product n.

$$n \times 1 = n$$

FIGURE 12.8 Equivalent Fractions $\frac{1}{2}$ and $\frac{2}{4}$ Using a Common Whole Unit

To show that the fraction $\frac{2}{2} = 1$, make two identical squares from tagboard and cut one in half. Demonstrate that, when the two halves are recombined, they are congruent to the whole square. Multiplying any number by $\frac{2}{2}$, therefore, is equivalent to multiplying it by 1. Have the children cut geometric figures to show that $\frac{3}{3}$, $\frac{4}{4}$, and $\frac{5}{5}$ also equal 1. An understanding of fraction multiplication is needed in order to construct a family of equivalent fractions using this procedure. For example:

$$\frac{1}{2} \times 1 = \frac{1}{2}$$

$$\frac{1}{2} \times \frac{2}{2} = \frac{2}{4}$$

$$\frac{1}{2} \times \frac{3}{3} = \frac{3}{6}$$

$$\frac{1}{2} \times \frac{4}{4} = \frac{4}{8}$$

$$\frac{1}{2} \times \frac{5}{5} = \frac{5}{10}$$

Since each of these exercises involves multiplying the fraction $\frac{1}{2}$ by 1 (i.e., $\frac{2}{2}$, $\frac{3}{3}$, etc.), each result must remain equal to $\frac{1}{2}$. An infinite family of equivalent fractional representations is possible for any rational number ($\frac{1}{2}$, $\frac{2}{4}$, $\frac{3}{6}$, $\frac{4}{8}$, $\frac{5}{10}$, $\frac{6}{12}$, . . .).

Compute five equivalent fractions for each of the following values.

1. $\frac{1}{2}$

2. $\frac{2}{3}$

3. $\frac{3}{4}$

4. $\frac{2}{5}$

5. $\frac{3}{10}$

ADDITION AND SUBTRACTION OF COMMON FRACTIONS

Adding and Subtracting Fractions with Like Denominators

In the previous section, fractions were implicitly added when counting the number of fourths equivalent to one-half. Adding fractions with the same, or **like**, denominators parallels the procedure for adding whole numbers. Whole number operations involve counting whole units. For fraction operations, the units are always smaller than one whole (see Figure 12.9).

Fraction pies and strips are instructional materials based on an area model. These materials can be constructed by reproducing each shape in Figure 12.10 on different colored construction paper. The circles (or squares) are then cut into the

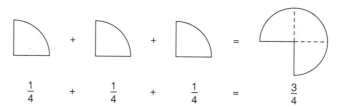

$$\frac{1}{4} \quad + \quad \frac{1}{4} \quad + \quad \frac{1}{4} \quad = \quad \frac{3}{4}$$

FIGURE 12.9   Sum of Three Fourths

fractional pieces 1, $\frac{1}{2}$, $\frac{1}{3}$, $\frac{1}{4}$, $\frac{1}{5}$, $\frac{1}{6}$, $\frac{1}{8}$, $\frac{1}{9}$, $\frac{1}{10}$, and $\frac{1}{12}$. Label each piece with the symbol (e.g., $\frac{1}{3}$) on the front and the fraction word (one-third) on the back. Each student can construct a personal fraction set.

Addition and subtraction exercises can be represented using pies (or strips). To add two fractions with like denominators, have children count out the number of pieces needed to represent each value, put the two piles together, and count the total number of identical pieces (see Figure 12.11).

Notice that the sum $\frac{5}{5}$ is also 1 whole. Sums larger than 1 whole can also be shown using strips and pies. Problems involving mixed numbers require two or more sets of materials (see Figure 12.12).

As with whole numbers, to subtract fractions with like denominators, it is necessary to remove from the minuend the number of identical pieces indicated by the subtrahend. The difference is the number represented by the remaining pieces (see Figure 12.13).

Using fraction pies or strips, carry out the necessary actions to compute the following sums and differences. Record the results using symbols.

1. $\frac{1}{2} + \frac{1}{2} = ?$ 6. $\frac{3}{4} - \frac{1}{4} = ?$

2. $\frac{2}{3} + \frac{1}{3} = ?$ 7. $\frac{2}{2} - \frac{1}{2} = ?$

3. $\frac{3}{10} + \frac{1}{10} = ?$ 8. $\frac{5}{8} - \frac{2}{8} = ?$

4. $\frac{3}{8} + \frac{1}{8} = ?$ 9. $\frac{7}{10} - \frac{5}{10} = ?$

5. $\frac{3}{5} + \frac{2}{5} = ?$ 10. $\frac{5}{8} - \frac{2}{8} = ?$

Finding Common Denominators

To add or subtract fractions with **unlike** denominators, children must first rename one or both fractions so their denominators are identical (i.e., are represented by the *same* sized fraction pie or strip pieces). The process does not change the size of the fractions involved, but simply substitutes one member of an equivalent family for another. Once the denominators are identical, sums and differences can be calculated by simple counting procedures.

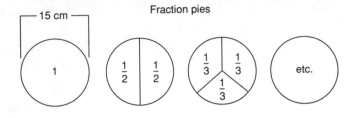

Fraction pies

15 cm

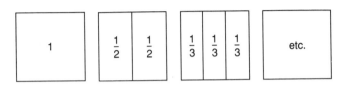

Fraction strips

FIGURE 12.10   Fractional Parts of a Whole

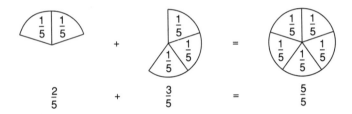

$$\frac{2}{5} \quad + \quad \frac{3}{5} \quad = \quad \frac{5}{5}$$

FIGURE 12.11

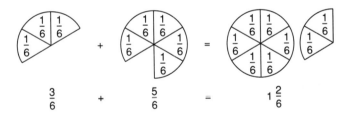

$$\frac{3}{6} \quad + \quad \frac{5}{6} \quad = \quad 1\frac{2}{6}$$

FIGURE 12.12

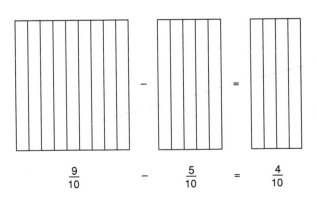

$$\frac{9}{10} \quad - \quad \frac{5}{10} \quad = \quad \frac{4}{10}$$

FIGURE 12.13

To compare the relative size of $\frac{1}{2}$ and $\frac{1}{3}$, children select the $\frac{1}{2}$ and $\frac{1}{3}$ strips (or pie sections) and place one over the other. Part of the $\frac{1}{2}$ strip sticks out beyond the $\frac{1}{3}$ strip, showing that $\frac{1}{2} > \frac{1}{3}$.

Common denominators can also be shown for two fractions using fraction pies or strips. For $\frac{1}{2}$ and $\frac{1}{3}$, have children try to cover the $\frac{1}{2}$ and the $\frac{1}{3}$ pieces exactly with several identical strips. In this case, three $\frac{1}{6}$ strips exactly cover the $\frac{1}{2}$ strip, and two $\frac{1}{6}$ strips exactly cover the $\frac{1}{3}$ strip. Therefore, as shown in Figure 12.14, sixths is a common denominator.

The common denominator for two fractions can also be found by observing fraction families. For example, to determine whether $\frac{3}{5}$ or $\frac{5}{7}$ is the larger value, children first list the family of equivalent fractions for each value until a member with the same denominator is found in each. The bold values have common denominators for the $\frac{3}{5}$ and $\frac{5}{7}$ families. The larger fraction value can be identified by directly comparing the numerators ($\frac{25}{35} > \frac{21}{35}$).

Three-fifths family— $\frac{3}{5}, \frac{6}{10}, \frac{9}{15}, \frac{12}{20}, \frac{15}{25}, \frac{18}{30}, \mathbf{\frac{21}{35}}, \frac{24}{40}$
Five-sevenths family— $\frac{5}{7}, \frac{10}{14}, \frac{15}{21}, \frac{20}{28}, \mathbf{\frac{25}{35}}$

In this case, 35 was the *smallest* denominator that was common to both families ($\frac{42}{70}$ and $\frac{50}{70}$ are the next larger set of common denominators). The smallest common denominator is called the **least common denominator** (LCD). The LCD for fifths and sevenths is thirty-fifths. The LCD can be found using the prime factor method for finding the least common multiple (LCM) discussed in chapter 11.

Children can use any of the following algorithms to find common denominators. The first does not require prior knowledge of fraction multiplication. Techniques 2

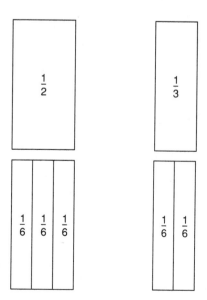

FIGURE 12.14 Common Denominator for $\frac{1}{2}$ and $\frac{1}{3}$

and 3 assume children have already been introduced to multiplication of fractions. The following examples show how to find the common denominator for $\frac{5}{6}$ and $\frac{3}{4}$.

Algorithm 1:

$$\frac{5}{6} = \frac{5+5}{6+6} = \frac{10}{12}$$

a. Repeatedly add respective denominators to each series below the line until the sums are equal.

$$\frac{3}{4} = \frac{3+3+3}{4+4+4} = \frac{9}{12}$$

b. Then add the same number of numerators above each line.

c. Total the values in the numerator and denominator.

Algorithm 2:

$$\frac{5}{6} \times \frac{2}{2} = \frac{10}{12}$$

a. Determine LCM of the denominators by computing the product of the maximum set of unique prime factors ($6 = 2 \times \mathbf{3}$; $4 = \mathbf{2} \times \mathbf{2}$; LCM $= \mathbf{3} \times \mathbf{2} \times \mathbf{2} = 12$).

$$\frac{3}{4} \times \frac{3}{3} = \frac{9}{12}$$

b. Determine the unit multipliers ($\frac{2}{2}$ and $\frac{3}{3}$) needed to give a denominator equal to the LCM (12).

c. Multiply the numerator and denominator by the unit multiplier.

Algorithm 3:

$$\frac{5}{6} \times \frac{4}{4} = \frac{20}{24}$$

a. Multiply the numerator and denominator of each fraction by the denominator of the other fraction ($\frac{4}{4}$ and $\frac{6}{6}$).

$$\frac{3}{4} \times \frac{6}{6} = \frac{18}{24}$$

Algorithms 1 and 2 will always give results in terms of the least common denominator. Algorithm 3 always gives a common denominator but *not* necessarily the smallest possible. Though Algorithm 3 is the easiest to apply, the fractions that result may yield sums or differences that are not in simplest form.

Using the three techniques shown, write each of the following pairs as fractions with common denominators. Show each example using fraction strips or pies.

1. $\frac{1}{2}, \frac{3}{4}$

2. $\frac{2}{3}, \frac{1}{6}$

3. $\frac{3}{5}, \frac{1}{10}$

4. $\frac{1}{4}, \frac{2}{3}$

5. $\frac{5}{6}, \frac{5}{12}$

Adding and Subtracting Fractions with Unlike Denominators

To add $\frac{1}{2} + \frac{1}{4}$, children must first find a common denominator. Using fraction strips, children can substitute $\frac{2}{4}$ for $\frac{1}{2}$ and solve the equivalent exercise

$$\frac{1}{2} + \frac{1}{4} \rightarrow \frac{2}{4} + \frac{1}{4} = \frac{3}{4}$$

As shown in Figure 12.15, it is also possible to substitute $\frac{4}{8}$ for $\frac{1}{2}$ and $\frac{2}{8}$ for $\frac{1}{4}$ giving the equivalent exercise

$$\frac{1}{2} + \frac{1}{4} \rightarrow \frac{4}{8} + \frac{2}{8} = \frac{6}{8}$$

Subtraction follows a similar process. For example, $\frac{3}{5} - \frac{1}{2}$ can be shown by first finding a common denominator using fraction strips, and then matching the appropriate number of minuend pieces to the subtrahend display. As shown in Figure 12.16, the pieces left over represent the difference.

Exercises must be carefully selected to ensure that the required denominators can be represented by the pies or strips available. For example, to add thirds and fifths (e.g., $\frac{1}{3} + \frac{2}{5}$), the required common denominator 15 is not represented in the fraction pies and strips. The pie and strip materials shown can be used to calculate problems involving the pairs of denominators shown in Table 12.1.

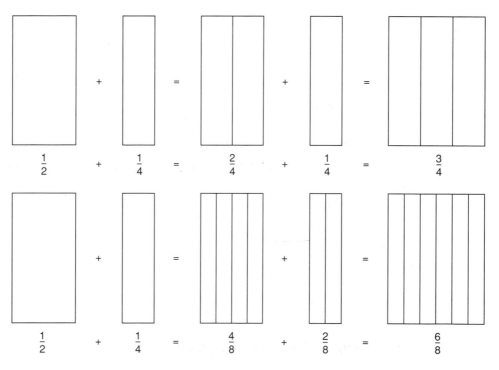

FIGURE 12.15

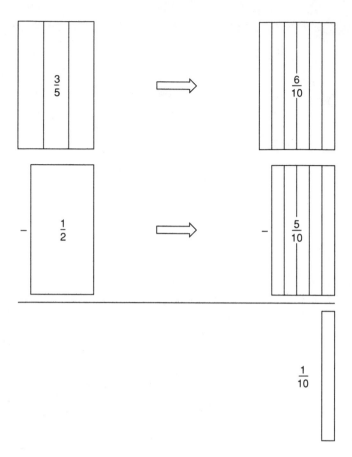

FIGURE 12.16

> Using fraction pies and/or strips, show the actions required to compute the following exercises.
>
> 1. $\frac{1}{2} - \frac{1}{3} = ?$
> 2. $\frac{1}{3} + \frac{5}{12} = ?$
> 3. $\frac{1}{4} + \frac{1}{3} = ?$
> 4. $\frac{3}{10} - \frac{1}{5} = ?$
> 5. $\frac{2}{3} - \frac{3}{6} = ?$

Algorithms for Adding and Subtracting Fractions

After children demonstrate facility calculating exercises using fraction pies and strips, symbolic algorithms can be introduced. Note that addition and subtraction of appro-

TABLE 12.1 Possible Fraction Pie/Strip Exercises

Unlike Denominator Pairs	Examples
2,3	$\frac{1}{2} - \frac{1}{3}$
2,4	$\frac{1}{2} + \frac{3}{4}$
2,5	$\frac{2}{5} + \frac{1}{2}$
2,6	$\frac{5}{6} - \frac{1}{2}$
2,8	$\frac{3}{2} + \frac{1}{8}$
2,10	$\frac{2}{10} + \frac{1}{2}$
2,12	$\frac{1}{2} + \frac{7}{12}$
3,4	$\frac{1}{3} - \frac{1}{4}$
3,6	$\frac{5}{6} + \frac{1}{3}$
3,9	$\frac{1}{3} + \frac{5}{9}$
3,12	$\frac{1}{3} - \frac{1}{12}$
4,8	$\frac{3}{4} - \frac{1}{8}$
4,12	$\frac{1}{4} + \frac{9}{12}$
5,10	$\frac{3}{10} - \frac{1}{5}$
6,12	$\frac{5}{6} + \frac{3}{12}$

priate fractions with like and unlike denominators can be displayed using fraction pies or strips with no prior knowledge of fraction multiplication. Similarly, Algorithm 1 below allows addition of fractions with *unlike* denominators with no prior knowledge of fraction multiplication. Algorithms 2 and 3 require a prior understanding of fraction multiplication.

The following three algorithms can be used for adding and subtracting fractions and mixed numbers (Algorithm 3 requires mixed numbers to be renamed as improper fractions).

Addition *Subtraction*

Algorithm 1:

$\frac{1}{2} = \frac{1+1}{2+2} = \frac{2}{4}$ $\frac{1}{2} = \frac{1+1}{2+2} = \frac{2}{4}$

$+\frac{1}{4} = +\frac{1}{4} = +\frac{1}{4}$ $-\frac{1}{4} = -\frac{1}{4} = -\frac{1}{4}$

$\frac{3}{4}$ $\frac{1}{4}$

Algorithm 2:

$$\frac{1}{2} \times \frac{2}{2} = \frac{2}{4}$$

$$+\frac{1}{4} \times \frac{1}{1} = +\frac{1}{4}$$

$$\frac{3}{4}$$

$$\frac{1}{2} \times \frac{2}{2} = \frac{2}{4}$$

$$-\frac{1}{4} \times \frac{1}{1} = -\frac{1}{4}$$

$$\frac{1}{4}$$

Algorithm 3:

$$\frac{1}{2} \times \frac{4}{4} = \frac{4}{8}$$

$$+\frac{1}{4} \times \frac{2}{2} = +\frac{2}{8}$$

$$\frac{6}{8}$$

$$\frac{1}{2} \times \frac{4}{4} = \frac{4}{8}$$

$$-\frac{1}{4} \times \frac{2}{2} = -\frac{2}{8}$$

$$\frac{2}{8}$$

Algorithm 1 uses repeated addition to calculate the common denominator for the pair of fractions involved. While often more time-consuming than other methods, this algorithm does not require children to initially specify the unit multiplier.

Algorithm 2 is generally accepted as the standard procedure for adding and subtracting fractions. While this procedure is efficient, many children find it difficult to accurately determine the unit multiplier (i.e., $\frac{2}{2}$) that generates the least common denominator.

Algorithm 3 removes the guess work since opposite denominators are automatically selected as the unit multiplier. Though all three algorithms can give results that are not in simplest form ($\frac{8}{15} - \frac{1}{3} = \frac{8}{15} - \frac{5}{15} = \frac{3}{15}$), procedures 1 and 2 minimize the amount of renaming required. Children should be given the opportunity to try each algorithm and practice using the one that suits them best.

Calculate each of the following exercises using the three alternative algorithms.

1. $\frac{1}{2} + \frac{3}{4} = ?$

2. $\frac{2}{3} - \frac{1}{6} = ?$

3. $\frac{3}{5} - \frac{1}{10} = ?$

4. $\frac{1}{4} + \frac{2}{3} = ?$

5. $\frac{5}{6} + \frac{5}{12} = ?$

Mixed Numbers

Though fractions are often thought of as numbers between 0 and 1, rational numbers can be larger than 1 and smaller than 0 (see Figure 12.17).

FIGURE 12.17 Fraction Number Line

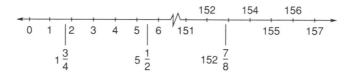

FIGURE 12.18

Fractional numbers larger than 1 (or smaller than $^-1$) are called **mixed numbers**. Mixed numbers can be thought of as the sum of a whole number and a fraction. On the number line, values such as $1\frac{3}{4}$, $5\frac{1}{2}$, and $152\frac{7}{8}$ are represented by the point three-fourths of the distance between 1 and 2, half the distance between 5 and 6, and seven-eights of the distance from 152 to 153, respectively. Grade 4–6 children can graph fractions on the number line as a way to compare the relative size of values (see Figure 12.18).

Draw a number line and graph the following rational numbers in their proper positions.

1. $\frac{1}{3}$

2. $1\frac{1}{5}$

3. $4\frac{3}{4}$

4. $3\frac{2}{3}$

5. $6\frac{7}{8}$

Mixed numbers are also rational numbers because they can be written in the form $\frac{a}{b}$. For example:

$$1\frac{3}{4} = \frac{4}{4} + \frac{3}{4} = \frac{7}{4}$$

$$5\frac{1}{2} = \frac{10}{2} + \frac{1}{2} = \frac{11}{2}$$

$$152\frac{7}{8} = \frac{1216}{8} + \frac{7}{8} = \frac{1223}{8}$$

When mixed numbers are written in $\frac{a}{b}$ form, the resulting fraction always has a numerator equal to or larger than the denominator. Though mathematicians regard them as perfectly fine values, such numbers are called **improper fractions**. Children can rename, or **simplify**, improper fractions as a mixed number by dividing the denominator into the numerator. The quotient is the number of whole units contained in the fraction, and the remainder indicates the number of fractional parts remaining. For example:

$$\frac{9}{2} \rightarrow \quad 2\overline{)9}^{\,\,4R\,1} \quad \rightarrow 4\frac{1}{2}$$

Justify this procedure by reversing the process that renames a mixed number as an improper fraction. For the previous example, since $4 = \frac{8}{2}$, the number $4\frac{1}{2}$ can be written as $\frac{8}{2} + \frac{1}{2}$, or $\frac{9}{2}$. An algorithm can be introduced once students understand the relationship between mixed numbers and improper fractions. Multiplying the denominator in a mixed number by the whole number, and adding the result to the numerator is an easy way to rename a mixed number as its $\frac{a}{b}$ equivalent. For example:

$$4\frac{1}{2} = \frac{(2 \times 4) + 1}{2} = \frac{9}{2}$$

Write each of the following mixed numbers as improper fractions. Then rename the result as a mixed number again.

1. $2\frac{1}{3}$

2. $3\frac{3}{5}$

3. $5\frac{3}{4}$

4. $2\frac{1}{7}$

5. $6\frac{3}{10}$

Figure 12.19 shows how to solve the exercise $3\frac{4}{5} + 1\frac{2}{5} = ?$ using fraction pies. This problem requires regrouping $\frac{6}{5}$ into $1\frac{1}{5}$ and adding the whole units ($3\frac{4}{5} + 1\frac{2}{5} = 4\frac{6}{5} = 4 + 1 + \frac{1}{5} = 5\frac{1}{5}$).

Addition and subtraction of mixed numbers with unlike denominators follows a procedure similar to that of other fractions. First, children must find a common denominator using fraction pies or strips. The strips are combined for addition or, for subtraction, those in the minuend are matched with those in the subtrahend, and the remainder becomes the difference. Subtraction exercises like $3\frac{1}{4} - 1\frac{1}{2} = ?$ require regrouping one whole unit in the minuend in order to subtract the fraction part of the subtrahend. To regroup, have children substitute the correct number of strips for one whole unit in the minuend (e.g., $\frac{4}{4}$ for 1). As shown in Figure 12.20,

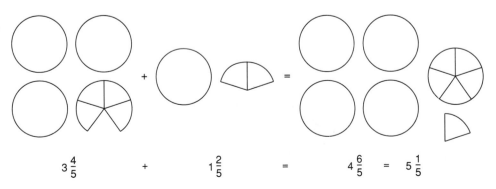

$$3\frac{4}{5} \quad + \quad 1\frac{2}{5} \quad = \quad 4\frac{6}{5} \quad = \quad 5\frac{1}{5}$$

FIGURE 12.19

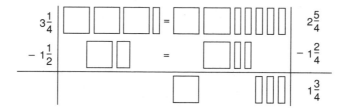

FIGURE 12.20 Subtraction with Unlike Denominators

this provides enough strips to match with the fraction part of the subtrahend, and the remainder is the difference. Children should note that substituting four-fourths for one whole does not change the overall value of the mixed number.

MULTIPLICATION AND DIVISION OF FRACTIONS

Multiplying two fractions between 0 and 1 *always* gives a product less than either of the two factors. Children may find this observation confusing when compared with the products of numbers greater than one. Fraction strips can be used to demonstrate how these small numbers behave when multiplied. The area of the whole square is considered as **one unit**. To multiply $\frac{1}{2} \times \frac{1}{3}$, children select the half and third strips and position them on the **unit square**. The shaded area in Figure 12.21 indicates where the two strips overlap. The dotted lines show that the shaded region comprises $\frac{1}{6}$ of the total unit area. The process can be verbalized as, One-half of one-third of the unit square equals one-sixth of the unit square.

$$\frac{1}{2} \times \frac{1}{3} = \frac{1}{6}$$

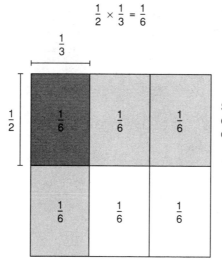

FIGURE 12.21

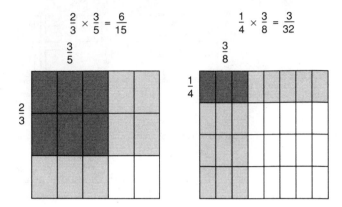

FIGURE 12.22

Different combinations of fraction strips can be used to show multiplication exercises. The product of two fractions is represented by the overlapping region (see Figure 12.22).

An algorithm for multiplying fractions can be introduced after children are successful at using fraction strips to solve multiplication exercises. To multiply two fractions, calculate the products of the numerators and denominators as shown in the following examples.

1. $\frac{1}{2} \times \frac{1}{3} = \frac{1 \times 1}{2 \times 3} = \frac{1}{6}$

2. $\frac{2}{3} \times \frac{3}{5} = \frac{2 \times 3}{3 \times 5} = \frac{6}{15}$

3. $\frac{1}{4} \times \frac{3}{8} = \frac{1 \times 3}{4 \times 8} = \frac{3}{32}$

While mixed numbers can be multiplied using procedures similar to those of whole numbers, it is generally more efficient to first rename each mixed number as an improper fraction, then use the standard fraction multiplication algorithm. For example:

Vertical Algorithm	*Improper Fraction Algorithm*

$$1\frac{2}{3}$$
$$\underline{\times\ 4\frac{1}{2}}$$

$$1\frac{2}{3} \times 4\frac{1}{2} \rightarrow \frac{5}{3} \times \frac{9}{2} = \frac{45}{6} = 7\frac{3}{6} = 7\frac{1}{2}$$

$$\frac{2}{6} = \frac{1}{2} \times \frac{2}{3}$$

$$\frac{1}{2} = \frac{1}{2} \times 1$$

$$\frac{8}{3} = 4 \times \frac{2}{3}$$

$$\underline{+\ 4\ = 4 \times 1}$$

$$7\frac{1}{2} = 4 + \frac{8}{3} + \frac{1}{2} + \frac{2}{6}$$

Use fraction strips to compute the following products.

1. $\frac{1}{2} \times \frac{1}{4} = ?$
2. $\frac{1}{2} \times \frac{1}{3} = ?$
3. $\frac{1}{2} \times \frac{3}{4} = ?$
4. $\frac{1}{3} \times \frac{1}{4} = ?$
5. $2\frac{2}{5} \times 3\frac{1}{2} = ?$

Reciprocals

A special pair of fractional factors called **reciprocals** always have the product 1, the **multiplicative identity**. For example, $\frac{3}{4}$ and $\frac{4}{3}$ are reciprocals because their product equals 1.

$$\frac{3}{4} \times \frac{4}{3} = \frac{12}{12} = 1$$

The reciprocal of a number is also called its **multiplicative inverse**. Several numbers and their multiplicative inverses (reciprocals) are listed in Table 12.2.

To introduce children to reciprocals, give them a set of missing-factor exercises such as the following.

$$\frac{2}{3} \times ? = \frac{6}{6}$$

$$\frac{4}{5} \times ? = \frac{20}{20}$$

$$\frac{7}{3} \times ? = \frac{21}{21}$$

$$5 \times ? = \frac{5}{5}$$

$$1\frac{3}{4} \times ? = \frac{28}{28}$$

Discuss what is special about all of the answers for this type of exercise (the missing factor is the reciprocal of the first factor). Have the children make up miss-

TABLE 12.2

Number	Reciprocal
$\frac{3}{4}$	$\frac{4}{3}$
3	$\frac{1}{3}$
10	$\frac{1}{10}$
$2\frac{1}{2}$	$\frac{2}{5}$

ing-factor and missing-product exercises involving reciprocals and various types of numbers.

$$\frac{2}{3} \times \frac{3}{2} = ?$$

$$? \times \frac{10}{12} = \frac{144}{144}$$

$$? \times \frac{10}{12} = 1$$

$$4 \times ? = 1$$

$$2\frac{1}{3} \times \frac{3}{7} = ?$$

Dividing Fractions

As the *inverse* of multiplication, division by a number between 0 and 1 always gives a quotient *larger* than the dividend. This observation may make fraction division confusing for children. For example, $4 \div \frac{1}{2}$ can be thought of as, How many halves are contained in four?

$$4 \div \frac{1}{2} = 8$$

Children may also confuse *division by one-half* with dividing something *in half*. For example, dividing 4 in half means to multiply by one-half, or divide by 2.

$$4 \times \frac{1}{2} = 4 \div 2 = 2$$

Give oral examples of problems using very simple fractions to help children establish the difference between these English phrases. For example:

A recipe calls for 4 cups of flour. How much flour is needed if we cut the recipe in half?

(4 × $\frac{1}{2}$ = 2 cups of flour)

Plans for a bookcase call for $\frac{1}{2}$-foot shelves. How many shelves can be cut from an 8-foot board?

(8 ÷ $\frac{1}{2}$ = $\frac{16}{2}$ ÷ $\frac{1}{2}$ = 16 shelves)

How many pieces of cheese will there be if we cut a $5\frac{1}{2}$-inch block into $\frac{1}{4}$-inch slices?

($5\frac{1}{2}$ ÷ $\frac{1}{4}$ = $\frac{11}{2}$ ÷ $\frac{1}{4}$ = $\frac{22}{4}$ ÷ $\frac{1}{4}$ = 22 slices)

Fraction division can be introduced by initially selecting exercises with whole number dividends *and* quotients (e.g., $1 \div \frac{1}{2} = 2$). Using fraction strips, have children show $2 \div \frac{1}{2}$ as pictured in Figure 12.23. Children should solve several carefully selected exercises that have whole number quotients ($3 \div \frac{1}{6} = ?$; $2 \div \frac{2}{5} = ?$; etc.) and record the solutions using symbols.

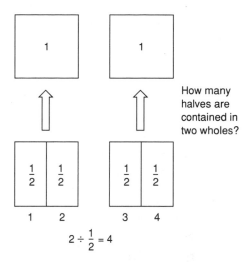

$$2 \div \frac{1}{2} = 4$$

FIGURE 12.23 Division Using Fraction Strips

Common Denominator Algorithm

An intuitive algorithm for fraction division is the inverse of the multiplication algorithm. To find the quotient, divide the numerators and divide the denominators. The algorithm works fine for exercises that give whole number quotients for both the numerator and the denominator of the answer. For example:

$$\frac{2}{8} \div \frac{1}{2} = \frac{2}{4} = \frac{1}{2}$$

$$\frac{10}{12} \div \frac{2}{3} = \frac{5}{4} = 1\frac{1}{4}$$

For fraction pairs that do not give whole number numerators and denominators in the quotient, children can first rename one or both fractions so they have a common denominator. This procedure always generates a 1 in the denominator of the quotient, so the final answer can be easily determined by observing the numerator. For example:

$$\frac{2}{1} \div \frac{1}{2} = \frac{4}{2} \div \frac{1}{2} = \frac{4 \div 1}{2 \div 2} = \frac{4}{1} = 4$$

$$\frac{2}{3} \div \frac{1}{2} = \frac{4}{6} \div \frac{3}{6} = \frac{4 \div 3}{6 \div 6} = \frac{4 \div 3}{1} = 4 \div 3 = \frac{4}{3} = 1\frac{1}{3}$$

Reciprocal Algorithm

The standard fraction division algorithm, **invert and multiply**, uses a property of reciprocals that relates multiplication and division. Demonstrate to the class that dividing by any number gives the same answer as multiplying by its reciprocal.

Therefore, to divide one fraction by another, we can simply multiply the dividend by the *reciprocal* of the divisor. For example:

$$6 \div 2 = 6 \times \tfrac{1}{2} = 3$$

$$6 \div \tfrac{1}{3} = 6 \times 3 = 18$$

$$\tfrac{1}{6} \div \tfrac{2}{3} = \tfrac{1}{6} \times \tfrac{3}{2} = \tfrac{3}{12} = \tfrac{1}{4}$$

Have the children practice dividing fractions using this procedure. Have them check the results of several exercises using the common denominator algorithm and, for simple examples, show the results using fraction strips. After they become proficient using the reciprocal algorithm, give the class a worksheet of mixed practice to check their ability to use the correct algorithm for each exercise.

Renaming Fractions in Simplest Form

One member of each family of equivalent fractions has the smallest denominator. For example, in the one-half family ($\tfrac{1}{2}$, $\tfrac{2}{4}$, $\tfrac{3}{6}$, etc.), $\tfrac{1}{2}$ has the smallest denominator. The process of finding the family member with the smallest denominator is called **simplifying**.

To rename a fraction in simplest form, we employ the principle that dividing by one does not alter the size of the original number. Therefore, a fraction divided by any unit (e.g., $\tfrac{2}{2}$, $\tfrac{3}{3}$, $\tfrac{4}{4}$), gives a quotient in the *same* fraction family. Simplifying fractions using this technique requires prior understanding of fraction division. For example:

$$\tfrac{2}{4} \div \tfrac{2}{2} = \tfrac{1}{2}$$

To rename a fraction in simplest form, children must first find a common divisor for the numerator and denominator. For numerators and denominators to 120, children only need to check the prime divisors 2, 3, 5, and 7. For example, to simplify $\tfrac{8}{12}$, check to see if 2 is a common divisor for 8 and 12. Since both values are even, they are divisible by 2. Dividing $\tfrac{8}{12}$ by the unit $\tfrac{2}{2}$ gives $\tfrac{4}{6}$, a member of the same fraction family. The process is repeated until no common divisors remain.

$$\tfrac{8}{12} \div \tfrac{2}{2} \quad = \quad \tfrac{4}{6} \div \tfrac{2}{2} = \tfrac{2}{3} \quad - \quad \text{Fully simplified}$$

First common divisor (2)   Second common divisor (2)

$$\tfrac{8}{12} \div \tfrac{4}{4} = \tfrac{2}{3} \quad - \quad \text{Fully simplified}$$

Greatest common factor (4)

Notice that when the greatest common factor (GCF) was selected for the unit divisor in the second example, the resulting quotient was fully simplified in one step.

The term *simplified* (and especially the outdated term *reduced*) may give the mistaken impression that the value of the fraction has changed. It is important to point out to children that the change in the numerator exactly compensates for the change in the denominator, so the fraction value remains constant.

This raises a question as to the value of teaching fraction simplification at all. While $\frac{1}{2}$ may appear to be a simpler description of the fraction concept of *halfness*, other representations of one-half can be useful as well (e.g., $\frac{5}{10} = 0.5$ in decimal notation). As an old boat builder once said, "I never put away my tools since it is just as likely I will need my hammer where I last used it, as on the workbench where it belongs." Perhaps the best advice is to simplify fractions only when necessary (when adding $\frac{1}{3}$ and $\frac{8}{12}$, it may be easier to simplify the $\frac{8}{12}$ to $\frac{2}{3}$ and find the sum rather than renaming $\frac{1}{3}$ as $\frac{4}{12}$) and, otherwise, leave them as you find them.

The careful use of concrete models like fraction pies and strips offers children the opportunity to develop fractional number and operation concepts in a meaningful way. As with whole numbers, the concrete displays can support learning during initial concept development. The use of concrete materials can be reduced as students gain facility with the more flexible symbolic representations and algorithms.

Cancellation

An application of fraction simplification offers a short-cut procedure for multiplying and dividing fraction expressions. Multiplying and dividing some fractions can be simplified by dividing the numerator and denominator of a fraction by the same value. This process is commonly called *canceling*. For example, in $\frac{3}{8} \times \frac{12}{15}$, the 3 in the numerator of the first fraction and the 15 in the denominator of the second can both be canceled by dividing each by 3. Similarly, the 8 and the 12 can be canceled by dividing each by 4.

$$\overset{1}{\underset{2}{\frac{3}{8}}} \times \overset{3}{\underset{5}{\frac{12}{15}}} = \frac{1}{2} \times \frac{3}{5} = \frac{3}{10}$$

With practice, the canceling procedure can simplify multiplication and division problems involving fractions. Caution children not to attempt to cancel across the numerators and denominators of fractions being added or subtracted.

> Explain why cancellation enables multiplication and division exercises to be simplified. Why doesn't cancellation work for addition and subtraction exercises?

RATES, RATIOS, AND PROPORTIONS

Real world situations sometimes require us to make systematic comparisons among measures or sets of objects. Often, these comparisons are expressed as rates, ratios, or proportions.

TABLE 12.3 Equivalent Ratios

30:1
60:2
90:3
120:4

Rates and Ratios

To describe how fast you ride a bicycle requires consideration of two measures, the elapsed time and the distance traveled. It might take two hours to ride twenty miles. More commonly, speed is reported as a standard **rate** of the number of miles traveled each hour (miles per hour). For example, suppose a triathalon racer could ride her bicycle at a steady rate of thirty miles per hour. This rate could be written as a ratio of 30:1 (every thirty miles takes one hour). If the athlete is able to maintain this level of effort for several hours, a table of equivalent ratios can be constructed that will predict the distance traveled for any number of hours (see Table 12.3).

A **ratio** is a way to describe the size relationship between two or more groups. For example, the normal ratio of persons-to-shoes is one person for every two shoes, or one-to-two. This ratio can be written 1:2 or (1, 2). The ratio 1:2 seems to give the same information as the fraction $\frac{1}{2}$. However, the ratio 1:2 establishes a more powerful relationship between persons and shoes than does the fraction $\frac{1}{2}$. The ratio 1:2 tells us that *every* person has two shoes. The fraction $\frac{1}{2}$ only indicates that there are half as many persons as shoes. A fraction does not guarantee an even distribution of objects. For example, one person may wear two pairs at the same time (e.g., boots over shoes), while another may be barefoot. There are still half as many persons as shoes, but not in a ratio of 1:2. Figure 12.24 shows some possible arrangements for ratio and fraction distributions.

Children often misinterpret ratios, especially when a comparison is being made between a set of objects and a subset. For example, say a pencil sharpener breaks the lead one time for every five times that it functions properly (a ratio of 1:5); if all thirty children in the class sharpen their pencils, how many will break their lead? Children often give the incorrect answer of six broken pencil leads for this problem. As shown in Figure 12.25, this comparison involves class inclusion. There is one broken pencil lead for each six sharpenings. Therefore, there would be five broken leads in thirty sharpenings.

The ratio of people to shoes is 1:2
(each person wearing 2 shoes)

$\frac{1}{2}$ as many people as shoes
(2 people with 4 shoes and 2 shoeless people)

FIGURE 12.24

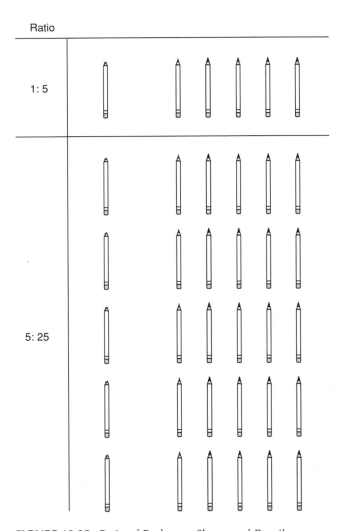

FIGURE 12.25 Ratio of Broken to Sharpened Pencils

Ratios can not be added, subtracted, multiplied, or divided like fractions. To combine two different ratios, we add the number of events in each category. For example, suppose another class has a better sharpener, and their ratio of broken to sharpened pencils is 2:15. The combined performance of the two sharpeners is (1 + 2) : (5 + 15), or 3:20.

Proportion

Two ratios are said to be **proportional** if their corresponding fraction representations are equivalent. For example, the ratios 1:2 and 2:4 are proportional because $\frac{1}{2} = \frac{2}{4}$.

Proportional relationships are *not* identical relationships; a 1:2 ratio assumes a different *distribution* than a 2:4 relationship. For example, suppose the 1:2 ratio stipulates that each person receives two hamburgers. The 2:4 ratio stipulates that each two persons must receive four hamburgers. As shown in Figure 12.26, it is possible that, with a ratio of 2:4, someone will have nothing to eat.

Proportions can be written using three different notations. Notation 3 below is used frequently for solving problems because corresponding fraction operations can be used. The symbol **::** means **is proportional to**.

1. 1:2 :: 2:4
2. $(1, 2) :: (2, 4)$
3. $\frac{1}{2} :: \frac{2}{4}$

Exploiting proportional relationships may help children to solve a wide range of practical problems. In chapter 6 we learned that the sides of *similar triangles* are proportional. Using this information, we can calculate the height of a flag pole by comparing the length of its shadow to the shadow of a meter stick. As these two shadows form similar triangles, we can compute the unknown height by solving the proportion

$$\frac{\text{Pole height (?)}}{\text{Pole shadow length (known)}} = \frac{\text{Meter stick height (1 m)}}{\text{Meter shadow length (known)}}$$

For example, suppose the meter stick cast a shadow of three meters, and the pole cast a shadow of twenty-four meters. The pole height P can be calculated from the proportion

$$\frac{P}{24} :: \frac{1}{3}$$

This proportion can be solved by inspection. For the two ratios to be equivalent, P must be one-third of twenty-four, or eight meters. To solve more complex proportions, the ratios are written using fraction notation, and an equal sign is substituted for the proportion symbol. Like a pan balance, we can add to, subtract from, multiply, or divide both sides of the equation by the same value without upsetting the balance. To solve this proportion, crossmultiply both sides of the equation by the denominator (24) of the missing term P. Simplifying the right side of the equation gives the solution: $P = 8$.

Ratio of persons to hamburgers is:

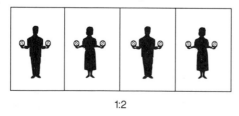  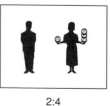

1:2   2:4   2:4

FIGURE 12.26

$$\frac{P}{24} = \frac{1}{3}$$

$$P = \frac{1 \times 24}{3}$$

$$P = 8$$

Therefore, the flag pole is eight meters high. This algorithm is based on the following steps.

a. $\frac{P}{24} = \frac{1}{3}$ Initial proportion

b. $\frac{24 \times P}{24} = \frac{1 \times 24}{3}$ Multiply both sides of equation by 24.

c. $P = 8$ Simplify

Mature Grade 5–6 children can use this procedure to solve simple proportions. When the missing term in a proportion is in the denominator, it is more convenient to invert both ratios before solving the equation (e.g., $\frac{1}{n} = \frac{5}{8} \rightarrow \frac{n}{1} = \frac{8}{5}$). Try to justify that the proportion relating two ratios is equivalent to the proportion for the reciprocal ratios. In chapter 13, crossmultiplying is discussed in more detail as a technique for solving percentage problems.

Solve the following proportions for n using crossmultiplication. Make up a word or picture problem to go with each example.

1. $\frac{4}{20} :: \frac{n}{80}$

2. $\frac{15}{3} :: \frac{n}{12}$

3. $\frac{n}{10} :: \frac{7}{35}$

4. $\frac{12}{7} :: \frac{36}{n}$

5. $\frac{2}{10} :: \frac{n}{25}$

CLASSROOM ACTIVITIES

Sharing Equally

Some children may be satisfied with unequal divisions when constructing fractional parts of a set or whole. For example, when asked to draw lines to partition a circle into thirds, a child might make the divisions shown in Figure 12.27.

Young children can practice making the equal divisions necessary for defining fractional parts through small group sharing experiences. Using easily partitioned objects (clay, bread, oranges) or sets (dried beans, raisins, peanuts), K–1 students can work in pairs to practice halves, groups of three for thirds, and so on. In each group, one child makes the first partition into the required number of groups. The second

FIGURE 12.27 Unequal Thirds Partition

adjusts the partitioning as desired. The process continues in turn until everyone has had an opportunity to adjust the partitioning. In the same order that the partitions were made, each child gets to choose a pile. For example, to divide an orange into thirds, Child 1 makes the first partition. Children 2 and 3 adjust the division in turn. Then Child 1 chooses first, followed by 2, and then 3. The last child to adjust the partitions also chooses last.

This process, known to every parent, encourages children to focus on the *equivalence* among partitions when making fractional groupings.

Equivalent Fraction Chart

To help children visualize equivalent families of fractions, construct a bulletin board display of the equivalent fraction families chart shown in Figure 12.28. Using nine different colors of construction paper, cut one 5 cm × 100 cm (2 in × 36 in) strip of each color. By measuring or folding, mark each strip to show one of the following fraction partitions: $\frac{1}{2}, \frac{1}{3}, \frac{1}{4}, \frac{1}{5}, \frac{1}{6}, \frac{1}{8}, \frac{1}{9}, \frac{1}{10}, \frac{1}{12}$. The final strip represents one whole. Children can position a stick vertically to show equivalent fraction families. The stick shown in Figure 12.28 crosses the strips to show the equivalent fraction family: $\frac{2}{3}, \frac{4}{6}, \frac{6}{9}$, and $\frac{8}{12}$.

Cuisenaire Rods

Cuisenaire Rods can be used to display fraction values and operations (see Figure 12.29).

To show $\frac{1}{2}$, first select a rod to represent a whole unit. This rod must be selected such that two identical rods laid end to end exactly match its length. As shown in Figure 12.30, five rods can be chosen as a unit to show $\frac{1}{2}$ (red, purple, dark green, brown, orange). The light green, dark green, or blue rod can be used to show $\frac{1}{3}$ (Figure 12.31).

To show the addition problem $\frac{1}{2} + \frac{1}{3}$, have children select the unit rod that allows the representation of both halves and thirds. Selecting the correct unit rod (the dark green rod in this case) is equivalent to finding a common denominator. As shown in Figure 12.32, the dark green rod can also be partitioned into sixths, represented by white rods.

To show $\frac{1}{2} + \frac{1}{3}$, place the $\frac{1}{2}$ rod (light green) and $\frac{1}{3}$ rod (red) end to end. Since the blocks are of different lengths, their total length cannot be eas-

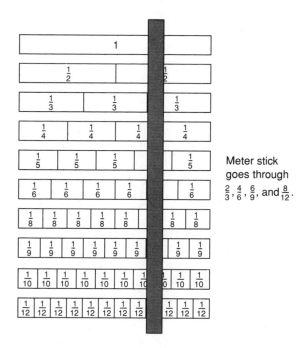

FIGURE 12.28 Equivalent Fraction Family

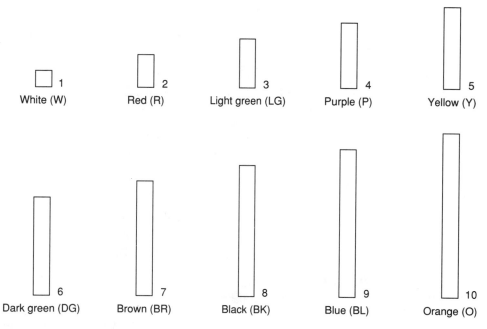

FIGURE 12.29 Cuisenaire Rods

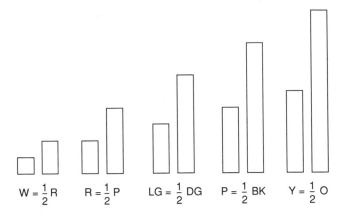

$$W = \frac{1}{2}R \qquad R = \frac{1}{2}P \qquad LG = \frac{1}{2}DG \qquad P = \frac{1}{2}BK \qquad Y = \frac{1}{2}O$$

FIGURE 12.30 Representations of One-Half Using Cuisenaire Rods

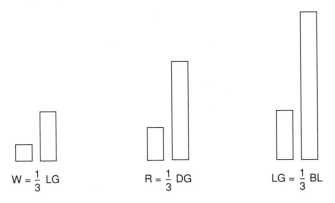

$$W = \frac{1}{3} LG \qquad R = \frac{1}{3} DG \qquad LG = \frac{1}{3} BL$$

FIGURE 12.31 Three Representations of $\frac{1}{3}$

ily described. As shown in Figure 12.33, after subdividing each fraction into sixths (white rods), the total can be found by counting the five white rods, each representing $\frac{1}{6}$ ($\frac{1}{6} + \frac{1}{6} + \frac{1}{6} + \frac{1}{6} + \frac{1}{6} = \frac{5}{6}$).

Sums and differences involving halves, thirds, fourths, fifths, sixths, eighths, ninths, and tenths can be readily displayed using Cuisenaire rods. The references at the end of this chapter provide additional information on the use of these materials.

PROBLEM SOLVING INVOLVING RATIONAL NUMBERS AND RATIOS

With a partner, work through several of the following problem-solving experiences. Keep a record of your solutions to share with other members of your class.

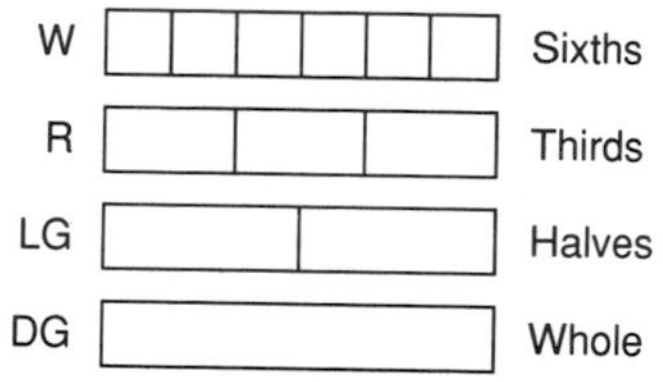

FIGURE 12.32 Fractional Parts of Dark Green Rod

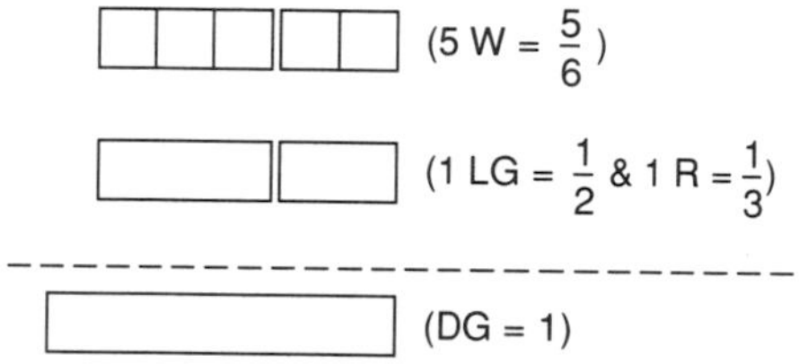

FIGURE 12.33 $\frac{1}{2} + \frac{1}{3} = \frac{5}{6}$

Billiard Table Math

In which corner will the ball end up if it is shot from the lower left corner at a 45° angle? Figure 12.34 shows *interesting* paths traced by the ball on several rectangular tables of different dimensions. Figure 12.35 shows other tables that display *boring* paths in which some squares are not crossed by the ball.

Using centimeter squared paper, carry out several billiard table experiments and make a list of interesting tables (where all the squares are crossed by the path of the ball) and boring tables (those that contain squares not crossed by the path of the ball). Record the results as in Table 12.4. Look for a pattern based on the dimensions of the billiard table that will predict whether a table will generate an interesting or boring display.

When the edge lengths of *interesting* tables are written as fractions ($\frac{2}{3}$, $\frac{3}{4}$, $\frac{1}{5}$), notice that they are renamed in simplest terms. *Boring* tables are represented by fractions that can be further simplified ($\frac{2}{4}$, $\frac{4}{6}$, $\frac{2}{6}$). In Figure 12.36, two billiard tables that represent fractions in the same family display proportionally shaped paths. Will simplifying the fraction represented by a boring table always give an interesting table that ends at the *same* corner?

The ball can end in one of three corners: upper left (UL), upper right (UR), and lower right (LR). Using only interesting games, organize a table by the ending position of the ball. Table 12.5 shows the ending corner for several interesting tables. Observe the dimensions of each table and try to find a pattern that accurately predicts the ending corner based on the table dimensions.

See the computer software *Problem Solving Strategies* (MECC) for a useful simulation of *Pooling Around* to help solve these problems.

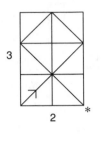
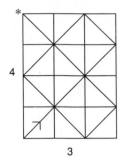
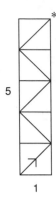
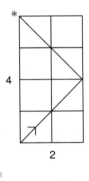

FIGURE 12.34 Interesting Tables

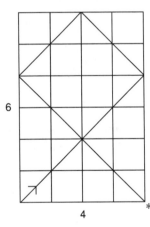
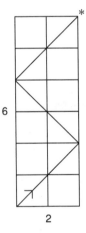

FIGURE 12.35 Boring Tables

TABLE 12.4 Billiard Table Math

Interesting Tables	Boring Tables
(2,3)	(2,4)
(3,4)	(4,6)
(1,5)	(2,6)
(3,8)	(4,8)
(7,10)	(4,12)

Note: The first coordinate represents the length along the horizontal axis, the second coordinate the length along the vertical axis.

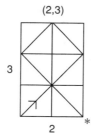
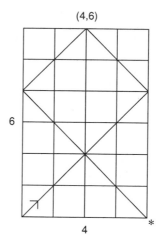

FIGURE 12.36 Similar Pool Table Paths

TABLE 12.5 Billiard Table

Path Finishes in Corner		
UL	**UR**	**LR**
(3,4)	(3,5)	(2,3)
(5,8)	(1,1)	(4,9)
(7,10)	(7,9)	(6,11)

Graphing Proportions

A family of equivalent ratios can be graphed on the coordinate plane. For example, suppose six arcade tokens to play electronic games cost seventy-five cents. This relationship can be written as the ratio 6:75. Twelve tokens would cost $1.50, giving the ratio 12:150. Table 12.6 lists the set of equivalent ratios for this example.

This table describes a proportional relationship that can be graphed on the coordinate plane. Each ratio is treated as a coordinate pair and is graphed on the coordinate plane (see Figure 12.37).

The line connecting the points on the graph represents the cost of tokens if they could be purchased in quantities other than groups of six. The graph shows the functional relationship between the tokens (T) and cost (C). Use the *finite difference* method discussed in chapter 11 to verify the function:

$$12.5 \times T = C$$

TABLE 12.6 Equivalent Ratio Table

Tokens	Cost (Cents)
0	0
6	75
12	150
18	225
24	300
.	.
.	.
.	.

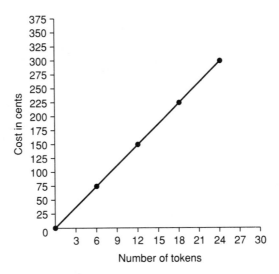

FIGURE 12.37 How Do Travel Costs Grow?

COMPUTER APPLICATIONS

Fraction Bars

The *Fraction Bars Computer Program* (Scott Resources) is a series of programs that uses graphics of fraction bars, number lines, blocks, and dots to introduce the concepts of fractional part, fraction operations, and ratio. Fraction bars are identical strips of cardboard (approximately 1 in × 6 in) that are marked off in fractional parts. Common fractions are shown by shading the appropriate part of each bar. For example, two-thirds would be shown by shading two of the three equal segments on a bar. Each of the seven disks covers one of the following fraction topics: basic concepts, equality, inequality, addition, subtraction, multiplication, and division. Each topic begins with graphic representations of the concepts and moves to symbolic exercises (see Figure 12.38). By typing *HELP*, children receive step-by-step tutorial

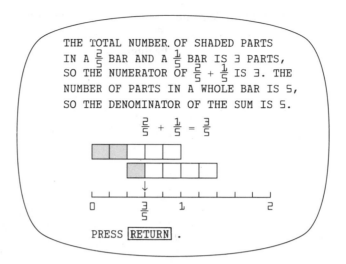

THE TOTAL NUMBER OF SHADED PARTS
IN A $\frac{2}{5}$ BAR AND A $\frac{1}{5}$ BAR IS 3 PARTS,
SO THE NUMERATOR OF $\frac{2}{5} + \frac{1}{5}$ IS 3. THE
NUMBER OF PARTS IN A WHOLE BAR IS 5,
SO THE DENOMINATOR OF THE SUM IS 5.

$$\frac{2}{5} + \frac{1}{5} = \frac{3}{5}$$

PRESS RETURN.

FIGURE 12.38 Fraction Bars. *Note:* From "Fraction Bars Computer Program" [Computer program] by Scott Resources, 1984. Copyright 1984 by Scott Resources, Inc. Reprinted by permission.

assistance on the current exercise. Word problems are included on each disk to provide practice using newly learned fraction concepts and operations. Additional practice is provided in a card game format for each topic. Each disk contains an achievement test to help children determine when to move on to the next topic and what concepts need review.

ADAPTING INSTRUCTION FOR CHILDREN WITH SPECIAL NEEDS

Operations with fractions can often be very difficult for special-needs children to accomplish. Early work with concrete models and clear graphic representations of fraction concepts are very important for children with visual discrimination and figure–ground problems. The use of colored materials such as Cuisenaire Rods may help such children's understanding. For example, have the child use like-colored crayons to record the numeral for fractions represented by Cuisenaire Rods (one-half can be represented as a red rod over a purple rod and recorded by writing the value 1 in red and 2 in purple in the number $\frac{1}{2}$).

Finding Error Patterns

Observing children's error patterns can be helpful when they are learning to use the fraction algorithms. Recognizing such patterns early may make it easier to correct

TABLE 12.7 Common Errors among Elementary Children

	Problem	**Error**
Addition:	$\frac{2}{3} + \frac{3}{4} = \frac{5}{7}$	Adding numerators and denominators.
Subtraction:	$\frac{4}{5} - \frac{2}{3} = \frac{2}{2}$	Finding difference between numerators and denominators.
	$\frac{1}{2} - \frac{5}{8} = \frac{4}{6}$	
	$3\frac{3}{5} - 1\frac{2}{5} = \frac{1}{5}$	Ignoring whole numbers when subtracting mixed numbers.
Multiplication:	$\frac{3}{8} \times \frac{5}{8} = \frac{15}{8}$	Failure to multiply denominators when they are the same value.
	$\frac{2}{3} \times \frac{3}{5} = \frac{2}{3} \times \frac{5}{3} = \frac{10}{9}$	Inverting and multiplying.
Division:	$\frac{3}{4} \div \frac{1}{8} = \frac{3}{4} \times \frac{1}{8} = \frac{3}{32}$	Failure to invert before multiplying.

misconceptions before they become routinized. The errors described in Table 12.7 are common among elementary children.

Assigning Mixed Practice

Special-needs children often display more rigid thinking patterns than other students. This tendency may be unnecessarily reinforced by the practice of making assignments that focus on the practice of one subskill at a time. For example, after completing several problems containing examples of addition of fractions with like denominators, the child may not recognize the need to find common denominators when faced with a set of mixed exercises.

While focused skill practice seems to be particularly important for the special-needs learner, limiting practice to a *single* type of exercise may be misleading and cause unintended reinforcement of incorrect procedures.

To help students develop more flexible thinking patterns, prepare practice worksheets with mixed examples. Before children begin to calculate, have them evaluate which examples require the application of the procedure being practiced and systematically cross out those that do not. They then solve only those exercises not crossed out. For example, if a child needs to practice finding common denominators, mix unlike-denominator addition and subtraction exercises with like-denominator exercises. Have the child evaluate which examples require finding common denominators and cross out all the others. They then work only the unlike-denominator exercises. This procedure gives children practice thinking about the underlying characteristics of the exercises to be solved in addition to the needed drill on specific skills.

Inventing Algorithms

Gifted and talented children can explore fraction, rates, ratio, and proportion concepts to extend their understanding of the algorithms and their problem-solving applications. Mathematically mature children should be encouraged to invent their own algorithms and explain why they work. As described earlier, Algorithm 1 for adding and subtracting fractions with unlike denominators uses a skip-counting technique to locate a common denominator. Have children practice using this procedure. By comparing it with Algorithm 2, see if they can explain why it works. As a hint, have them think about how skip counting is related to multiplication. Similarly, children can explore the common denominator algorithm for dividing fractions.

SUMMARY

Common fraction operations are introduced in the upper elementary grades. While the fraction number concept is introduced as early as Kindergarten, systematic work with operations involving common fractions and work with ratios and proportions constitute a significant component of the mathematics curriculum in Grades 5 and 6. Conservation of length, area, and quantity are associated with the meaningful understanding of fractional part because models representing fractions are commonly based on these attributes. The concept of fractional part involves partitioning a whole unit, or set, into a given number of equal parts, or groups. Other meanings for common fractions include the division of the numerator by the denominator and the representation of a ratio. Instructional materials such as fraction pies, fraction strips, and Cuisenaire Rods can be used to introduce fraction concepts and operations. Fractions can be classified into families of equivalent fractions. The family member with the smallest denominator is said to be in simplest terms. Families of fractions can be used to find common denominators for two or more fractions. Adding and subtracting fractions with unlike denominators requires finding equivalent members of the respective families that have the same, or common, denominator. Several algorithms are available to facilitate adding and subtracting common fractions. The most efficient algorithm requires finding least common multiples and multiplying each fraction by an appropriate unit multiplier. Multiplication of common fractions can be modeled using fraction strips. Division is introduced as the inverse of fraction multiplication. Proportional thinking is an important problem-solving skill, particularly in geometry and measurement work with scales. Primary children often have difficulty understanding proportional reasoning. Young children can approach proportional situations using graphic and concrete representations, and mature children can be introduced to solving proportional equations using crossmultiplying. Special-needs children often exhibit rigid thinking patterns. Providing sets of mixed exercises may promote more flexible thinking in such children. Gifted and talented children benefit from experience inventing their own algorithms and should be encouraged to try to justify why common fraction algorithms work.

COURSE ACTIVITIES

1. Construct a complete set of fraction pies and strips. Practice showing representations of specific fractions, developing families of equivalent fractions, simplifying fractions, adding and subtracting like and unlike fractions and mixed numbers, and multiplying and dividing selected fractions and mixed numbers. Give a demonstration for your peers. If possible, write a lesson plan and try out at least one activity with a group of students in a classroom setting.

2. Using squared paper, draw sketches of rectangular billiard tables (corner pockets only) with sides of various whole number lengths. Work out a rule, based on the length and width of the table, that predicts how many times the ball will bounce off the edge (see Figure 12.39). Always shoot the ball at 45° from the lower left corner. Make a table of results, look for patterns, and uncover a function.

3. Read two articles listed in the references that discuss the role of standard fractions in the mathematics curriculum over the next decade. Be sure to include articles about the possible effects of increased use of calculators, microcomputers, and metrics. List ways standard fractions will continue to be used over the next decade and in what way their importance is likely to diminish.

4. Include examples of fraction, ratio, and proportion lessons in your idea file. Organize them according to appropriate grade levels.

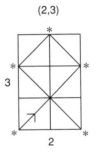

FIGURE 12.39

MICROCOMPUTER SOFTWARE

Edu-Ware Fractions Tutorial on fraction concepts and practice with operations (Edu-Ware Services, Inc.).

Fraction Bars Computer Program: Basic Concepts, Equality, Inequality, Addition, Subtraction, Multiplication, Division A series of seven disks that use fraction bars as a model for introducing fraction concepts and operations (Scott Resources).

Fractions—Basic Concepts Practice with fraction number and operation skills (Sterling Swift Publishing Company).

Problem Solving Strategies Set of four problem-solving situations, including *Pooling Around*, that employ the solution strategies: making a table, looking for patterns, and guess-and-test (MECC).

REFERENCES AND READINGS

Bennett, A., & Davidson, P. (1978). *Fraction bars*. Fort Collins, CO: Scott Resources.

Carpenter, T., Corbitt, M., Kepner, H., Lindquist, M., & Reys, R. (1981). Decimals: results and implications from national assessment. *Arithmetic Teacher, 28*(8), 34–37.

Cathcart, W. (1977). Metric measurement: Important curricular considerations. *Arithmetic Teacher, 24*(2), 158–160.

Chiosi, L. (1984). Fractions revisited. *Arithmetic Teacher, 31*(8), 46–47.

Coxford, A., & Ellerbruch, L. (1975). Fractional numbers. In J. Payne (Ed.), *Mathematics learning in early childhood: 1975 yearbook* pp. 192–203. Reston, VA: National Council of Teachers of Mathematics.

Davidson, P. (1969). *Using the Cuisenaire rods: A photo/text guide for teachers*. New Rochelle, NY: Cuisenaire Co. of America.

Davidson, P. (1977). *Idea book for Cuisenaire rods at the primary level*. New Rochelle, NY: Cuisenaire Co. of America.

Edge, D. (1987). Fractions and panes. *Arithmetic Teacher, 34*(8), 13–17.

Firl, D. (1977). Fractions, decimals and their futures. *Arithmetic Teacher, 24*(2), 238–240.

Kalman, D. (1985). Up fractions! Up n/m! *Arithmetic Teacher, 32*(8), 42–43.

Litwiller, B., & Duncan, D. (1985). Pentagonal patterns in the addition table. *Arithmetic Teacher, 32*(8), 36–38.

Quintero, A. (1987). Helping children understand ratios. *Arithmetic Teacher, 34*(9), 17–21.

Steiner, E. (1987). Division of fractions: Developing conceptual sense with dollars and cents. *Arithmetic Teacher, 34*(9), 36–42.

Sweetland, R. (1984). Understanding multiplication of fractions. *Arithmetic Teacher, 32*(1), 48–52.

13

Decimals, Percents, and Integers

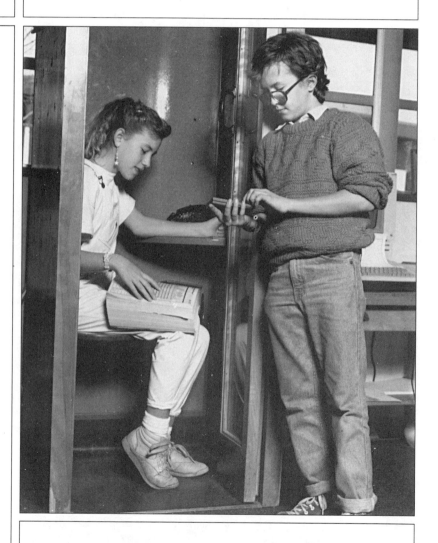

I never could figure where to put that damned spot!
—Winston Churchill

Upon completing Chapter 13, the reader will be able to

1. Describe the relationship between common and decimal fractions and suggest appropriate classroom activities.

2. Describe how decimal operations can be introduced using concrete materials and plan appropriate classroom activities.

3. Model integer operations on the number line and describe applications in measurement and coordinate geometry.

4. Describe scientific notation and give examples of its uses.

5. Describe how to solve three types of percent problems using the proportion method and give examples of appropriate classroom activities.

6. Describe methods for adapting decimal operation instruction for special-needs children.

7. Describe activities involving decimal fractions that offer extension experiences for gifted and talented children.

The term **decimal** is based on the Latin word *deci*, meaning ten. Decimals are simply fractions that can be written with powers of 10 as denominators ($\frac{1}{10}$, $\frac{1}{100}$, etc.). Using the indicated division meaning, any fraction can be written as a decimal with a finite number of decimal places ($\frac{1}{10} = 0.1$ or $\frac{1}{2} = 0.5$) or as a decimal with a pattern of digits that repeats indefinitely ($\frac{1}{3} = 0.333\ldots$, or $\frac{2}{11} = 0.1818\ldots$).

INTRODUCING DECIMAL FRACTIONS

Decimals as Special Fractions

One reason decimals are useful is because they can be manipulated by adapting the whole number algorithms. Because decimal fractions are extensions of the base-10 place value system, many concepts and procedures developed for whole numbers carry over to work with them. It is important, however, to introduce decimal fraction concepts using clear concrete models to give meaning to the decimal point and the decimal fraction place values (Hiebert, 1987). Most children do not recognize the relationship between common fractions and decimals (Bell, Swan, & Taylor, 1981; Hiebert & Wearne, 1985). Also, the complex rules used to manipulate decimals are quickly confused without meaningful instruction. For many children, skill in working decimal algorithms does not improve with time (Carpenter, Corbitt, Kepner, Lindquist, & Reys, 1981; Grossman, 1983).

Familiar base-10 materials can be used to extend the base-10 relationship to numbers between 0 and 1. The block previously associated with the hundreds place

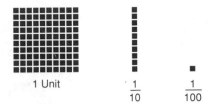

FIGURE 13.1 Base-10 Block Decimal Representation

(100-block) is redefined as the 1-unit. As shown in Figure 13.1, the 10-rod becomes one-tenth ($\frac{1}{10}$), and the small cube becomes one-hundredth ($\frac{1}{100}$).

Introduce children to decimal fraction place values by having them construct concrete representations for digits in the tenths and hundredths place values. At first, values composed of only one nonzero digit should be used. Children record the numbers in place value tables as shown in Figure 13.2.

Next, problems composed of digits in the ones place and either the tenths or hundredths place values are introduced (Figure 13.3).

Reading Decimal Values

After the tenths and hundredths place values are introduced separately, values composed of digits in the ones, tenths and hundredths place values can be introduced. Values such as 0.23 are *not* normally read as "two-tenths and three-hundredths." The fraction part of the decimal is read as a single value rather than as separate fractions. As two-tenths is equivalent to twenty-hundredths ($\frac{2}{10} = \frac{20}{100}$), it is customary to state the sum of the two place values ($\frac{23}{100}$—"twenty-three hundredths") when verbalizing the number 0.23.

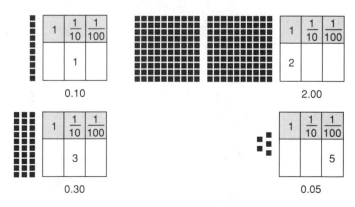

FIGURE 13.2

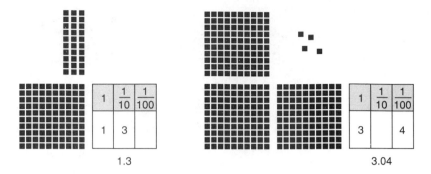

FIGURE 13.3

The base-10 system fortunately offers a built-in shortcut to this process. In English, the decimal fraction is read as if it were a separate whole number value, and the place value of the final digit is appended. Reversing the process allows students to construct decimal representations from orally presented values (see Figure 13.4).

Because the denominators can be accurately predicted from the relative position of each digit, they are omitted from the decimal symbol. For example:

$$42 + \frac{2}{10} + \frac{3}{100} = 42.23$$

$$36 + \qquad \frac{6}{100} = 36.06$$

The convenience of not having to write the denominators (tenths, hundredths, etc.) when using decimals may create a problem for some students when the 0 digit is involved. If a zero place holder is overlooked in a number such as 2.001, a serious error will occur. Children should have considerable practice concretely constructing such numbers as 0.04 and 1.03 using base-10 materials, recording them in place value tables, and expressing each number orally.

FIGURE 13.4

Using base-10 materials, construct the following numbers and record the results in place value tables. Say each value orally and have a peer write the value using decimal symbols. Compare this value with the original.

1. 1.24

2. 11.2

3. 2.04

4. 10.02

5. 12.04

The Decimal Point

It is customary to include at least one whole number digit (it may be zero in the ones column) when writing a decimal fraction. The **decimal point** is simply a convention to separate the whole number from the fraction component of the decimal. We write

0.14 not .14

Normally, any trailing zeros in the decimal fraction are dropped. Some specific applications, such as currency or rounded measures, require that extra zeros be retained:

0.2 not 0.20

$0.20 not $0.2

$12.00 not $12

When rounding 2.58 m
to the nearest decimeter: 2.60 m not 2.6 m

Note that whole numbers are also decimals. Decimal numbers are made up of a whole number component and a decimal fraction component (either of which may be zero), separated by a decimal point. When writing *only* the whole number component of a decimal number, it is customary to omit the decimal point:

234 not 234.

Money, Metrics, and Calculators

In addition to the base-10 system of numeration, **money**, **metric measures**, and **calculators** offer excellent opportunities to introduce decimal concepts and skills. Each offers an alternative representation of decimal fraction place values (see Figure 13.5).

Children who can readily recognize pennies, dimes, and dollars can be introduced to decimal notation and operations through comparing, recording, and reading decimal values. Cash register tapes, price tags, menus, and price lists show how

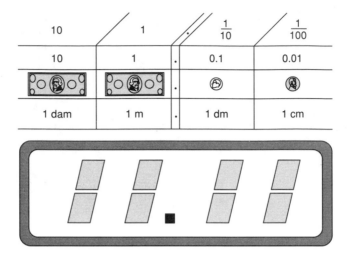

10	1	.	$\frac{1}{10}$	$\frac{1}{100}$
10	1	.	0.1	0.01
		.		
1 dam	1 m	.	1 dm	1 cm

FIGURE 13.5 Decimal Place Value Representations

decimals are used in everyday life. It is important for children to understand the relationship between $ sign and ¢ sign usage (i.e., $0.42 = 42¢).

A useful activity to help children see the two ways money can be represented is to compare item pricing and cash register pricing. First, go to a grocery store and buy eight to ten needed items. Keep the register tape for these items in a safe place. As the items are used at home, open the cans and boxes so that the prices remain visible (cut the bottom out of the cans instead of the top). When the empty packages and cans are available, take them to school and make an overhead slide of the register tape. Show the class each priced item (e.g., a can of soup for 65¢) and have them identify its value printed on a register tape ($0.65). Using a place value table on the overhead, the price of each item in dollars, dimes, and pennies (play money) can be recorded in the ones, tenths and hundredths columns. The can of soup, for example, can be recorded as six dimes in the tenths column and five pennies in the hundredths column, which matches the decimal representation on the register tape (see Figure 13.6).

Collect menus and price lists from fast food restaurants, auto parts stores, record shops, and newspaper ads to use in class. Plan activities in which students use play

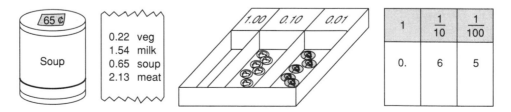

FIGURE 13.6 Price on Can, Tape, Till and Place Value Table

money to represent decimal amounts and record the values in place value tables. For example:

☐ Have each student plan a birthday party by listing the items needed, the cost of each, and the total.
☐ Have small groups of students compute the cost of items needed for a back-packing trip.
☐ Have each student design a private classroom study area and compute the cost of the required building materials.

One of the important advantages of the SI metric system is that it is based on the decimal system. Working with metric measures can provide practice with reading, writing, and comparing decimals. For example, the meter is the basic unit of metric length. Its subunits include the decimeter (0.1 m) and centimeter (0.01 m). When measuring a length, the result is written in whole and decimal parts of a meter (see Figure 13.7).

Use an accurate graduated cylinder to introduce work with decimals by measuring the volume of common containers in liter units (see Figure 13.8). Decimal calculations involving weight are less convenient because measuring parts of a gram requires accurate scales not generally found in elementary schools.

Comparing Decimals

Calculators are excellent tools for experimenting with decimals. As a beginning activity, have children work in pairs each with their own calculator. Have one child press the CLEAR and the DECIMAL POINT key and, with eyes closed, enter a two- or three-digit decimal fraction on the calculator. The second child does the same. Both

FIGURE 13.7

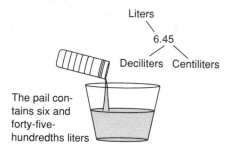

The pail contains six and forty-five-hundredths liters

FIGURE 13.8

values are recorded in a place value table and compared. Any disagreement about which is larger can be resolved by constructing each number using base-10 materials or play money. If desired, each pair of children can keep track of who accumulates the highest number of larger (or smaller) decimal values in ten rounds.

DECIMAL OPERATIONS

Introducing Addition and Subtraction

Finding sums and differences of decimal fractions is simplified in that the corresponding digits in two decimal numbers (digits having the same place value) *always* have common denominators. Finding sums and differences of decimal fractions, therefore, is simply an application of addition and subtraction of common fractions with like denominators. Because decimal fractions are an extension of the whole number place value system, operations can be carried out using the standard algorithms.

Children must take care in lining up the place value columns so that the corresponding values will have common denominators. Table 13.1 compares addition using common fraction and decimal fraction notation. Note that when the decimal points are lined up, the corresponding decimal fraction place values always have the same denominator. The zero place holder in the number 36.06 makes it clear that the 6 digit is in the hundredths column.

TABLE 13.1 Comparing Common and Decimal Fraction Addition

Common Fraction Notation		Decimal Fraction Notation
$42 + \frac{2}{10} + \frac{3}{100}$		42.23
$+\ 36\ + \qquad \frac{6}{100}$	$\rightarrow$	$+\ 36.06$
$78 + \frac{2}{10} + \frac{9}{100}$		78.29

TABLE 13.2 Comparing Common and Decimal Fraction Subtraction

Common Fraction Notation		Decimal Fraction Notation
$42 + \frac{2}{10} + \frac{3}{100} = 42 + \frac{1}{10} + \frac{13}{100}$		42.23
$- 36 + \qquad \frac{6}{100} = 36 + \qquad \frac{6}{100}$	$\rightarrow$	$- 36.06$
$6 + \frac{1}{10} + \frac{7}{100}$		6.17

Similarly, Table 13.2 compares subtraction using common fraction and decimal fraction notation.

Adding Decimals Using Base-10 Materials

Children can use base-10 materials to display the actions involved in decimal addition. The procedures parallel those for adding whole numbers.

Have children display the values on a decimal addition mat as shown for 3.42 + 1.89 in Figure 13.9. The ones column is represented by the flat, the tenths column by the rod, and the hundredths column by the small cube. The hundredths column is combined first (Figure 13.9a). There are more than 9-hundredths in the sum, so 10-hundredths are regrouped into 1-tenth and placed in the tenths column, leaving 1-hundredth in the hundredths column. Each column is combined in turn and regrouped as required, and each step is recorded using place value tables or standard decimal notation.

Calculate the following decimal sums using base-10 materials and record the results in place value tables.

1. 2.47 + 1.72 = ?

2. 1.34 + 2.85 = ?

3. 2.06 + 1.64 = ?

4. 0.78 + 1.7 = ?

5. 2.81 + 1.09 = ?

Subtracting Decimals Using Base-10 Materials

The actions involved in decimal subtraction also parallel the whole number algorithm. Figure 13.10 shows the steps for the exercise 4.13 − 1.26. Because there are not

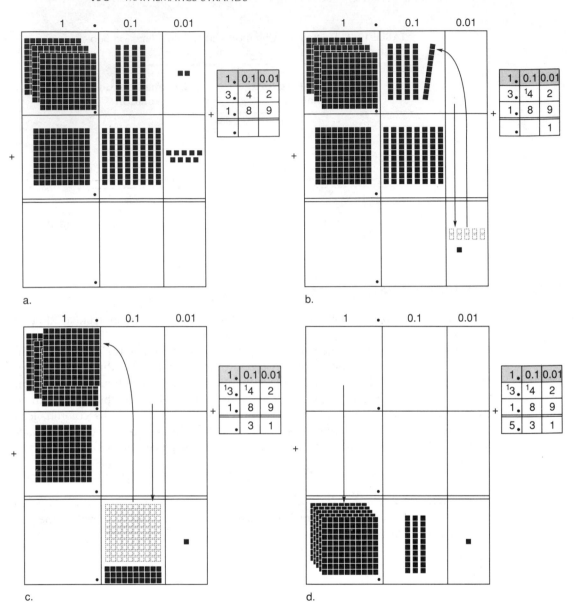

FIGURE 13.9

enough hundredths in the minuend to subtract 6-hundredths, 1-tenth is regrouped into 10-hundredths and placed in the hundredths column. We can then subtract the 6-tenths (Figure 13.10b) and move the difference to the answer box. Since there are 0-tenths remaining in the minuend, 1-one must be regrouped into 10-tenths and placed in the tenths column (Figure 13.10c). We can then subtract 2-tenths from the 10-tenths and move the difference to the answer box. Finally, the ones column can be subtracted (Figure 13.10e).

Calculate the following decimal differences using base-10 materials and record the results in place value tables.

1. $2.67 - 1.23 = ?$

2. $3.62 - 1.27 = ?$

3. $3.42 - 2.69 = ?$

4. $2.04 - 1.27 = ?$

5. $3.2 - 2.04 = ?$

Introducing Multiplication and Division

As with common fractions, concrete displays of the multiplication and division of decimal fractions do not exactly parallel procedures for whole numbers. Because the amounts involved increase and decrease substantially when dividing or multiplying by a decimal fraction, the use of physical models tends to get unwieldy. Fortunately, when decimal fraction multiplication and division are introduced in Grades 5–6, children are generally able to extend the whole number algorithms to work with decimals.

Multiplication of Decimals Using Fraction Strips

Fraction strips or base-10 materials can be used to show multiplication of decimal fractions. When children are working with fraction strips, have them use only the whole square and $\frac{1}{10}$ strips. To show 0.3×0.4 using strips, have the children first represent 0.3 ($\frac{3}{10}$) and 0.4 ($\frac{4}{10}$) and position them on the whole square as shown in Figure 13.11 (draw lines on the whole square to show 100 small squares in 10 rows and columns). As with common fraction multiplication, the area of the overlapping strips (12 hundredths) is the product: $0.3 \times 0.4 = 0.12$. Have the class do several exercises involving tenths $\times$ tenths.

Products of two decimals such as 2.3×1.2 can be shown using base-10 materials where 1 is represented by a flat, 0.1 by a rod, and 0.01 by a small cube. Following the same procedures as for whole numbers, have the children show each factor and construct an array with 2.3 rows and 1.2 columns (Figure 13.12). Each of the four subproducts is represented by a component of the array. Several exercises of this type can be used to show how decimal multiplication can be accomplished using the same algorithm as for whole number products.

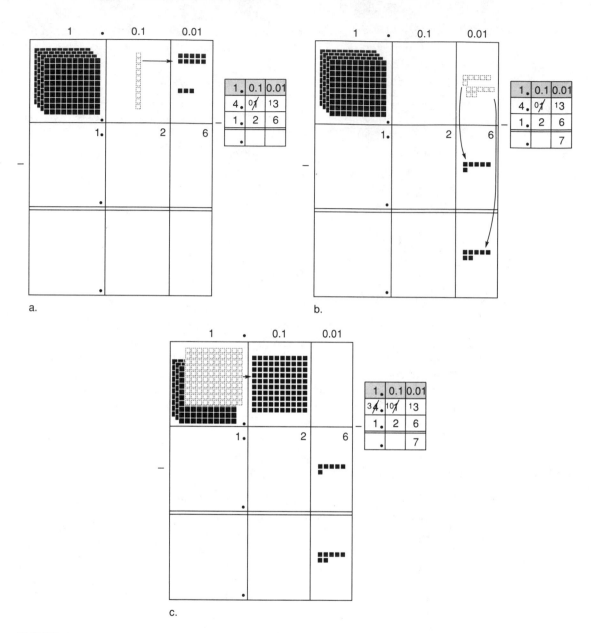

a.

b.

c.

FIGURE 13.10

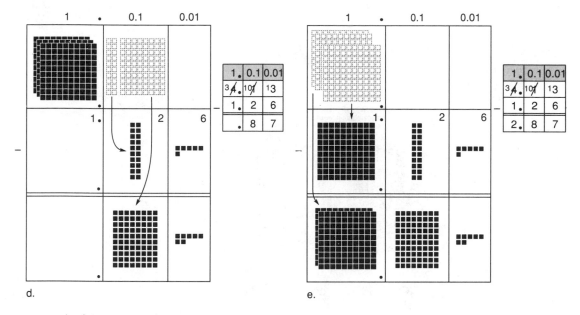

d.   e.

FIGURE 13.10   *continued*

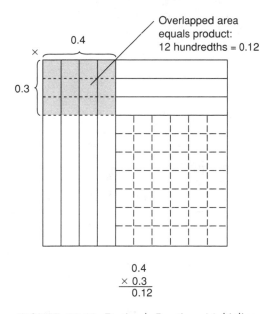

FIGURE 13.11 Decimal Fraction Multiplication

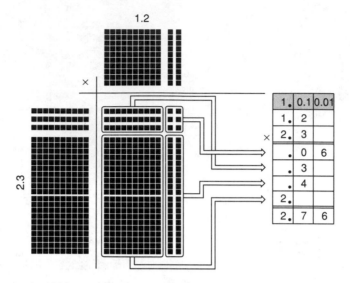

FIGURE 13.12 Decimal Multiplication Using Base-10 Materials

Calculate the following decimal products using base-10 materials and record the results using place value tables.

1. $2.1 \times 1.4 = ?$

2. $1.3 \times 1.6 = ?$

3. $2.4 \times 2.1 = ?$

4. $1.5 \times 2.5 = ?$

5. $2.4 \times 3.6 = ?$

Decimal Multiplication Algorithm

After children are confident with these exercises, introduce more complex decimal exercises. For complex examples, there is no need to keep track of the place value of each subproduct. Point out that, by initially ignoring the decimal point, the factors can be multiplied using the standard algorithm and the decimal point positioned in the answer. The number of places to the right of the decimal point can be determined by mentally calculating the decimal place value of the product. For example, tenths × tenths = hundredths, so the decimal point in Example 1 should be placed so that the decimal fraction part of the product shows hundredths (i.e., so that the decimal point has two place values to its right):

1.	0.5	(1 decimal fraction place value)
	× 0.6	(1 decimal fraction place value)
	0.30	(2 decimal fraction place values)

2.   0.31   (2 decimal fraction place values)
 × 0.2   (1 decimal fraction place value)
   0.062 (3 decimal fraction place values)

3.   2.43   (2 decimal fraction place values)
 × 0.17   (2 decimal fraction place values)
   1601
   243
 0.4031   (4 decimal fraction place values)

Eventually, children should discover that the number of decimal places in the product is the *sum* of the numbers of decimal places in each factor.

You can create a useful exercise for children by making a worksheet that shows several possible answers for each decimal multiplication problem. The children estimate the decimal fraction place values and circle the correct answer. For example:

3.9 × 12.4 =   4836   483.6   48.36   4.836   0.4836

12.04 × 0.204 =   24561.6   2456.16   245.616   24.5616   2.45616

Decimal Division

A number line can be used to show division of a decimal fraction by a decimal fraction. For example, if a group of friends is planning to row a boat 2.5 miles around a lake, and each person can row 0.5 miles, how many people are needed to complete the trip? This model uses the repeated subtraction model of division discussed in chapter 10. Figure 13.13 shows how this division problem can be solved using a number line.

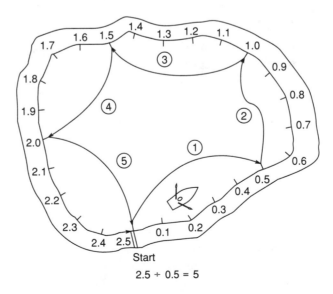

Start
2.5 ÷ 0.5 = 5

FIGURE 13.13 Decimal Division on the Number Line

$$0.24 \div 3 = 0.08$$

FIGURE 13.14 Division of Decimal Fraction by Whole Number Using Money

Work with money can be used to show division of a decimal fraction by a whole number. For example, to show $0.24 \div 3$, have children start with two dimes (20¢ = $0.20) and four pennies (4¢ = $0.04). Partitioning these coins into three equal groups requires regrouping the dimes into twenty pennies as shown in Figure 13.14. Each group contains eight pennies, so $0.24 \div 3 = 0.08$. Have children work several examples using play money.

Calculate the following decimal quotients using a number line or money and record the results using standard symbols.

1. $0.36 \div 3 = ?$

2. $0.45 \div 5 = ?$

3. $2.2 \div 0.2 = ?$

4. $1.6 \div 0.4 = ?$

5. $8.4 \div 4 = ?$

Decimal Division Algorithm

Division of a decimal fraction by a whole number uses the same algorithm as whole number division except the decimal point must be positioned above the decimal point in the dividend.

$$
\begin{array}{r}
0.58 \\
4\overline{)2.32} \\
\underline{20} \\
32 \\
\underline{32} \\
0
\end{array}
\qquad
\begin{array}{r}
3.18 \\
8\overline{)25.44} \\
\underline{24} \\
14 \\
\underline{8} \\
64 \\
\underline{64} \\
0
\end{array}
$$

Remainders are generally not written as part of the quotient for decimal fraction division as they may not be whole numbers. Instead, when greater accuracy is required, additional zeros are appended to the dividend, and the division is carried out to further decimal places.

For example, to compute ($14.24 \div 6$) to the *nearest hundredth* (two decimal places), children append one 0 to the dividend, carry out the division three decimal places (thousandths place value), and round off to the nearest hundredth.

$$
\begin{array}{r}
2.373 \approx 2.37 \\
6\overline{)14.240} \\
\underline{12} \\
22 \\
\underline{18} \\
44 \\
\underline{42} \\
20 \\
\underline{18} \\
2
\end{array}
$$

To position the decimal point correctly, division exercises involving decimal fraction divisors require an additional step prior to carrying out the standard algorithm. Because it is more difficult to divide by a fraction than by a whole number, have children first multiply the divisor by the power of 10 (10, 100, 1000, etc.) that will convert it to a whole number. Then multiply the dividend by the same power of 10 to maintain an equivalent proportion between the divisor and the dividend. Dividing these two new values gives the same quotient as in the original exercise, because both are enlarged proportionally ($8 \div 4 = 80 \div 40 = 800 \div 400$).

After children have completed several examples using this method, ask if anyone sees a pattern that simplifies the process (moving the decimal point an equal number of positions in both the divisor and dividend has the effect of multiplying each by the same power of 10).

When applying this rule, have students insert a **caret** ($\wedge$) to stand for the new decimal point in the divisor that makes it a whole number. Place a caret in the dividend the same number of places to the right of its decimal point. Solving the problem marked with the caret in the following examples will give the same answer as in the original exercise. Since each of the divisors in the transformed exercises is a whole number, the division algorithm is identical to that of whole numbers. The decimal point, of course, must be placed in the quotient directly above its *new* position in the dividend, identified by the caret. Each of those exercises has a remainder so the quotion is approximate. Students can check the results of their work using a calculator.

$$
2.4\overline{)42.68} \;=\; 2.4\wedge\overline{)42.6\wedge 8} \quad \begin{array}{c} 17.7 \end{array}
$$

(position caret one place value to the right in divisor and dividend)

$$
.34\overline{)5.276} \;=\; .34\wedge\overline{)5.27\wedge 6} \quad \begin{array}{c} 15.5 \end{array}
$$

(position caret two place values to the right in divisor and dividend)

$$\begin{array}{r} 9\ 1. \\ .71\overline{)64.9} \quad = \quad .7\ 1\overline{)6\ 4.9\ 0} \end{array}$$ (position caret two place values
to the right in divisor and
dividend by appending a 0)

As with whole number operations, the effective use of concrete materials can assist children in developing decimal fraction concepts and skills. Systematically reducing reliance on perceptual supports as facility with decimal operations increases will improve conceptual understanding and the ability to apply skills to problem-solving situations.

USING PERCENTS

A useful application of decimal fractions is working with percents, or **rates**. The term **percent** (%) literally means *for each hundred*. For example, if 10 percent of the students in a school were absent with the flu, an average of 10 out of each 100 students was sick. If there were 400 students in the school, a ratio of 10:100, or 40 students, was ill. The total number of objects considered (student population) is called the *base*, and the result (sick students) is called the *percentage*.

To introduce percents, give each child one base-10 flat and a supply of one unit-cubes (10 × 10 squares and single squares made from centimeter squared paper can be substituted). Ask the class to put ten unit-cubes on the flat and write a fraction that describes how much of the flat is covered by the cubes (i.e., $\frac{10}{100}$). Repeat the process with 6, 18, 25, 40, 70, and 85 cubes. Finally, have the students cover the flat with cubes and record the fractional part covered ($\frac{100}{100}$). Have each child write each fraction as a decimal ($\frac{10}{100} = 0.10$). Next, explain that people often write these special *hundredths* fractions another way and call them *percents*. Percent means per hundred, so we write $\frac{10}{100}$, or 0.10, as 10 percent, or 10%. Have the class write each decimal they recorded as a percent. Have them make up other problems using the flat and cubes and record the results as a fraction, decimal, and percent (Figure 13.15).

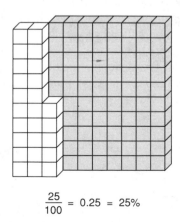

$$\frac{25}{100} = 0.25 = 25\%$$

FIGURE 13.15 Finding Percents

Other fractions can also be written as percents if they can be renamed as hundredths. For example, $\frac{1}{2}$ can be renamed as $\frac{50}{100} = 0.50 = 50\%$. Have children use flats and cubes to find the percent equivalents of common fractions such as $\frac{1}{4}$, $\frac{3}{4}$, $\frac{2}{5}$, and $\frac{3}{10}$ ($\frac{1}{4} = \frac{25}{100} = 0.25 = 25\%$).

Such fractions as $\frac{1}{3}$ cannot be exactly renamed as hundredths. However, using the division meaning of fractions, children can rewrite $\frac{1}{3}$ as a decimal approximation and a percent ($\frac{1}{3} \approx 0.33 = 33\%$). Have children use calculators to help them write the following fractions as approximate percent equivalents: $\frac{1}{3}$, $\frac{2}{3}$, $\frac{5}{6}$, and $\frac{1}{8}$ (note that some fractions like $\frac{1}{8}$ can be written as exact percents if we allow for fractional parts of a percent: $\frac{1}{8} = 0.125 = 12.5\%$).

Finding the Percentage

Twenty-five percent is equivalent to $\frac{25}{100}$, or 0.25. To calculate $\frac{1}{2}$ of 6, we multiply $\frac{1}{2} \times 6$. Similarly, to find twenty-five percent of 60, multiply 0.25×60 (or $\frac{25}{100} \times 60$).

$$25\% \text{ of } 60 = 0.25 \times 60 = 15$$
$$50\% \text{ of } 150 = 0.50 \times 150 = 75$$
$$90\% \text{ of } 35 = 0.90 \times 35 = 31.5$$
$$100\% \text{ of } 85 = 1.00 \times 85 = 85$$

Stores often have sales that advertise discounts in terms of percents. The *sale* price is the original price minus a discount. If items are reduced thirty percent, the sale price is computed by multiplying the regular price by 0.30 and subtracting the resulting **percentage** from the original price. Have children cut out ads from the newspaper and make their own sale by marking everything down by twenty-five percent. Have them calculate the sale price for each item and record their results as shown in Table 13.3.

Sales tax can also be computed. Have children calculate a sales tax of six percent on each item and add it to the sale price (Table 13.4).

Finding the Rate

A bank loans money and charges the borrower a fee for its use. A $100 loan might require a repayment of $110 at the end of one year. The extra $10, called **interest**, is often expressed as a percent of the amount borrowed. In this case, the percent, or **rate**, of interest is $10 per $100 per year, or ten percent per year. Have children

TABLE 13.3 Finding Percentage

Item	Discount	Percentage	Sale Price
$11.95	30%	$3.59	$ 8.36
$ 8.50	30%	$2.55	$ 5.95
$22.75	30%	$6.83	$15.92

TABLE 13.4   Calculating Sales Tax

Sale Price	Sales Tax	Percentage	Total Cost
$ 8.36	6%	$0.50	$ 8.86
$ 5.95	6%	$0.36	$ 6.31
$15.92	6%	$0.96	$16.88

compute the percent of interest (rate) by using a calculator to divide the yearly interest fee (percentage) by the total amount borrowed (base). The resulting decimal is renamed as a percent by dividing the percent by 100 ($0.25 = \frac{25}{100} = 25\%$). Point out that dividing a number by 100 has the effect of moving the decimal point in the dividend two places to the left (Table 13.5).

Another application of this procedure is finding the rate of return, or **profit margin**, for an investment. Suppose a storekeeper purchased an item wholesale for $1.92 and sells the item for $2.15. Her profit is $0.23. Make up a list of items with the purchase and sale prices and have children use a calculator to compute rate of return by dividing the profit by the original price. Have them record the results as shown in Table 13.6.

Finding the Base

Some applications ask that the **base** amount be computed given the percentage and rate. For example, suppose someone was willing to spend $15 in interest on the purchase of a bicycle. If the current rate of interest is ten percent, how expensive a bicycle can be purchased if it is financed for one year? Since the percentage and rate are known, the base amount can be found by dividing the percentage by the rate.

$$\text{rate} \times \text{base} = \text{percentage}$$

$$\text{base} = \frac{\text{percentage}}{\text{rate}}$$

$$0.10 \times b = 15$$

$$b = 15 \div 0.10 = \$150.00$$

TABLE 13.5   Calculating Interest

Interest Fee (Percentage)		Amount Borrowed (Base)		Interest Rate (Rate)	
$42	÷	$500	=	0.084	= 8.4%
$12	÷	$ 95	=	0.126	= 12.6%
$25	÷	$295	=	0.085	= 8.5%

TABLE 13.6 Calculating Rate of Return

Profit (Percentage)		Original Price (Base)		Profit Margin (Rate)
$0.23	÷	$1.92	=	0.119 ≈ 12%

Have children use a calculator to compute the amount that could be borrowed for various interest rates and record the results as shown in Table 13.7. Discuss why the amount that can be borrowed increases as the interest rate decreases (i.e., as the divisors get smaller, the quotients get larger). How much money can be borrowed if the interest rate drops to zero percent? (Note that as the rates gets closer to 0, the amount borrowed increases rapidly; at zero percent, the amount is undefined because we cannot divide by 0.)

Proportion Method for Working Percents

A **proportional algorithm** can be used to solve all three types of percent problems. The three values—rate, percentage, and base—can be expressed as a proportion. A rate expressed as a percent (e.g., 25%) must be divided by 100 to transform it to its decimal equivalent.

$$\frac{r}{100} = \frac{p}{b}$$

Given any two of the values r, p, and b, the third can be calculated by crossmultiplying and simplifying. The following three examples show how to solve for each unknown.

Example 1   Find the percentage p given $b = 160$ and r equals 25%.

$$\frac{25}{100} = \frac{p}{160}$$

TABLE 13.7 Computing the Base

Percentage (Interest)	Rate (Percent)	Base (Price)
15	0.13	$ 115.39
15	0.10	$ 150.00
15	0.08	$ 187.50
15	0.06	$ 250.00
15	0.01	$1500.00
15	0.00	?

$$\frac{160 \times 25}{100} = p$$

$$40 = p$$

Example 2   Find the rate r given $b = 160$ and $p = 40$.

$$\frac{r}{100} = \frac{40}{160}$$

$$r = \frac{40 \times 100}{160}$$

$$r = 25\%$$

Example 3   Find the base b given $p = 40$ and $r = 25\%$.

$$\frac{25}{100} = \frac{40}{b}$$ Reciprocals of equivalent

$$\frac{100}{25} = \frac{b}{40}$$ ratios are equivalent

$$\frac{40 \times 100}{25} = b$$ $(\frac{1}{2} = \frac{2}{4} \rightarrow \frac{2}{1} = \frac{4}{2})$.

$$160 = b$$

Solve the following percent problems using the proportion method.

1. How much sales tax (6%) should be charged for the purchase of a $4500 automobile?

2. What is the rate of return on a $1000 investment that pays $1135 at the end of one year?

3. How much money could be borrowed for one year at a 10% interest rate by making a single interest payment of $25 at the end of the year?

INTEGER OPERATIONS

Graphing Integers on the Number Line

In chapter 7, negative whole numbers were introduced as a way to record temperatures below 0. The set of whole numbers and their negative counterparts constitutes the set of **integers**. Negative values used as *ordinal* labels on a scale such as temperature can be introduced to young children. *Cardinal* applications involving integer operations such as balancing a checking account are generally introduced as a pre-algebra topic in Grade 6 or later. Scientific notation introduced later in this chapter also depends on knowledge of integer operations.

Children can be introduced to comparison of positive and negative values using a number line. As with whole numbers, for any two integers on the number line,

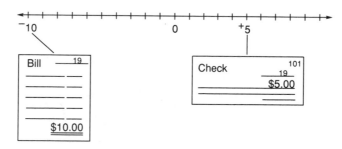

FIGURE 13.16

the value farthest to the *left* is smaller. As a way to introduce the notion that (⁻10 < ⁺5), have children plot the values on a number line, as shown in Figure 13.16, and select the value farther to the left. It is also useful to employ a real-world application of positive and negative values. For example, receiving a check for $5.00 (positive integer) benefits the recipient more than a bill for $10.00 (negative integer).

Other real-world uses of integers include measuring distances above and below sea level. For example, the Dead Sea between Israel and Jordan has an elevation of ⁻1312 feet, and Mt. Everest between China and Nepal has an elevation of ⁺29,028 feet. Businesses also use positive and negative values to indicate profit and loss.

The Cartesian coordinate plane was introduced in chapter 6. Children can specify (graph) each point in the coordinate plane using ordered pairs. The first value in the ordered pair represents the distance in the direction of the horizontal axis and the second indicates the distance in the direction parallel to the vertical axis. Point (⁺2, ⁺3) is located by moving two units to the right and then up three units. Similarly, children can use integer values to show points to the left and below the origin. For example, (⁻2, ⁻4) is graphed by moving two units to the left, then four units down. Have children graph sets of positive and negative ordered pairs that create a simple picture when connected with lines, such as the square shown in Figure 13.17.

Integer Addition

Exercises involving addition of integers can be introduced using the number line. When adding, a negative sign attached to a value indicates a move to the left, and a positive sign indicates a move to the right. To compute ⁻3 + ⁺2 = ?, begin at 0 on the number line, move 3 units to the left and then 2 units to the right, giving the sum ⁻1. To solve the exercise ⁺3 + ⁻2 = ? on the number line, begin at 0, move 3 units to the right, then 2 units to the left, ending at the point ⁺1. Similarly, for the exercise ⁻3 + ⁻2 = ?, begin at 0, move 3 units to the left followed by 2 more units to the left, giving the sum ⁻5 (Figure 13.18).

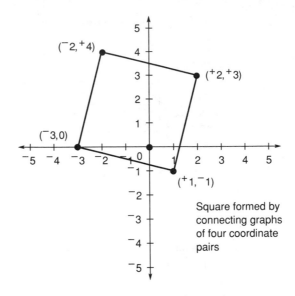

FIGURE 13.17 Cartesian Coordinate Plane

Compute the following sums using a number line.

1. $^-4 + ^-2 = ?$
2. $^+3 + ^-4 = ?$
3. $^+5 + ^+3 = ?$
4. $^-8 + ^+9 = ?$
5. $^-5 + ^+5 = ?$

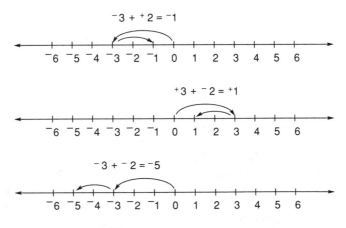

FIGURE 13.18 Adding Integers

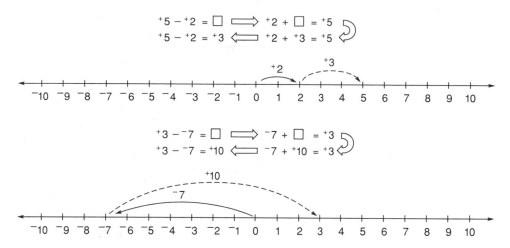

FIGURE 13.19 Subtracting Integers

Integer Subtraction

Since subtraction can be written as a missing addend exercise, it is convenient to rewrite all integer subtractions in this form before solving them on the number line. To solve the integer subtraction problem $^+5 - ^+2 = ?$, first rewrite the exercise as $^+2 + ? = ^+5$. Then begin at 0, move 2 units to the right and count the number of units needed to arrive at $^+5$ (answer $^+3$). Similarly, for $^+3 - ^-7 = ?$, rewrite the exercise as $^-7 + ? = ^+3$. Then begin at 0, move 7 units to the left and count the number of units to the right needed to arrive at $^+3$ (answer $^+10$) (Figure 13.19).

> Rewrite the following subtraction exercises as missing addend examples and solve each using a numberline.
>
> **1.** $^+4 - ^+6 = ?$
> **2.** $^+5 - ^-3 = ?$
> **3.** $^-7 - ^+5 = ?$
> **4.** $^-6 - ^-10 = ?$
> **5.** $^-8 - ^+12 = ?$

Integer Multiplication

Some integer multiplication exercises can be shown on the number line by using repeated addition. For example, to show $^+3 \times ^-4 = ?$, have children first write the exercise as an equivalent repeated addition example $^-4 + ^-4 + ^-4 = ?$ and show the sum $^-12$ on the number line, as in Figure 13.20. Examples such as $^-5 \times ^+4 = ?$ can be rewritten as $^+4 \times ^-5 = ?$ using the commutative property before being calculated as $^-5 + ^-5 + ^-5 + ^-5 = ?$.

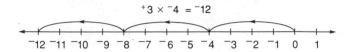

$$^+3 \times {}^-4 = {}^-12$$

FIGURE 13.20 Multiplying Integers

When both factors are negative, as in the example $^-3 \times {}^-4 = ?$, the solution cannot be directly displayed on the number line. The following algebraic justification can be used to show mathematically mature children why a negative integer times a negative integer gives a positive integer product.

Equation	*Justification*
$^-3 = {}^-3$	Identity relation
$^-3 \times ({}^+1 + {}^-1) = {}^-3 \times ({}^+1 + {}^-1)$	Identity relation
$({}^-3 \times {}^+1) + ({}^-3 \times {}^-1) = {}^-3 \times 0$	Distributive property
$^-3 + ({}^-3 \times {}^-1) = 0$	Multiplication of negative $\times$ positive integer and multiplication by 0
$^+3 + {}^-3 + ({}^-3 \times {}^-1) = {}^+3 + 0$	Adding same value to both sides of an equation
$({}^-3 \times {}^-1) = {}^+3$	Therefore negative $\times$ negative = positive

Integer Division

When modeling integer division on the number line, children can use the partitioning meaning of division whenever the divisor is positive. For example, $^-15 \div {}^+3 = ?$ can be shown on the number line by first plotting the dividend $^-15$, then partitioning this distance on the number line into three equal segments (the divisor). As shown in Figure 13.21, the quotient $^-5$ is the value of each equal segment.

For exercises with negative dividends *and* divisor, such as $^-15 \div {}^-3 = ?$, the repeated subtraction meaning of division can be used. For example, Figure 13.22 shows how the value $^-3$ can be subtracted from $^-15$ five times. Therefore, $^-15 \div {}^-3 = {}^+5$.

Exercises that involve positive dividends and negative divisors such as $^+12 \div {}^-4 = ?$ cannot be directly modeled on the number line because it is not possible to construct negative partitions, or to count the number of times $^-4$ can be subtracted from $^+12$. To solve this type of exercise, it is necessary to employ the principle that division and multiplication are *inverse* operations. For example, $^+12 \div {}^-4 =$

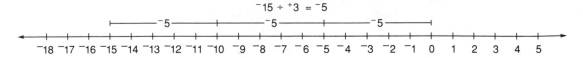

$$^-15 \div {}^+3 = {}^-5$$

FIGURE 13.21 Dividing Integers

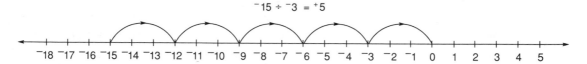

FIGURE 13.22 Integer Division Using Repeated Subtraction

? is another way of writing $? \times {}^-4 = {}^+12$. As ${}^-3 \times {}^-4 = {}^+12$, then ${}^+12 \div {}^-4 = {}^-3$.

Integer Operation Rules

Children often experience difficulty carrying out integer operations, especially subtraction exercises involving negative minuends. After preliminary number line exercises with simple integers, children can be introduced to rules for adding, subtracting, multiplying, and dividing integers. A useful simplification is to convert *all* subtraction exercises immediately into addition exercises by changing the **operation sign** and the **value sign** of the minuend. This rule reflects the fact that subtracting a value n from any number has the same effect as adding its negative value ${}^-n$. For example:

$$
{}^-4 - {}^+9 = {}^-4 + {}^-9 = {}^-13
$$
$$
{}^-4 - {}^-9 = {}^-4 + {}^+9 = {}^+5
$$
$$
{}^+4 - {}^+9 = {}^+4 + {}^-9 = {}^-5
$$
$$
{}^+4 - {}^-9 = {}^+4 + {}^+9 = {}^+13
$$

A simple rule for integer multiplication and division is that the answer is positive when the value signs are *alike* (both negative or both positive). When the value signs are *different* (one negative and the other positive), the answer is always negative.

Convert the following subtraction examples into equivalent addition exercises and use a number line to convince yourself that both forms give the same answers.

1. ${}^+14 - {}^-9 = ?$

2. ${}^-5 - {}^+20 = ?$

3. ${}^+7 - {}^-10 = ?$

4. ${}^-12 - {}^+9 = ?$

5. ${}^+6 - {}^-15 = ?$

Solve the following integer exercises using a number line.

6. ${}^-25 + {}^-9 = ?$

7. ${}^+14 - {}^-10 = ?$

8. ${}^-4 \times {}^-5 = ?$

9. ${}^-7 \times {}^+4 = ?$

10. ${}^+18 \div {}^-9 = ?$

TABLE 13.8 Decimal Place Values Written in Scientific Notation

100,000	10,000	1000	100	10	1	.	0.1	0.01	0.001	0.0001	0.00001
1.0×10^5	1.0×10^4	1.0×10^3	1.0×10^2	1.0×10^1	1.0×10^0		1.0×10^{-1}	1.0×10^{-2}	1.0×10^{-3}	1.0×10^{-4}	1.0×10^{-5}

SCIENTIFIC NOTATION

People often refer to very small and very large numbers without truly understanding the magnitude of the values involved. Have children try to imagine one-millionth of a second (how far would light travel in this period?), or one billion dollars (how long would it take to count it?). Many people work with these numbers every day in the electronics and finance industries, yet it is difficult, indeed, to visualize the magnitude of these measures.

Scientific notation is a way of representing very small and very large decimals by writing a number as a product of two values. The first factor is a decimal value < 10 and the second factor is a power of 10. For example, using scientific notation, the number 200 is written: 2×10^2. Decimal place values can be written in scientific notation as shown in Table 13.8.

Table 13.9 gives several examples of special decimal numbers written in scientific notation.

Numbers written in scientific notation that have the *same* power of 10 are said to be of the same **order of magnitude**. This is another way of saying they are about the same size relative to other very big numbers. For example, give Grade 5–6 children a list of the planets and their distance from the sun. Have them write each distance in scientific notation and classify the planets according to the magnitude of their orbits. Table 13.10 shows that the orbits of the nine planets can be classified into three orders of magnitude. Ask the children what information order of magnitude provides (i.e., any values of the same order of magnitude differ by no more than a multiple of 10).

The results in Table 13.10 do not mean that Mercury, Venus, and Earth are in exactly the same orbit. In fact, the distance from the sun to Mercury is 3.6×10^7 miles; Venus is 6.7×10^7 miles; and Earth is 9.3×10^7 miles. Saying that the distances of

TABLE 13.9 Representing Large Values

Decimal Notation		Scientific Notation
330 m/sec	=	3.3×10^2 m/sec (approximate speed of sound)
928,000,000 km	=	9.28×10^8 kilometers (length of earth orbit)
299,800,000 m/sec	=	2.998×10^8 m/sec (approximate speed of light)
1,440,000,000,000,000,000 t	=	1.44×10^{18} metric tons (mass of water in oceans)

TABLE 13.10   Distance of Planets from Sun

Planets	Order of Magnitude
Mercury, Venus, Earth	7
Mars, Jupiter, Saturn	8
Uranus, Neptune, Pluto	9

these planets from the sun are of the same order of magnitude indicates that their orbits are close together when compared with the other planets.

As shown in the place value table, decimal fractions can also be written using scientific notation. To show fractions, powers of 10 smaller than one are written using **negative** powers. This notation is simply another way that mathematicians write fractions. For example:

$$10^{-1} = \tfrac{1}{10^1} = 0.1$$
$$10^{-2} = \tfrac{1}{10^2} = \tfrac{1}{100} = 0.01$$
$$10^{-3} = \tfrac{1}{10^3} = \tfrac{1}{1000} = 0.001$$

Table 13.11 shows several very small numbers written in scientific notation.

CLASSROOM ACTIVITIES

Decimal Chip Trading

Decimal operations can be performed using place value mats and colored chip trading counters. The procedures are similar to those for working with whole numbers. The counters, or chips, are used like money and are assigned an exchange rate of ten-to-one for each of four colors (red, green, blue, and yellow). Initially, let the green chips represent the one-unit, or $1.00. Other colors represent larger and smaller decimal place values (red = $10.00, blue = $0.10, yellow = $0.01) (see Figure 13.23).

Children play the *Banker's Game* in groups of four. Each group needs a pair of different colored dice, one representing tenths (dimes) and the other hundredths

TABLE 13.11   Representing Small Numbers

Decimal Notation		Scientific Notation
0.0001 m	=	1.0×10^{-4} m (diameter of a hair)
0.000000000000000000000167 gm	=	1.67×10^{-24} gm (mass of one hydrogen atom)
0.00000000000000000000000000091 gm	=	9.1×10^{-28} gm (mass of one electron)

Red 10	Green 1	Blue 0.1	Yellow 0.01

FIGURE 13.23 Chip Trading Place Value Mat

(cents), and each child needs a mat and supply of chips. Players roll the dice in turn and add the number of yellow and blue chips shown to their mats, regrouping as necessary. The first to get ten dollars wins the round. All players record their final amounts in place value tables, remove their chips, and resume play. After five rounds, the player with the largest total wins.

The game can be played in reverse by beginning with a $10.00 red chip on each mat. On each roll of the dice, the amount shown is removed from the mat by regrouping the $10 chip as necessary. When one player runs out of chips, the round ends. The player with the largest total at the end of five rounds wins the *Tax Game*.

Rounding Off to the Nearest 1, $\frac{1}{10}$, and $\frac{1}{100}$

Rounding off was introduced previously as a technique for comparing numbers and performing mental arithmetic. Rounding off decimal fractions is a practical skill for work with money and measures.

Give children a list of meter measures with one digit circled or written in bold. Point out to the class that, when rounding off to the indicated place value, they should first look at the digit in the next smaller place value. If it is 5, 6, 7, 8, or 9, the bold digit is increased by 1, and the remaining digits to the right are dropped; if it is smaller than 5, the bold digit stays the same, and the remaining digits to the right are dropped. Note that if the digit 9 is rounded up 10, as in the second example below, the result must be regrouped into the next larger place value and a zero left in the tenths column to indicate that the value was rounded to the tenths column. Similar procedures are followed for rounding gram and liter measures.

To the nearest cm (0.01):   2.0**4**2 m = 2.04 m

35.7**9**9 m = 35.80 m

To the nearest dm (0.1):   2.**0**42 m = 2.0 m

35.**7**99 m = 35.8 m

To the nearest m (1):   **2**.042 m = 2 m

3**5**.799 m = 36 m

Prices are often rounded to the nearest dollar to enable quick estimates of values and totals. In this case, rounding to the nearest dollar is the same as rounding a decimal fraction to the nearest 1-unit. Give children a copy of a cash register tape showing the cost of ten to fifteen items without the total. Have them mentally round each value to the nearest dollar and find the sum to see if a twenty-dollar bill would be enough money to buy the items. Similarly, have children use copies of actual menus and price lists to create an order or shopping list and use rounding to estimate the total bill.

$$\$12.95 \approx \$13.00$$
$$\$5.39 \approx \$5.00$$
$$\$124.50 \approx \$125.00$$

Round off $162.48 for each place value indicated below.

1. Tenths (or dimes)
2. Hundreds (or hundred dollars)
3. Ones (or dollars)
4. Tens (or ten dollars)
5. Hundredths (or pennies)

PROBLEM SOLVING INVOLVING DECIMALS AND INTEGERS

Foreign Travel

Most countries print their own currency. When traveling to another country, people from the United States exchange their dollars for the local currency. Suppose one U.S. dollar could be exchanged for $1.30 in Canadian dollars. Have children use a calculator to compute the amount of money in U.S. dollars that a family would have to take to pay for the travel expenses listed in Canadian dollars in Table 13.12 (e.g., C$150 ÷ 1.30 = US$115.39). Children should first **predict** whether the cost will be more or fewer U.S. dollars than Canadian dollars for each item.

TABLE 13.12 Travel Expenses

Expense	Cost in Canadian $	Cost in U.S. $
Food	150	115.39
Gasoline	85	?
Motel rooms	225	?
Other	175	?
Total	$635	?

Suppose the family decided to go to Mexico the following summer. One U.S. dollar can be exchanged for about 2500 Mexican pesos. Have the class calculate the number of Mexican pesos the family would need if they planned to spend the same amount of U.S. dollars for their expenses. Again, have children predict whether there will be more or fewer U.S. dollars than Mexican pesos.

As a follow-up activity, have each child select a country to visit from an exchange rate table in the business section of the newspaper. List the expenses the family will have during the visit and convert the amounts to the local currency using a calculator. Record the results in a table.

Using Proportions

Problems involving conversions between measures can also be solved using **proportions**. For example, suppose a family traveling in Canada bought gasoline that cost $1.50 in Canadian money per Imperial gallon (one Imperial gallon equals about 1.2 U.S. gallons). If gasoline at home cost $1.25 per U.S. gallon in U.S. money, was the gasoline more or less expensive in Canada? One way to approach this problem is to assume that the family bought, say, ten Imperial gallons at a cost of $15 in Canadian money. How many U.S. gallons would they have bought? The proportion $1.2{:}1 = \square{:}10$ describes the relationship between the Imperial and U.S. gallons. Solving the proportion using crossmultiplication gives

$$\frac{1.2}{1} = \frac{\square}{10}$$

$$\frac{10 \times 1.2}{1} = \square$$

$$10 \times 1.2 = \square$$

$$12 = \square$$

The proportion $1{:}1.3 = \square{:}15$ describes the relationship between the U.S. and Canadian dollar. Solving the proportion gives

$$\frac{1}{1.3} = \frac{\square}{15}$$

$$\frac{1 \times 15}{1.3} = \square$$

$$\frac{15}{1.3} = \square$$

$$11.54 = \square$$

Therefore, ten Imperial gallons purchased for $15 Canadian money is equivalent to twelve U.S. gallons purchased at $11.54 U.S. money. The price of Canadian gasoline in U.S. dollars was $11.54 \div 12 = \$0.96$ per U.S. gallon. The gasoline was less expensive in Canada. Have children make up similar problems for Mexico, where gasoline might cost 650 pesos per liter (1 liter $\approx$ 0.25 gallons).

Checks and Bills

To practice adding and subtracting integers, children can use simulated **checks and bills** to represent positive and negative values. Prepare sets of green and red file cards with various whole dollar amounts ($1–$20). The green cards represent checks received for work done, and the red cards are bills received for amounts owed. When you receive a check for $5, the transaction is recorded as adding a positive ($+ {}^+5$) and when you give someone else a check for $5, it is recorded as subtracting a positive ($- {}^+5$). When someone gives you a bill, it is recorded as adding a negative ($+ {}^-5$) and when you give someone a bill, the transaction is recorded as subtracting a negative ($- {}^-5$). Children can use checks and bills to solve word problems such as those shown in Figures 13.24–13.26.

One day, Sharon earned $5 for mowing the lawn, and afterwards her sister sold her a book for $2. How much money did she have left at the end of the day?

Checks and bills	5	+	2	=	
	(Green)		(Red)		

| Number sentence | ${}^+5$ | + | ${}^-2$ | = | ${}^+3$ |

FIGURE 13.24

Dana owed Danny $8 yesterday. Today Dana worked in the garden and earned $5. How much money would Dana have if he paid Danny the $5?

Checks and bills	8	+	5	=	
	(Red)		(Green)		

| Number sentence | ${}^-8$ | + | ${}^+5$ | = | ${}^-3$ |

FIGURE 13.25

Brandon had $3 and was paid back a loan for $10 he made to Jenny. How much money does Brandon have now?

Checks and bills	3	−	10	=	3	+	10	=	
	(Green)		(Red)		(Green)		(Green)		

| Number sentence | ${}^+3$ | − | ${}^-10$ | = | ${}^+3$ | + | ${}^+10$ | = | ${}^+13$ |

FIGURE 13.26

Each of these problems can be solved using the checks and bills. Have children make up their own problem situations, exchange, and solve them. Have them try to invent multiplication and division problems involving integers that can be solved using checks and bills.

CALCULATOR AND COMPUTER APPLICATIONS

Calculator Activities

Exploring Repeating Decimals. Calculators can be used to help children rename common fractions as decimals. Using the division meaning of fractions, fractions can be written in decimal form by dividing the denominator into the numerator. For example

$$\tfrac{1}{2} = 1 \div 2 = 0.5$$

To help children verify this result, point out that $\tfrac{1}{2} = \tfrac{5}{10}$, which is just another way of writing 0.5. Verifying the decimal equivalent for other fractions is not as convenient. For example, $\tfrac{1}{3}$ can never be *exactly* renamed as a fraction with a power of 10 as a denominator, because no number times 3 equals any power of 10 (the digital root of all powers of 10 is 1, and 1 is not evenly divisible by 3). However, by using a calculator a decimal *approximation* for $\tfrac{1}{3}$ can be computed:

$$\tfrac{1}{3} = 1 \div 3 = 0.33333333\ldots$$

Though the calculator is unable to continue dividing forever, we append three dots, called an *ellipsis*, to the sequence of threes to indicate that the pattern continues indefinitely (the calculator does not show the ellipsis). For most applications, decimal approximations are computed to the nearest hundredth or thousandth. Banks carry out their computations to five or more decimal places to maintain accurate records of interest calculations.

Computer Software

Shark Estimation Games. This software provides games for practicing decimal value estimation by simulating the harpooning of sharks (InterLearn, Inc.). Children try to hit a shark fin displayed on the screen by estimating corresponding decimal coordinates on a number line or coordinate plane (see Figure 13.27). After each unsuccessful attempt, hints are given to remind the user to select larger or smaller values. The end points of the two axes can be labeled as randomly selected whole numbers, integers, or decimals, facilitating a wide range of uses. There are four related games available, each of which automatically increases the level of difficulty as the user becomes a more successful harpooner.

Decimal Estimation Games. Computer simulations can provide motivating practice for students in identifying and comparing decimal numbers. *Decimals: Practice*

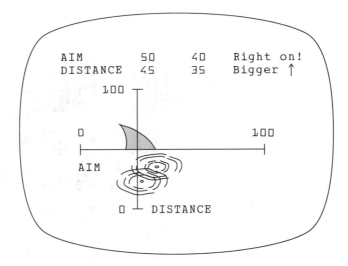

FIGURE 13.27   Shark Estimation Games. *Note*: From "Shark Estimation Games" [Computer program] by InterLearn, Inc., 1984.   Copyright 1984 by InterLearn, Inc.   Reprinted by permission.

(Control Data Publishing Co.) presents students with a dart game in which they try to burst balloons arranged at random along a number line by identifying the corresponding decimal locations. There are two lesson programs, each with multiple levels of difficulty. The first type uses integers as end points on the number line, and the second has decimal end points. Students make an estimate of the decimal position on the number line and refine their choices in successive selections until they burst the balloon.

Decimal Squares Computer Games.   This series of eight activities provides practice representing and recognizing decimal values using base-10 displays (Scott Resources, Inc.). Children use their knowledge of decimal equality, inequality, place value, addition, and subtraction to win games such as miniature golf, tug-of-war and blackjack (Figure 13.28). The programs offer several levels of difficulty to accommodate grade levels 3–6.

ADAPTING INSTRUCTION FOR CHILDREN WITH SPECIAL NEEDS

Using Money to Introduce Decimals

Learning to work with decimal numbers and operations is often difficult for special-needs children. After children have had extensive experience counting money and making change, decimal numbers and operations can be introduced using money as a model. Limiting decimal fractions to two decimal places (i.e., tenths—pennies;

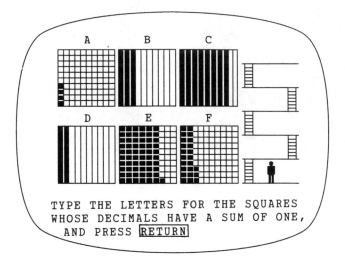

FIGURE 13.28 Decimal Squares Computer Game. *Note*: From "Decimal Squares" [Computer program] by Scott Resources, Inc., 1985. Copyright 1985 by Scott Resources, Inc. Reprinted by permission.

hundredths—dimes) simplifies the process of lining up the decimal point when adding and subtracting. Relating base-10 materials to pennies, dimes, and dollars helps children visualize the place value relationship between the decimal fraction and whole number component of numbers. Figure 13.29 shows a set of printed base-10 materials that uses coins to show graphically the place value relationship. One set of these materials can be constructed from coins, xeroxed, glued to cardboard

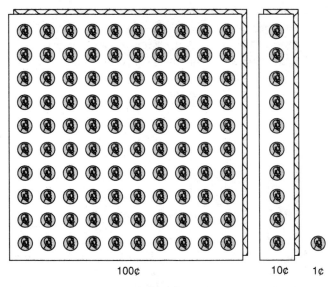

FIGURE 13.29 Coin Base-10 Materials

squares, and used like base-10 materials by special-needs children when solving decimal addition, decimal subtraction, and percent exercises.

Exploring Decimal Patterns

Gifted and talented children can extend their understanding of decimal notation by exploring the patterns of digits in common fractions renamed as decimals. Some fractions, such as $\frac{1}{2}$, repeat zeros after one or more nonzero digits. These are called *terminating decimals*. The fraction $\frac{1}{9}$ written as a decimal equals 0.1111.... Some fractions repeat pairs of digits (e.g., $\frac{5}{11} = 0.454545...$), repeat in cycles of three ($\frac{1}{27} = 0.037037...$), or longer cycles ($\frac{1}{13} = 0.076923076923...$). Children can use calculators to rename fractions as decimals and classify them according to the number of digits in their repeating cycles. (Note: Some calculators round to the last digit and show $\frac{1}{6} = 0.166666667$. Tell children using such calculators to ignore the last digit or cover it with tape.)

$$\frac{2}{3} = \quad \begin{array}{r} .6666 \\ 3\overline{)2.0000} \end{array} \qquad \text{Repeats every digit}$$

$$\frac{1}{7} = \quad \begin{array}{r} .142857142857 \\ 7\overline{)1.000000000000} \end{array} \qquad \text{Repeats in six-digit cycles}$$

$$\frac{1}{2} = \quad \begin{array}{r} .5000 \\ 2\overline{)1.0000} \end{array} \qquad \text{Repeats every digit (after first)}$$

To help children understand why fractions always generate repeating decimals, have them carry out the long division exercise for $\frac{1}{7}$. Because we are dividing by 7, the only possible remainders are the values 0–6. Notice that the cycle begins to repeat at that point where a remainder repeats. If the remainder is 0, the decimal terminates.

Have children find the prime factors of the denominator for all the fractions that generate terminating decimals (repeat zeros). If the fractions are renamed in lowest terms (e.g., $\frac{1}{2}$, not $\frac{3}{6}$), they should discover that these denominators all have sets of prime factors that include only 2s and 5s. All other fractions generate nonterminating, repeating decimal equivalents (i.e., the set of prime factors may include 2s and 5s but must contain at least one other prime as well). Why is this true?

Use a calculator to compute the decimal equivalent of each of the following fractions. Carry out the division to sufficient decimal places to verify a repeating pattern.

1. $\frac{1}{3}$

2. $\frac{1}{5}$

3. $\frac{1}{4}$

4. $\frac{2}{7}$

5. $\frac{1}{13}$

Many numbers cannot be written as a fraction or a repeating decimal and, therefore, are not rational numbers. These values are called **irrational** numbers. Numbers that cannot be written as fractions include π, $\sqrt{2}$, and 0.10110111011110.... No repeating pattern of digits can be found for these values regardless of the number of decimal places calculated. Though infrequently encountered in elementary mathematics, such irrational numbers are important for the study of algebra and geometry. Together, the set of rational (repeating decimals) and irrational (nonrepeating decimals) numbers constitutes the set of **real numbers**.

> Write five rational and five irrational numbers. Make a sketch of plans for a real number bulletin board showing examples and key features of rational and irrational numbers.

SUMMARY

Decimal fractions are another way to write common fractions having only powers of 10 for denominators. Each component of a decimal fraction is displayed in its own place value, separated from the whole number place values by the decimal point. Operations with decimals never require finding common denominators, because all decimal denominators are powers of 10. Concrete materials used to develop whole number operations and algorithms can be adapted to introduce decimal operations. The equivalence of two ratios is called a proportion. An important application of proportional reasoning is work with percents. Scientific notation can be used to record and compare very large and small values. Addition and subtraction of integers is required to understand scientific notation. Special-needs children may require additional work with concrete materials to extend work with whole numbers to operations with decimals. Decimals, percents, and proportions offer excellent opportunities to extend the mathematics understanding and problem-solving abilities of gifted and talented children.

COURSE ACTIVITIES

1. Using base-10 or chip trading materials, practice representing, adding, and subtracting decimal fractions involving tenths and hundredths. Give a presentation to your peers and, if possible, write a lesson plan and introduce decimal addition to a group of Grade 5–6 students in a classroom setting. Keep track of points students have difficulty understanding and discuss possible teaching solutions with your peers.

2. Most calculators are unable to display more than eight to ten digits. Think of a way to use the calculator to carry out division to more decimal places than can be displayed. Use this procedure to help you find fractions that repeat in groups of 5, 6, or more

digits. Make a list of fractions that repeat in groups of 1, 2, 3, 4, 5, and 6 digits.

3. Rename the six fractions ($\frac{1}{7}$, $\frac{2}{7}$, $\frac{3}{7}$, $\frac{4}{7}$, $\frac{5}{7}$, and $\frac{6}{7}$) as decimals carried out to twelve decimal places. Look for patterns in the decimal representations. As an extended activity, do the same with other series of fractions (eg., $\frac{1}{13}$, $\frac{2}{13}$, . . .) and look for repeating patterns. Discuss how this activity could be incorporated into a lesson on decimals for Grades 5 and 6. Write a lesson plan that uses this activity.

4. Design a bulletin board to display the proportional relationship involved in percent. Try to arrange the display so that the viewer must engage in some activity. List ways bulletin boards can be integrated into learning experiences rather than used primarily as decoration.

5. Include examples of decimal and percent lessons in your idea file. Organize them according to appropriate grade levels.

6. Read the *Arithmetic Teacher* article by L. Chang listed in the reference section. Write a brief report summarizing the main ideas of the article and describe the advantages of each instructional model for teaching integer operations.

MICROCOMPUTER SOFTWARE

Base Ten on Basic Arithmetic Practice working with multiples of ten and decimal numbers (MECC).

Decimals: Practice Practice identifying decimals on a number line in order to burst balloons in a simulated dart game (Control Data Publishing, Co.).

Decimal Skills and Mixed Number Skills Tutorial on decimal and fraction operations (Milton Bradley).

Decimal Squares Computer Games Eight computer games that provide practice with decimal place value and operations with decimals using base-10 representations of decimal fractions (Scott Resources, Inc.).

Integers A set of five programs that provides tutorials and practice working integer operations using a number line model (JHM Software).

Maxit Provides practice adding integers in a game environment (California State Department of Education, TIC Materials).

Shark Estimation Games Practice estimating the position of a shark fin on the coordinate plane using integer and decimal values (InterLearn, Inc.).

REFERENCES AND READINGS

Battista, M. (1983). A complete model for operations on integers. *Arithmetic Teacher*, *30*(9), 26–31.

Bell, A., Swan, M., & Taylor, G. (1981). Choice of operation in verbal problems with decimal numbers. *Educational Studies in Mathematics*, *12*, 399–420.

Carpenter, T., Corbitt, M., Kepner, H., Lindquist, M., & Reys, R. (1981). Decimals: Results and implications from national assessment. *Arithmetic Teacher*, *28*(8), 34–37.

Cathcart, W. (1977). Metric measurement: Important curricular considerations. *Arithmetic Teacher*, *24*(2), 158–160.

Chang, L. (1985). Multiple methods of teaching the addition and subtraction of integers. *Arithmetic Teacher*, *33*(4), 14–20.

Clason, R. (1986). How our decimal money began. *Arithmetic Teacher*, *33*(5), 30-33.

Cole, B., & Weissenfluh, H. (1974). An analysis of teaching percentage. *Arithmetic Teacher*, *21*(3), 226–228.

Dirks, M. (1984). The integer abacus. *Arithmetic Teacher*, *31*(7), 50–54.

Grossman, A. (1983). Decimal notation: An important research finding. *Arithmetic Teacher*, *30*(9). 32–33.

Hiebert, J. (1987). Decimal fractions. *Arithmetic Teacher*, *34*(7), 22–23.

Hiebert, J., & Wearne, D. (1986). Procedures over concepts: The acquisition of decimal number knowledge. In J. Hiebert (Ed.), *Conceptual and procedural knowledge: A case for mathematics* (pp. 199-223). Hillsdale, NJ: Erlbaum.

Lichtenberg, B., & Lichtenberg, D. (1982). Decimals deserve distinction. In L. Silvey & J. Smart (Eds.), *Mathematics for the middle grades: 1982 Yearbook* (pp. 142–152). Reston, VA: National Council of Teachers of Mathematics

Payne, J., & Towsley, A. (1987). Ideas: Dollars and cents, tenths and hundredths, comparing and ordering decimals, estimating with decimals. *Arithmetic Teacher*, *34*(7), 26–28.

Priester, S. (1984). SUM 9.9: A game for decimals. *Arithmetic Teacher*, *31*(7), 46–47.

Zawojewski, J. (1983). Initial decimal concepts: Are they really so easy? *Arithmetic Teacher*, *30*(7), 52–56.

14

Graphing, Statistics, and Probability — Exploring Uncertainty

The true logic of this world is in the calculus of probabilities.
—James Maxwell

Upon completing Chapter 14, the reader will be able to

1. Describe six types of elementary graphs and give examples of appropriate classroom activities.
2. Describe three measures of central tendency and give examples for each.
3. Define the slope of a line graph and give examples of problem-solving applications.
4. Describe three systematic counting procedures and give an example of each.
5. Define the probability of an event and give an appropriate introductory lesson.
6. Describe the difference between independent and dependent events and give an example of each.
7. Define random sample and prepare a sample lesson to introduce the concept to children.
8. Describe the difference between experimental and theoretical probability and give an example showing how they are related.
9. Describe problems involving statistics and probability and give a sample lesson for each.

The study of statistics and probability arises from a need to make informed judgments about uncertain events. To establish an appropriate inventory, the manager of a running shoe store must decide which sizes and brands to restock. The weather bureau must issue reports that indicate the relative likelihood of rain, sunshine, hurricanes, tornados, and other weather conditions. Consumers are bombarded with often conflicting claims about product quality on which they must make informed decisions.

Learning to interpret statistical summaries, such as **frequency** (number of occurrences of an event), the **mean** (average), and the **range** (difference between the smallest and largest values observed), is fundamental to making informed decisions at work and in the marketplace. These statistics are often presented as *graphs*, *charts*, and *tables*. The ability to interpret graphs is becoming an increasingly important skill in our society. Television and print advertising rely heavily on the convincing arguments embodied in well-designed graphs.

Graphs can be misleading, however, if key information necessary for making an unbiased comparison is omitted from the display. For example, a recent advertisement compared two headache remedies claiming that Brand 1 was more effective because it has more relief-giving medication. The two analgesics were graphically compared by pouring the powdered medicine into two clear tubes (see Figure 14.1). Sure enough, Brand 1 rose higher than Brand 2. The informed consumer might question whether both powders contained the same concentration of medications. Was the claimed effectiveness due to the *quantity* of medication shown in the graph; or was it a question of the *quality* of the ingredients, which could be verified only by clinical evaluation?

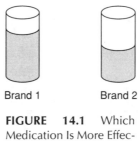

FIGURE 14.1 Which Medication Is More Effective?

USING STATISTICS
Introducing Graphs

Many of the activities presented in this book include suggestions that children record the results of their experiments using individual or class graphs. The following section reviews the construction and application of several types of elementary graphs.

The study of statistics initially involves organizing and summarizing data collected from experiments. Often data is displayed in tables, charts, and graphs so that relationships become clear. The amount of information presented can be reduced by using summary descriptions of data such as the **mean, median, mode,** and **range**.

Empirical Graphs.   Even Kindergarten children can participate in meaningful graphing activities by using objects themselves as a graphing medium. Such *empirical* graphing activities provide direct experience in constructing and interpreting graphs. For example, Figure 14.2 shows children standing in two lines that form an empirical graph of the number of lunch buyers and brown baggers. Empirical graphs enable teachers to introduce graphing concepts to young children. For example, the children must stand in straight lines, and the lines must start at the same **baseline** if accurate comparisons are to be made. The same conditions are also true for representational graphs children will construct later.

In Figure 14.3, children line up to find out if there are more shoes with laces in the classroom than without.

FIGURE 14.2 Are There More Brown Baggers or Buyers?

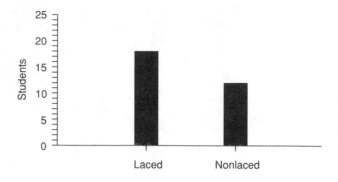

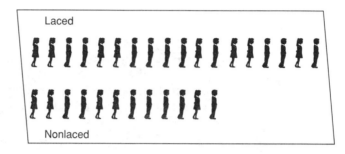

FIGURE 14.3 What Is the Most Popular Shoe Type?

Representational Graphs. A slightly more *representational* attendance graph can be constructed using pictures of each student with their names written on the back. The children turn their pictures face-out when they arrive in the morning, making it easy to take attendance. The pictures can be displayed in two rows to show clearly the relationship between the number of students present and absent. Graphs can be displayed using either vertical or horizontal columns. All columns (or rows) must begin at a common baseline, and the elements in each must be uniformly separated. This allows direct comparison of the number of elements in each category (Figure 14.4).

The next stage is to have children substitute uniform objects, such as blocks or colored squares on graph paper, to represent each item or event graphed. Because the objects are discrete, the graph can be reorganized without a record of the change if students change their minds or make a mistake. For example, all children with blue eyes can select a blue Unifix Cube and construct a column on the chalk tray representing blue-eyed children. Those with brown and green eyes can construct columns using like-colored cubes (Figure 14.5). Representational graphs provide a useful intermediary step for children learning to construct graphs.

Picture Graphs. A picture graph, or **pictograph**, is constructed using sketches or pictures to represent objects or events. Initially, each element in a picture graph can represent *one* object. Later, to save space, each element might represent several objects or events. A scale, or **key**, must be included to show the ratio of real objects

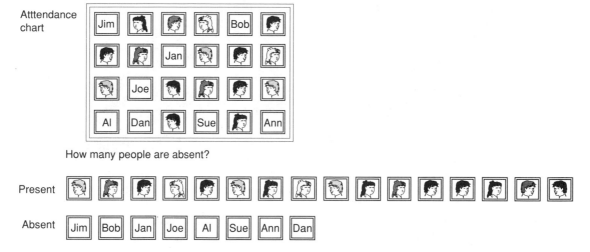

FIGURE 14.4

to those represented in a pictograph so that the graph can be accurately intepreted. The pictographs shown in Figure 14.6 indicate the number of calculators used in a classroom and the number of calculators sold by a department store over a five-year period.

Have children graph the area of their hands using their own thumbprints. Suppose it takes twenty-two thumbprints for a child to cover a cutout of her hand. Have her stamp twenty-two new thumbprints in a column above her name on a class graph to record the results. The whole class should record the area of their hands on the graph. To reduce the number of thumbprints needed to show the handprint area,

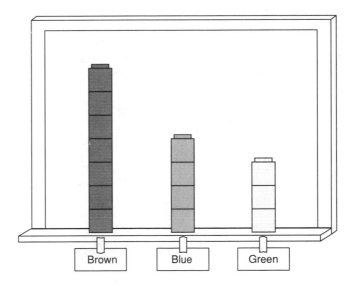

FIGURE 14.5 Are There More Brown or Blue Eyes?

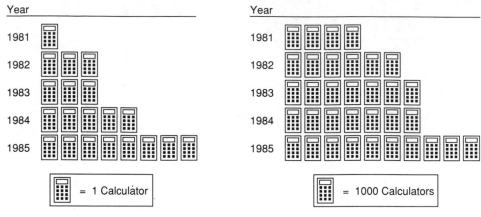

a. Calculators in Room 10

b. Calculators sold by Al's Calc

FIGURE 14.6

children can use one print on the graph to represent ten actual thumbprints. A key showing the ratio should be included on the graph to ensure accurate interpretation, as shown in Figure 14.7.

> Find out how many thumbprints it takes to cover your footprint. Make a class graph of the results.

Bar Graphs. Representational graphs and pictographs are often displayed in a simplified form called the **bar graph**. The bars, positioned horizontally or vertically, represent the number of items in each category. A scale is generally displayed along

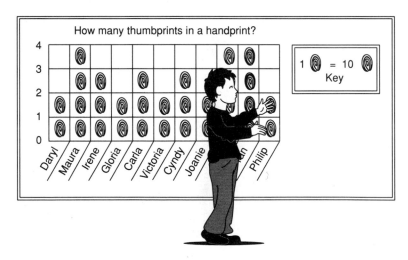

FIGURE 14.7

one axis, and the categories are listed along the second. These scales and labels facilitate the interpretation of the graph. A scale must be selected to accommodate the smallest and largest values to be graphed. Graphs used in school are generally titled with a question to stimulate viewer involvement (Figure 14.8).

Line Graphs.   Connecting the midpoints of the ends of each bar on a graph gives a simplified graphic display called a line graph, or **frequency polygon**. For example, children can make a line graph showing their favorite pets. First, identify the possible

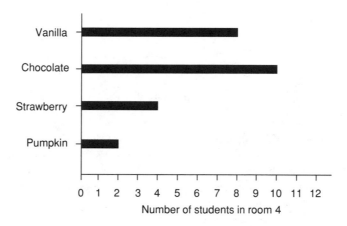

a.   What is the favorite ice cream flavor?

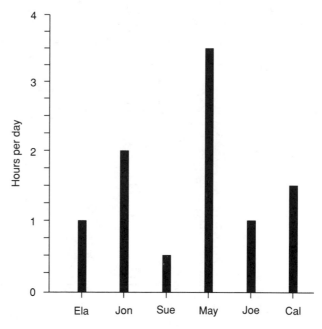

b.   How much T.V. watching time?

FIGURE 14.8

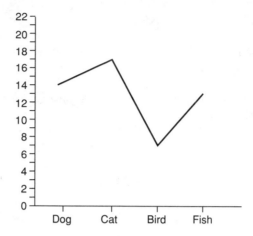

FIGURE 14.9 What Is Your Favorite Pet?

categories of pets (dogs, cats, birds, fish). Then, call out the categories and have the children vote for their favorite. Graph the results as shown in Figure 14.9. How would the results change if children could vote twice, or as often as they wished? Discuss different ways to collect the opinions and construct the graph (e.g., have everyone number the categories 1, 2, 3, 4, indicating least to most favorite pet, and graph the average class results for each pet).

Line graphs are often used to show continuous data rather than discrete categories. For example, in situations where intermediate values exist, such as temperature, speed, money, and length, a line graph generally gives a more accurate picture of the relationship than does a bar graph. Figure 14.10 shows the relationship

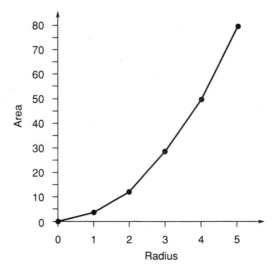

FIGURE 14.10 How Do Circles Grow?

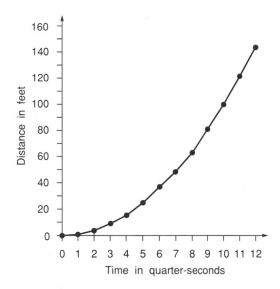

FIGURE 14.11 How Fast Do Things Fall?

between the radius and area of several circles. Figure 14.11 shows the relationship between time and distance for a falling object. Both relationships have line graphs called *parabolas*.

Coordinate Graphs.   Line graphs were used to show the *relationship* between two events. **Coordinate graphs**, which were introduced in chapter 6, can also be used for this purpose. A coordinate graph has a horizontal and a vertical scale. Two related events can be graphed at the same time by plotting one event on the horizontal scale and the other on the vertical scale. For example:

- ☐ The years 1980–1990 and the number of people living in Kansas.
- ☐ The ages 0–90 and the average height at each age.
- ☐ The grade levels K–12 and the number of students at each level.
- ☐ The heights 50–200 cm and the average shoe size for each height.
- ☐ The amount of money in children's pockets and their grade levels.
- ☐ The months in the year January–December and the average number of colds.

Notice that the scales do not have to be represented by numbers. However, each element in the sequence must have an easily interpreted, fixed position relative to all other members (e.g., October must follow September and precede November).

Grade 5–6 students can use a coordinate graph to investigate the relationship between grade level and the number of parents who attend open house. The children first collect the open house attendance data and organize the information as shown in Table 14.1.

Children can then construct a coordinate graph by writing the grade levels along the horizontal axis and a scale to record the number of parents along the vertical axis (Figure 14.12).

TABLE 14.1 Parent Attendance Data

Grade Level	Parents
K	23
1	19
2	18
3	12
4	6
5	4
6	2

The number of parents in attendance can be graphed by looking at each row in the table above. Each row can be written as an **ordered pair**. For example, the Kindergarten level can be represented by the ordered pair (K, 23). To graph this ordered pair, remind children to first move horizontally along the grade level scale to the point marked Kindergarten (K). Then count up twenty-three spaces in the direction parallel to the vertical axis. Place a dot at the point (K, 23). The remaining pairs can be graphed, or **plotted**, and the points connected in order with straight line segments (Figure 14.13).

The graph shows that fewer parents attended open house from the upper grades than the lower grades. There could be several reasons to account for this trend. Possibly, there were more primary than upper grade children. Because parents of older children have on the average been married longer, there may have been fewer parents in the upper grades due to divorce. Reduced parental interest in older children's schooling might also account for the variation. Graphs are an ex-

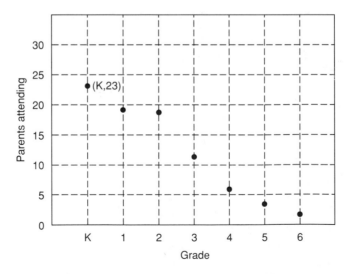

FIGURE 14.12 Did You Attend Open House?

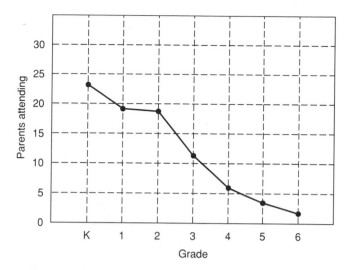

FIGURE 14.13 Open House Attendance Line Graph

cellent way for students to organize information and look for patterns to solve problems.

Circle Graphs. Another popular graph format uses fractional parts of a circle to show the amount each component contributes to making up the whole unit. The way a child spends her or his allowance can be shown as pie-wedge-shaped regions of a silver dollar (see Figure 14.14).

As an introduction to circle graphs, have children keep a diary of their activities for a complete day organized into five categories: studying, eating, playing, watching television, and sleeping. Have them make a table showing the total amount of time (to the nearest half-hour) spent in each category. Using a compass, each child draws a circle with an 8-inch (or 16 centimeters) diameter (4-inch radius) and marks off

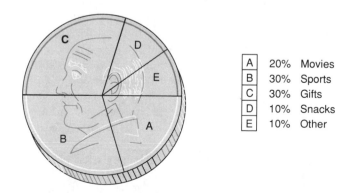

FIGURE 14.14 How Do You Spend Your Allowance?

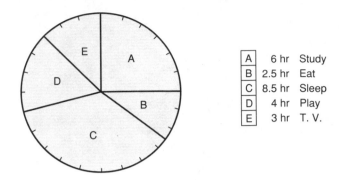

A	6 hr	Study
B	2.5 hr	Eat
C	8.5 hr	Sleep
D	4 hr	Play
E	3 hr	T. V.

FIGURE 14.15 What Do You Do All Day?

twenty-four equal segments (approximately 1-inch or 2-centimeter arcs) around the edge. Each segment equals one hour. Children shade an appropriately sized wedge-shaped section for each category. Younger children may have difficulty constructing pie graphs because of the dexterity involved in using a compass and marking off units around the circumference. Prepare circles with the circumference marked off in appropriate units to assist such children in constructing circle graphs (Figure 14.15).

Interpreting Graphs

Measures of Central Tendency. To communicate with other people, we often need to describe a collection of information in a simple way. The children in a classroom are all different ages unless two were born at exactly the same moment. Yet a third grade teacher, when asked what age children he or she teaches, might answer 8-year-olds. A value that is used to stand for a collection of numbers or objects is called a **measure of central tendency**. Three common measures of central tendency are the *mean*, *median*, and *mode*. Each of these statistics gives a single value, or category, that can be used to represent an entire set of observations.

The **mean**, or average, is the sum of a set of values divided by the number of values. To introduce this idea to children, use three towers of Unifix Cubes (e.g., 4 red, 7 black, 11 yellow) set on the chalk tray or table. Tell the class to imagine that the towers represent the number of minutes three children took to ride their bicycles to school. Ask the class to estimate the *average* amount of time it took for the trip. Figure 14.16 shows two ways to solve the problem using the cubes. First, children can move cubes from the tall tower to the shorter ones and adjust as necessary until all three towers are as near the same length as possible. Or, the towers could be dismantled and divided into three equal piles (with a possible remainder). Give several examples using three to five small values and have each child compute the mean using Unifix towers. Point out that dividing the total number of cubes into equal sets is the same as adding the values and dividing by the number of sets. When mean is introduced in Grades 5–6, children should already possess the required computation skills.

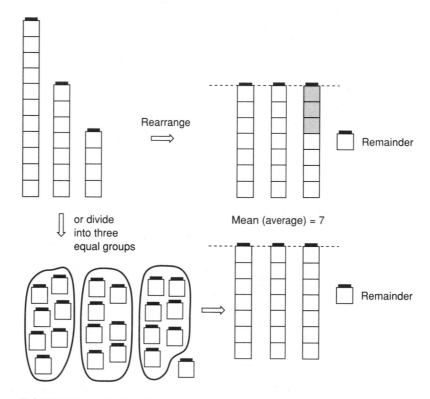

FIGURE 14.16 Finding the Mean

Have children calculate the mean height of the students in Grade 5 at their school by dividing the sum of the heights by the total number of fifth graders. Ask the class if they would expect the mean height for Grade 5 children to fall between that of Grade 4 and Grade 6. Why? How could they find out?

Sometimes the mean is not the best measure of central tendency. For example, a family moving to a new city may want to know the likely cost of a new home. If the real estate agent simply averaged the cost of all the homes sold in the past year, a small number of very expensive estates might increase the mean sufficiently to give a false impression of the true cost of housing. Real estate values are generally reported in terms of the **median** price of homes sold in the past year. This statistic is found by seriating the cost of all homes sold and selecting the value in the middle. (If there is an even number of values, the median equals the mean of the two middle values.) Notice that this technique ignores the size of the values once they are placed in order.

To introduce the idea of median to children, have them count the number of pencils they own. On the board or a prepared graph, as shown in Figure 14.17, have students record the number of pencils above their names. To find the median, have each child list the pencil counts in order from small to large and circle the value in the middle (if there is an even number of children, have them average the two

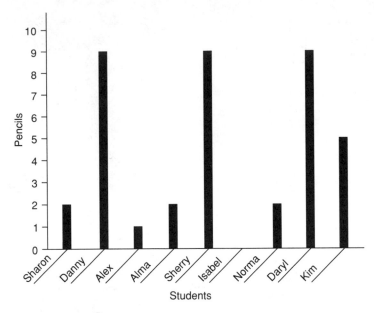

List: 0, 1, 2, 2, ②, 5, 9, 9, 9
Mean ≈ 4.4 pencils
Median (circled) = 2 pencils
Mode = 2 and 9 pencils

FIGURE 14.17 What Is the Average Number of Pencils Owned by Students?

middle values). Write the median on the graph. Also, compute the mean to compare it with the median.

A third measure of the central tendency is the **mode**. The mode is the value or category that occurs most frequently. It would not make sense to find the mean or median of favorite ice cream flavors. However, the fact that chocolate is the most frequently chosen flavor for a particular school is a useful statistic for the cafeteria manager. The most popular category (chocolate ice cream) is called the mode.

If a graph has two modes (two categories with equal frequencies larger than all others), we say the graph is *bimodal*. For example, in the pencil graph shown in Figure 14.17, three children have two pencils and three have nine pencils. This graph is bimodal.

The **range** of a set of observations is simply the numerical difference between the smallest and largest values. To introduce the concept of range, construct a graph of the height of each child in a class. Subtract the shortest value from the longest to compute the range (if the tallest child is four-feet six-inches and the shortest is four-feet one-inch, the range is five inches). Figure 14.18 shows the mean, median, mode, and range for a graph of the high temperature for each day in the month of June.

In general, if the range is relatively small, the mean is likely to be the best indicator of central tendency. As the range increases, the median or mode often gives

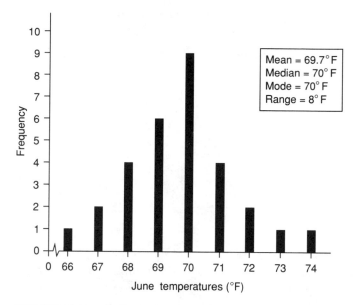

FIGURE 14.18 Daily High Temperature in June

a better indication of central tendency. For example, for the three sets of test scores listed in Table 14.2, the mean stays the same while the range and median change. Which measure seems to give the best indication of central tendency in each case? Notice that when the range is large, the mean does not represent the data as well as

TABLE 14.2 Test Scores with Same Mean and Different Median, Mode, and Range

	Test 1	Test 2	Test 3
	100	100	80
	100	100	80
	100	100	75
	100	80	75
	100	80	65
	100	80	65
	100	40	65
	0	40	65
	0	40	65
	0	40	65
Total	*700*	*700*	*700*
Mean	70	70	70
Median	100	80	65
Mode	100	40	65
Range	100	60	15

TABLE 14.3 Growth of Funds

Balance in Account at End of Year	Rate		
	5%	10%	15%
0	$100	$100	$100
1	105	110	115
2	110	120	130
3	115	130	145
4	120	140	160
5	125	150	175
6	130	160	190

the median or mode. Children should have several opportunities to work examples like these to help them develop an understanding of what each statistic descibes and its limitations.

> Construct a graph showing the following golf putting distances. Determine the mean, median, mode, and range for the set of measures. Which measure of central tendency best describes this set of data? (2m, 1m, 3m, 3m, 5m, 6m, 1m, 5m, 12m, 1m, 3m, 1m, 7m, 1m, 10m)

Slope. Graphs show how rapidly one variable changes in relation to a second variable. For example, to compare the effect of different interest rates on the value of a one hundred dollar bank deposit, children can construct a table showing the account balance at the end of each year. For simplicity, ignore any interest paid on previous years' earnings (i.e., no compounded interest). Table 14.3 shows the balance of the account at the end of each year for three different rates of interest: five percent, ten percent, and fifteen percent.

Plotting the three columns of account balances at the end of each year on the same graph shows the different rates of growth. Notice in Figure 14.19 that the vertical axis (balance) is interrupted with a broken line. This convention is used to indicate that a segment of the axis has been omitted to allow the graph to fit on the page. Remind children that when reading graphs, it is *always* good practice to check both axes to make sure they begin at 0 and to check whether the axes have a broken line to draw attention to any omission. Overlooking either of these factors might cause the reader to exaggerate the magnitude of the relationship displayed in the graph.

The steepness, or **slope**, of the line indicates that a fifteen percent rate of return generates the largest increases in balance and five percent the least. The slope is a measure of the amount of **rise** on the vertical axis for each unit of length on the horizontal axis. For example, the five percent line rises five units for each year on

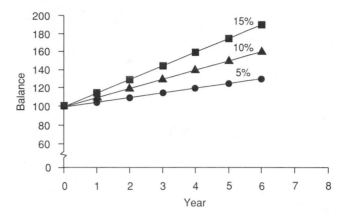

FIGURE 14.19 Growth of Funds

the horizontal axis. We say the five percent line has a slope of five. The ten percent line, therefore, has a slope of ten, and the fifteen percent line, a slope of fifteen. (Note: Banks frequently **compound** interest payments by including previous years' earnings in their computations. The slope of the graph increases more rapidly for each subsequent year when interest is compounded.)

> What would be the slope for twenty percent, fifty percent, and one hundred percent interest? Could the slope line for this problem ever be vertical? What percent return would be required? What would a zero percent graph look like? Could the line ever descend (hint—taxes!)?

EXPLORING PROBABILITY

Elementary probability concepts can be introduced in Grades 3–6. Grade 3–4 (and sometimes younger) children can make graphs of experiments with coins, dice, and spinners and informally predict the likelihood of specific events occurring in the future. Grade 5–6 children can be introduced to combinations and permutations, explore more complex probability experiments, and compare experimental and theoretical probabilities.

The probability that an event will occur is simply the ratio between the number of **desired outcomes** and the total number of **possible outcomes**. For example, the probability of heads showing when flipping a normal coin is 1:2, or $\frac{1}{2}$, because there is one desired outcome (a head) and two possible outcomes (head and tail).

To apply the concept of probability, children need a clear understanding of fractions, ratios, and proportions. Proportional reasoning is needed, for example, to understand that there is an equal likelihood of a red marble being drawn from each of the bags pictured in Figure 14.20.

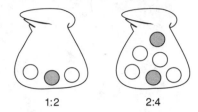

1:2   2:4

FIGURE 14.20 Ratio of Red Marbles to White Marbles

Children can be introduced to probability with carefully chosen experiments using familiar objects such as coins, dice, cards, spinners, marbles, and counters. Children compare the results of these experiments with the computed probability of each situation. These experiences will help children to understand that probability helps us predict the likelihood of the occurrence of some random event. In fact, it was just such an interest that prompted wealthy gamblers in the sixteenth and seventeenth centuries to support the research of prominent mathematicians to improve their understanding of games of chance.

Probability of an Event

Calculating the probability of an event requires two values:

☐ The number of ways a desired event can occur—favorable events (F)
☐ The total number of possible events—total events (T)

To be subject to the laws of probability, each event, or **outcome**, must have the same likelihood of occurring for each trial in an experiment. For example, to determine how likely it is that an ace will be drawn from a deck of cards, it is necessary to know

☐ the number of aces in the deck (desired events)
☐ the number of cards in the deck (total events)
☐ if the deck has been thoroughly shuffled (randomness of events)

When cards are selected one at a time from a standard deck, returned to the pack, and thoroughly shuffled after each draw, an average of 4 out of 52 cards drawn will be aces. Though it is highly unlikely, it is possible that the *same* ace will be drawn each time an ace appears.

The probability (P) of an event is defined as

$$\text{Probability} = \frac{\text{Favorable events}}{\text{Total events}} = \frac{F}{T}$$

A **certain** event has a probability of **1**, because the number of possible events equals the number of desired events (the probability of selecting a peanut from a jar containing n peanuts: $P = \frac{n}{n} = 1$). An **impossible** event has a probability of 0,

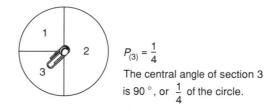

The central angle of section 3 is 90°, or $\frac{1}{4}$ of the circle.

$P_{(3)} = \frac{1}{4}$

FIGURE 14.21 Spinner

because there are zero desired events (the probability of catching a shark in a fresh water stream containing n fish: $P = \frac{0}{n} = 0$).

$$P_{(\text{certain})} = \frac{n}{n} = 1$$

$$P_{(\text{impossible})} = \frac{0}{n} = 0$$

The probability of a head showing when one coin is flipped is an example of two **equally likely** events.

$$P_{(\text{head})} = \frac{\text{Ways to show a head}}{\text{Possible events}} = \frac{1}{2}$$

Drawing an ace from a well-shuffled deck has a probability of

$$P_{(\text{ace})} = \frac{F}{T} = \frac{4}{52}$$

The probability of an even number showing on the roll of a standard die is

$$P_{(\text{even})} = \frac{3}{6}$$

The probability of getting the number three on the spinner in Figure 14.21 is found by determining the fractional part of the circle covered by the three-section. As this section is one fourth of the circle, the arrow should land on three, on the average, about one fourth of the time.

Compute the probability of selecting a white marble from the bag pictured in Figure 14.22. Do an experiment to verify your result.

Random Sampling

To determine the likelihood of a specific event, it is important to know if all events arise fairly. For an event to be **random**, each member of the set, or **sample space**, of all possible outcomes associated with a situation must have an *equal* opportunity to be selected. For a sample to be random, each event must have the same probability of being chosen for each selection, or **trial**.

FIGURE 14.22 Probability of Drawing a White Marble

Probability Experiments

Probability is introduced using experiments involving random samples. Flipping coins, rolling dice, using spinners, drawing from thoroughly shuffled cards, and selecting similar marbles of different colors from a bag all generate random samples.

Probability requires several random samples to be predictive. Sampling a single event gives little information about the likelihood of similar events occurring in the future. Suppose a coin is flipped only once, showing a head. This result indicates that a head is possible but does not indicate what is likely to occur on future flips. Normally, children conduct a random sample of thirty to one hundred trials to establish a reliable pattern of outcomes.

To test whether the computed, or **theoretical**, probability adequately predicts a pattern of outcomes, children should conduct experiments and compute experimental probabilities. An **experimental probability** is the ratio of the number of trials showing favorable events (F) and the total number of events (T):

$$P_{(\text{experimental})} = \frac{\text{Trials showing favorable events}}{\text{Total trials}} = \frac{F_{(\text{trials})}}{T_{(\text{trials})}}$$

As the number of trials gets large (thirty to one hundred or more), the experimental probability should approximate the theoretical probability. Have children carry out probability experiments such as flipping coins, drawing cards, tossing tacks, drawing marbles from a bag, and rolling dice. Have them keep a systematic record of their results in a table. Table 14.4 shows the results of flipping two coins fifty times.

TABLE 14.4 Flipping Two Coins (Heads/Tails)

H/H	H/T	T/H	T/T
/////	/////	/////	/////
/////	/////	/////	/////
//	/	////	///
12	11	14	13

The experimental and theoretical probability for *at least* one head showing can be written in decimal form to ease comparison.

$$P_{(experimental)} = \frac{12+11+14}{50} = \frac{37}{50} = 0.74$$

$$P_{(theoretical)} = \frac{3}{4} = 0.75$$

> Flip two coins one hundred times and record the results in a table. Compute the experimental probability that at least one head will show. Is the experimental probability result close to the theoretical probability? What is the experimental probability for at least one tail showing? What is the theoretical probability? How do the results compare with the experimental probability for exactly one head showing?

Fundamental Counting Principle

To calculate a probability, it is necessary to count the total number of possible outcomes. Several counting techniques can assist with this task. In a full deck of cards there are fifty-two possible events. Flipping one coin represents two possible events. There are six possible events associated with rolling a standard die.

When more than one object is involved, as when flipping two coins at the same time, counting the total number of *possible* events is more complex. The **fundamental counting principle** (FCP) is an efficient procedure for calculating the total number of possible outcomes for experiments that involve the combination of two or more independent events. The FCP states that, for two or more **independent** experiments (the outcome of one experiment is not influenced by the outcome of another), the total number of possible events for the combined experiment is the product of the number of events for each independent experiment. For example, when tossing two coins at the same time, each coin independently has two possible events (the outcome for one coin does not affect the outcome on the other). The FCP tells us that there are $2 \times 2 = 4$ possible outcomes. Table 14.5 shows the four possible outcomes when tossing two coins. Have children verify the FCP by listing the sample space (set of all possible outcomes) for rolling two dice ($6 \times 6 = 36$ outcomes).

TABLE 14.5 Fundamental Counting Principle

	Coin 1	Coin 2					
Event 1	H	H					
Event 2	H	T	Coin 1		Coin 2		Coin 1 and 2
Event 3	T	H	Events		Events		Events
Event 4	T	T	2	×	2	=	4

Many situations are not composed of independent events. For example, calculating the number of ways to deal two cards from a deck are **nonindependent** events. There are fifty-two possible first draws (events). After the first draw, there are fifty-one cards left from which to select the second card. The first event affects the probability of the second event. We can still apply the FCP to calculate the total number of ways two cards can be dealt from a deck. For nonindependent events, multiply the number of possibilities for the first event times the number of possibilities for the second event:

$$52 \times 51 = 2652 \text{ events}$$

Using the FCP, compute the number of possible events for each of the following experiments. Make a table showing the results.

1. Three coins flipped together.

2. Two dice rolled at the same time.

3. Three cards selected from a deck without replacing any cards as they are drawn.

Permutations

A second counting principle involves arrangements where *order matters*. Such arrangements are called **permutations**. Serious ice cream eaters know that the order of scoops on a cone makes a great deal of difference. A double-scooper with Jamoca Almond Fudge on top and German Chocolate Cake on the bottom is an entirely different epicurean experience than when the order is reversed. As shown in Figure 14.23, two-scoopers can be constructed in two ways, or permutations. For three-scoopers, there are three choices when selecting the first scoop, leaving two choices for the second and one choice for the third. Applying the FCP gives $3 \times 2 \times 1 = 6$ possible events. There are six permutations for three-scoop cones. (Note: For these

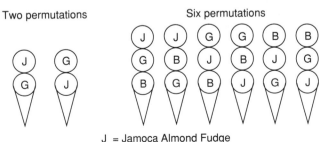

J = Jamoca Almond Fudge
G = German Chocolate Cake
B = Butterscotch Ripple

FIGURE 14.23

permutation problems, each scoop must be a different flavor. The "special-needs" ice cream eater wouldn't have it any other way!)

Four flavors of ice cream can be stacked $4 \times 3 \times 2 \times 1 = 24$ ways, or permutations. Note that, because order matters, each arrangement of a permutation is *unique* (the same flavors can occur in different arrangements, but each cone is considered different).

Calculating permutations involves finding the product of a sequence of counting numbers $(1, 2, 3, \ldots, n)$. This calculation is called a **factorial**. The previous ice cream examples can be written using the largest factor and the exclamation point ! to indicate a factorial calculation:

$$3! = 3 \times 2 \times 1 = 6$$
$$4! = 4 \times 3 \times 2 \times 1 = 24$$

Calculate the number of permutations for the following experiments.

1. Different types of ice cream cones with six scoops (all different flavors).

2. Number of ways a deck of seven different cards be arranged.

3. Number of ways eight race horses can be arranged in a starting gate.

4. Number of ways nine cars can line up at a toll booth.

Combinations

A third counting principle calculates the number of arrangements of two or more objects *without regard to order*. Such arrangements are called **combinations**. For example, suppose a chef has ten cookbooks but has room for only two at a time in a stove-top book rack. To determine the number of combinations, the FCP is first used to compute the total number of possible pairings. This result is divided by the number of ways each pair can be arranged, because the order in which the books are displayed does not matter to the chef.

Choosing two books from the same library are not *independent* events, as selecting the first book reduces the number of choices remaining for the second. In the chef's case, there are ten possible first choices and nine possible second choices. Applying the FCP gives $10 \times 9 = 90$ possible outcomes. There are two possible arrangements (the first book chosen can be placed on the left and the second on the right, or vice versa) and the chef is not concerned about their order, so there are $90 \div 2 = 45$ possible *combinations* of ten books chosen two at a time.

Using the procedure for calculating combinations, compute the number of ways four people can ride in pairs on a Ferris wheel. (Hint: How many choices does the first person have? The second person? and so on?)

Notice that for combinations and permutations, the same object is never used twice in a single arrangement (a book cannot be placed in the rack with itself; you do not shake hands with yourself; only amateurs eat two scoops of the same type of ice cream). The fundamental counting principle makes no such distinction (heads or tails can appear on both coins when flipping two coins). Selecting among these three counting methods depends on the application. Children must carefully consider the types of events and conditions of the experiment when selecting a counting method to help them work probability problems.

CLASSROOM ACTIVITIES

Opening Exercises

Daily graphing experiences arise naturally when updating daily attendance, lunch count, calendar, and weather information. A well-designed bulletin board display can turn the often tedious opening exercises into a valuable classroom learning experience for Grades K–3. In September, make a graph of everyone's birthday and leave it posted throughout the year as a reminder. Make attendance reports into a graphing activity by having children, as they arrive in the morning, turn their picture cards from the side showing their name to the side showing their picture. The children whose names are showing after class begins are absent. Record daily lunch and milk counts on a prepared monthly bar graph. Daily temperature, cloud cover, and precipitation conditions can also be displayed graphically. A monthly calendar and a time line showing the cumulative record of days from the beginning of school can be used to introduce the decade numbers, place value, and large numbers (the dates and time line can initially be recorded using base-10 materials cut from squared paper). The time line can also be used to post a memorable occasion for each day. A missing tooth, a new brother or sister, or a holiday can be noted on the daily time line as a concrete record of historical events and be reviewed at the end of the year (see Figure 8.3). Children also enjoy graphing their height and weight at the beginning of the year to compare with a graph at the year's end.

Surveys and Polls

A useful probability activity for elementary children is making a survey of people's opinions. Analyzing the results can help children make more objective decisions.

The public opinion **poll** has become an important tool in marketing and politics. For the results of a poll to be reliable, the persons surveyed must constitute a representative, random sample of the target population. Today, a national survey of about one thousand carefully selected individuals can indicate, with considerable accuracy, the probable outcome of an election, the most popular television programming, or the color preferences for new cars. These experiments generally produce results close to actual outcomes involving tens of millions of people.

Surveys differ from experiments involving coins and dice because an individual's personal preferences are unlikely to be random but, instead, are guided by some underlying value or belief. The task of carrying out a reliable survey requires selecting representative individuals whose responses are likely to agree with characteristic segments of the total population. If the population is small, one might simply poll everyone. However, as the population increases, it is much more efficient to survey a random sample within identifiable segments of the population.

Children might conduct a survey of food preferences by randomly selecting a sample of, say, three students from each grade level at a school and asking them to number the items on the school lunch menu in the order of their preference. Such a sample is said to be *stratified*, because subjects are selected from naturally occurring subcategories (grades) in the population. The results of this opinion poll can then be compared with the actual cafeteria lunch counts. A discussion of the results of the experiment could include opinions on the randomness of the sample, whether cafeteria sales really reflect student preference, and how these results might compare with the opinions of a second sample of students.

Charting the Market

Grade 4–6 children may enjoy trying their hand at investing in the stock market (Kelly, 1986). Detailed listings of thousands of stock prices are published daily in most local newspapers. Give each student an investment fund of one thousand dollars. Have students select a portfolio of five to ten stocks and record their purchases with their broker (the teacher). Once a week, supply a recent report of current prices from the newspaper and have the children compute the total current value of their portfolio. A line graph showing the performance of the portfolio should be maintained as well (see Figure 14.24). Have students write a quarterly financial report describing their investment performance.

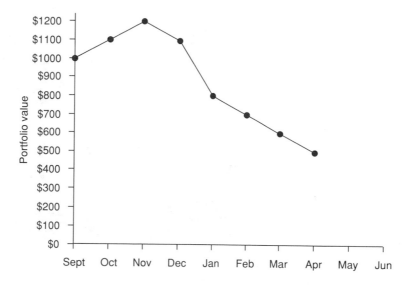

FIGURE 14.24 How Is My Portfolio Doing?

Letter Frequency

Which letter appears most frequently in written English? Grade 4–6 children can find out by selecting one paragraph from a textbook and making a bar graph of the frequency of each letter. Students' results can be compared to see if their experimental probabilities are similar (Figure 14.25).

The following is the list of letters in their order of frequency in written English:

E T A O N R I S H D L F C M U G Y P W B V K X J Q Z

The five most popular letters (E, T, A, O, and N) constitute over thirty-five percent of all letters used in English writing. The vowels make up about forty percent of all English words. The first nine letters in the frequency list constitute seventy percent of all the letters used in English words (Zim, 1975). Have children look at a commercial set of rub-on letters and see if the letters appear in proportions that agree with the class results. Compile the individual results in a class bar graph. Such a large sample should produce a reliable experimental probability.

Additional questions to explore include: How does the letter frequency in written English compare with that of other languages, such as Spanish? Does letter frequency vary among mathematics, basal readers, science, and social studies books? Have children construct class graphs to explore these questions. Have the class make up additional language-based probability experiments.

Tube Caps

Collect a class set of similar plastic caps from glue or toothpaste tubes. Have children compute the experimental probabilities of a cap landing on its big end, small end, and side by carrying out one hundred trials. Compare the children's results to see if the

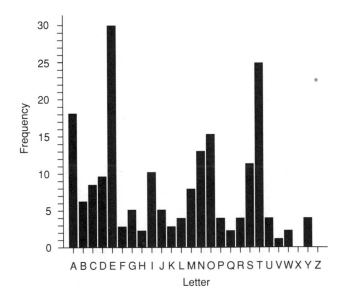

FIGURE 14.25 Which Are Popular Letters?

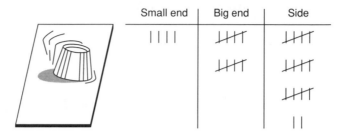

	Small end	Big end	Side
	∣∣∣∣	⫫⫫⫫	⫫⫫⫫
		⫫⫫⫫	⫫⫫⫫
			⫫⫫⫫
			∣∣

FIGURE 14.26   Tube Cap Probability

experimental probabilities are similar. Compare the results to the same experiment using thumb tacks or coins (see Figure 14.26).

Matching π

Children can approximate the number π (≈ 3.14) using a probability experiment. Trim ten match sticks or toothpicks to a length of five centimeters (or two inches). On a piece of unlined paper, draw a series of parallel lines five centimeters (or two inches) apart. Have each child lay the prepared sheet on a table and carefully drop ten sticks from a height of about twenty-five centimeters onto the paper, trying to keep the sticks within the edges of the paper. Have the children make a table recording the number of sticks touching a line (T) and the number lying completely between the lines (B) for thirty trials. Dividing the total number of sticks touching the line T by the number between the lines B gives a result approximately equal to π (Figure 14.27). Why do you think this is true?

Spinners and Quiet Dice

Spinners can easily be constructed from tagboard, paper fasteners, small washers, and paper clips. Constructed with various colored or numbered divisions, spinners can offer a range of experimental probability combinations (see Figure 14.28).

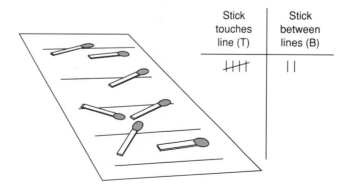

Stick touches line (T)	Stick between lines (B)
⫫⫫⫫	∣∣

FIGURE 14.27

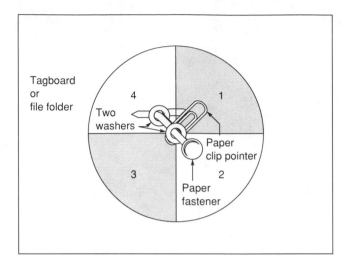

FIGURE 14.28 Plans for Constructing a Spinner

Quiet dice can be fashioned from thick sheets of foam rubber. Cut out cubes (equal edges) and use felt-tipped markers to write numbers or letters on the six faces. *Fair* dice with more faces (tetrahedrons, octohedrons, dodecahedrons, and icosahedrons) are available commercially (Figure 14.29).

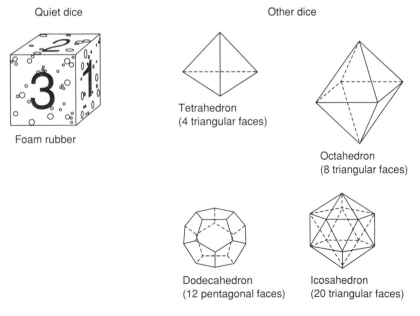

FIGURE 14.29

USING STATISTICS AND PROBABILITY TO SOLVE PROBLEMS

Popular Television Programs

To investigate the popularity of television programming, have each child list his or her favorite T.V. program. Construct a *class bar graph* showing the frequency of each program reported. Which is the most popular program (the mode)? Which programs are reported by only one student? To compute the experimental probability of the most popular program, divide the number of times the program was reported by the total number of students sampled. For the example shown in Figure 14.30, eight out of thirty-one students reported program *C*. The experimental probability that a similar class would select program *C* is

$$P_{(exp)} = \tfrac{8}{31} = 0.26$$

Have the class survey another class at the same grade level and compute the experimental probabilities for each program selected. How do the results compare with the previous example? If the experimental probabilities are similar, we say the results are *stable* across groups. Children can survey other grade levels to see if they get similar results. Is it likely that children of different ages will express different preferences? This experience points out the need to understand the population surveyed before generalizing results to other groups.

How Many Handshakes?

In chapter 4, we used a table to find out how many handshakes there would be in a party of ten persons. As shown in Figure 14.31, two persons require one hand-

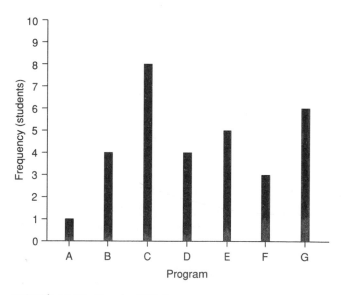

FIGURE 14.30 Popular T.V. Programs

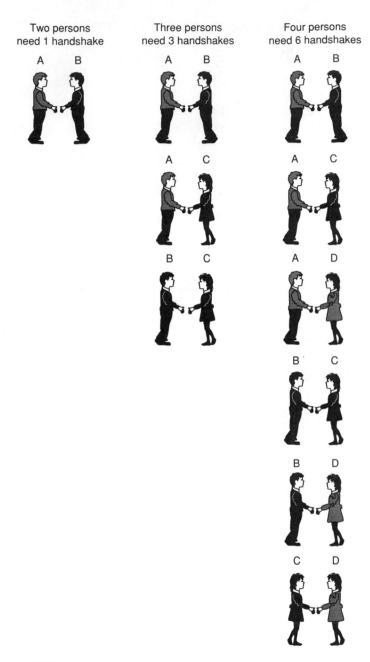

FIGURE 14.31 Handshake Problem

shake, three people require three handshakes, and four people require six handshakes.

For five people, there are five choices for the first person in the handshake couplet, leaving four choices for the second member. Using the fundamental counting

One three-person handshake

FIGURE 14.32   How Many Three-Person   Handshakes Would There Be for a Party of Ten People?

principle, there are 5 × 4 = 20 possible pairings. Since there are two equivalent ways for each pair to be arranged (first choice on the left and second choice on the right, or vice versa), the number of combinations (handshakes) is 20 ÷ 2 = 10. Using the rule for counting combinations, a formula can be written relating the number of people P and handshakes H:

$$\frac{P(P-1)}{2} = H$$

Therefore, ten people require

$$(10 \times 9) \div 2 = 45 \text{ handshakes}$$

Suppose, in some other culture, three persons are required for a proper handshake. For a party of ten, how many handshakes would be required for all possible groups of three to shake hands exactly once? Children can use the procedure for counting combinations. There are ten choices for the first member of the triplet, nine for the second, and eight for the third, giving 10 × 9 × 8 = 720 groupings. As there are six equivalent ways to arrange three persons (permutations of left, right, and middle positions), the number of handshakes required would be 720 ÷ 6 = 120 (see Figure 14.32). Children can try extending this problem to *four-way* handshakes.

Counting Fish

Wildlife biologists use probability to estimate the number of fish in a lake. First they net thirty or so fish, tag them with special paint, and return them to the lake. The next day they net a large number of fish and record the number of tagged and nontagged fish. A proportion can be written that relates the ratio of tagged fish netted to the number originally tagged with the ratio of the total number of netted fish to the original number of fish in the lake. Solving this proportion gives an estimate of the original number of fish in the lake. For example, suppose thirty tagged fish are

TABLE 14.6 How Many Fish in the Lake?

Tagged Fish	Unmarked Fish	Total Fish Netted
5	115	120

returned to the lake containing an unknown number of fish. The next day, 120 fish were netted, and five of them were tagged (see Table 14.6).

The following proportion can be written relating the ratio of tagged fish netted to the total number of tagged fish and the total number of fish netted to the total number of fish in the lake (F). Solving for F gives the estimated number of fish in the lake:

$$\frac{5}{30} = \frac{120}{F}$$

$$F = (30 \times 120) \div 5 = 720$$

Children can explore this procedure using goldfish crackers (Vissa, 1987). Put a box of goldfish crackers in a bowl and use a spoon to net several fish for tagging (dye them with green food coloring). Have the class estimate the fractional part of the total number of fish represented by the tagged fish. Return the tagged fish to the bowl, mix them thoroughly, and net a sample with a large spoon. Have the children record the results in a table and compute the proportion using a calculator. The process can be repeated several times and the results compared. Are the results always the same? Why not? Average the results of several trials to get a more stable estimate. Why do the fish have to be thoroughly mixed between trials? Could this explain why biologists wait a day after releasing the tagged fish before taking a sample?

> This estimation technique is used in medicine to determine the number of cancerous cells in patients by marking certain cells with small amounts of radioactivity. Wildlife biologists also use this technique to estimate populations of birds, rodents, insects, and other animals living in specific regions. Can you think of additional applications for this indirect counting technique?

CALCULATOR AND COMPUTER APPLICATIONS

Calculators

The development of statistics and probability concepts is often enhanced if children have access to calculators. The tedious calculations involved in computing means, decimal equivalents, probabilities, and proportions can divert attention from important concept development and problem-solving tasks. Ready access to a calculator allows children to focus more of their attention on the organization and interpretation of data while graphing probability experiments.

Probability Simulations and Tools

Computer simulations that model probability experiments can be helpful in extending concepts explored using manipulatives. Although it is possible to run an experiment of one thousand trials by dividing the task among the children in the class, a computer can simulate the experiment in a few seconds. Such software as *Probability* (MECC) allows students to flip simulated coins and roll simulated dice, predict outcomes, organize data, and compute experimental probabilities (Figure 14.33). By doing the same experiment several times using different numbers of trials, students gain understanding of the need for a large number of random samples in order for the laws of probability to be predictive. Students can make and test conjectures more efficiently using computer simulations.

Another software application useful for studying statistics and probability is a *graphing tool*. Examples of easy to use graphing tools are *Exploring Tables and Graphs* (Weekly Reader Family Software) and *MECC Graph* (MECC). These tools allow students to enter data and automatically produce high quality bar, line, or pie graphs (see Figure 14.34). Students design and scale the two axes and enter labels to ensure that the graph communicates information accurately. Scaling or labeling errors can be identified and corrected. For any given data set, several different graphic representations can be printed. This feature allows students to select graphs that best communicate the results of their statistics and probability experiments.

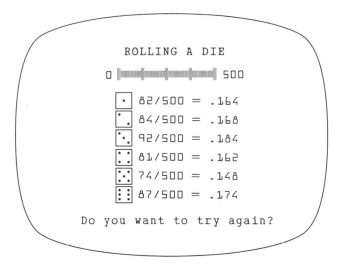

FIGURE 14.33 Probability. *Note:* From "Probability" [Computer program] by MECC, 1983. Copyright 1983 by MECC. Reprinted by permission.

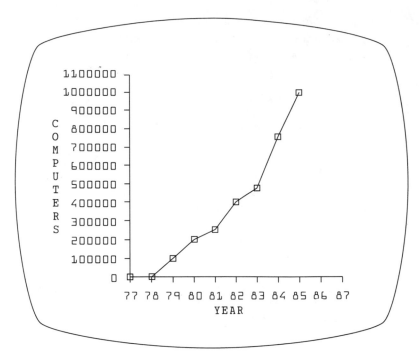

FIGURE 14.34 How Many Computers?

ADAPTING INSTRUCTION FOR CHILDREN WITH SPECIAL NEEDS

The graphing and probability concepts introduced in this chapter are important learning objectives for all elementary children. Children with **learning disabilities**, at a minimum, need to be able to construct and interpret simple **bar graphs** and **find averages**. Provide prepared axes for graphs drawn on one-inch squared paper, a supply of one-inch colored cubes, and like-colored felt pens to help such children construct bar graphs. Children first construct the graph (e.g., favorite pets) on the prepared graph using blocks with a different color for each bar. Each bar is then reproduced using a like-colored felt pen.

The topics introduced in this chapter are particularly appropriate as enrichment activities for **gifted and talented children**. Selecting appropriate measures of central tendency, interpreting slopes of graphs, calculating probabilities, and solving proportions are powerful mathematical ideas that have wide application in solving practical problems. The study of permutations and combinations offers an interesting introduction to number theory for mathematically mature Grade 5–6 children. Problems that can only be solved by using the laws of probability extend children's ability to apply proportional thinking. Many of the references listed at the end of this chapter include enrichment activities especially appropriate for gifted and talented children.

SUMMARY

Statistics and probability are mathematics topics that deserve more attention in the elementary curriculum. Designing and interpreting graphs can be integrated into all strands of mathematics. Objects can be arranged in lines starting with a common baseline to create an empirical graph. Representational bar graphs, pie graphs, and line graphs are used to display information graphically and help to solve problems. Coordinate graphs can display the relationship between two variables. The slope of a line graphed on the coordinate plane is a measure of how rapidly one variable changes relative to the other. Calculating the slope of a graph can be useful in solving some types of problems. The measures of central tendency—the mean, median, and mode—are statistics that are used to summarize large amounts of information. The theoretical probability of an event is the ratio of desired events to the total number of possible events. Independent events do not affect each other's chances of occurring, while dependent events do affect each other's chances of occurring. The fundamental counting principle states that, for two or more independent events, the total number of possible outcomes is the product of the number of outcomes for each separate event. Special applications of this principle make it possible to count the number of combinations (groupings where order does not matter) and permutations (groupings where order matters). These counting techniques are useful for computing theoretical probabilities. The experimental probability of an event is calculated from the ratio of the number of times a desired event occurs to the total number of events. If the sample is random and sufficient trials are recorded (thirty to one hundred), the experimental probability will approximate the theoretical probability. The laws of probability can be used to solve many nonroutine problems. While special-needs children should, at a minimum, learn how to interpret graphs and compute means, statistics and probability offer opportunities for instruction that are particularly appropriate for gifted and talented children.

COURSE ACTIVITIES

1. Make a list of twenty objects and events that could be used for primary level empirical graphing experiences. Try a class graphing activity with a group of children or peers.
2. Figure 14.35 shows a graph of the probability of two persons in various sized groups having the same birthday (date only, not year). In a class of thirty students, how likely is it that two will have the same birthday? How large a·group is needed to have a fifty percent chance that two birthdays will coincide? How many people would have to be present to be certain ($P = 1$) that two birthdays would coincide? Try this experiment

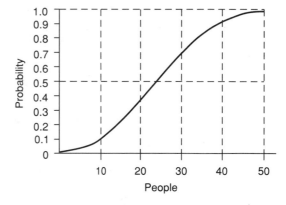

FIGURE 14.35 Chances of Having the Same Birthday

TABLE 14.7 Pascal's Triangle

$$
\begin{array}{rcl}
1 + 1 = 2 \text{ events} & \rightarrow & 1 \quad\quad 1 \leftarrow \text{ one coin} \\
1 + 2 + 1 = 4 \text{ events} & \rightarrow & 1 \quad 2 \quad 1 \leftarrow \text{ two coins} \\
1 + 3 + 3 + 1 = 8 \text{ events} & \rightarrow & 1 \quad 3 \quad 3 \quad 1 \leftarrow \text{ three coins} \\
1 + 4 + 6 + 4 + 1 = 16 \text{ events} & \rightarrow & 1 \quad 4 \quad 6 \quad 4 \quad 1 \leftarrow \text{ four coins} \\
32 \text{ events} & \rightarrow & 1 \quad 5 \quad 10 \quad 10 \quad 5 \quad 1 \leftarrow \text{ five coins} \\
64 \text{ events} & \rightarrow & 1 \quad 6 \quad 15 \quad 20 \quad 15 \quad 6 \quad 1 \leftarrow \text{ six coins}
\end{array}
$$

with your peers and, if possible, with a class of children.

3. Pascal's Triangle is a useful tool for computing probabilities of situations such as flipping coins. The first row in Table 14.7 represents the two events associated with flipping one coin (H,T). The second row shows the four possible events from flipping two coins (HH, HT, TH, TT).

Using Pascal's Triangle, compute the probabilities of getting at least one head when flipping one coin, two coins, three coins, four coins, five coins, and six coins. To calculate the numerator of each probability, note how many arrangements contain at least one head and look for a pattern in Pascal's Triangle. Enter the results in Table 14.8.

4. The sum of the possible values showing on two dice ranges from two to twelve. Con-duct an experiment of fifty trials using two dice and construct the bar graph of the sums as shown in Figure 14.36. Compute the experimental probabilities for each sum two through twelve. Which sum is most likely to occur (the mode)? Which is least likely to occur? Try the same activity with commercially available eight-, twelve-, or twenty-sided dice.

5. Invent a board game for children that requires an understanding of simple probability to win. Implement the game with a group of peers and, if possible, students in an actual classroom setting. Evaluate the effectiveness of the game as a teaching aid. Swap game ideas with your peers.

6. Include examples of statistics and probability activities in your idea file.

TABLE 14.8 Probability of at Least One Head

Coins	Probability
1	$P_{(\text{head})} = \frac{1}{(1+1)} = \frac{1}{2}$
2	$P_{(\text{head})} = \frac{(1+2)}{(1+2+1)} = \frac{3}{4}$
3	$P_{(\text{head})} = ?$
4	$P_{(\text{head})} = ?$
5	$P_{(\text{head})} = ?$
6	$P_{(\text{head})} = ?$

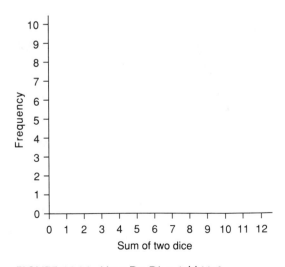

FIGURE 14.36 How Do Dice Add Up?

7. Read one of the *Arithmetic Teacher* articles listed in the reference section. Write a brief report summarizing the main ideas of the article and describe how the recommendations for instruction might apply to your own mathematics teaching.

MICROCOMPUTER SOFTWARE

AppleWorks Flexible spreadsheet, data base, and word processing tool integrated into a single, easy to learn package (Apple Computer Corporation).

Exploring Tables and Graphs: Levels 1 and 2 Two programs that teach children how to construct and interpret tables and graphs (Weekly Reader Family Software).

The Graphics Department Makes it easy to generate bar, line, and pie graphs from student-entered data (Sensible Software, Inc.).

MECC Graphing Primer Tutorial that provides instruction in analyzing bar, line, and pie graphs (MECC).

MECC Graph Easy to use graph generator *(MECC)*.

Probability Simulates tossing coins and rolling dice as probability experiments (MECC).

REFERENCES AND READINGS

Bruni, J., & Silverman, H. (1986). Developing concepts in probability and statistics—and much more. *Arithmetic Teacher, 33*(6), 34–37.

Burns, M. (1983). Put some probability in your classroom. *Arithmetic Teacher, 30*(7), 21–22.

Christopher, L. (1982). Graphs can jazz up the mathematics curriculum. *Arithmetic Teacher, 30*(1), 28–30.

Hoffer, A. (1978). *Statistics and information organization*. Palo Alto, CA: Creative Publications.

Horak, V., & Horak, W. (1983). Take a chance. *Arithmetic Teacher, 30*(9), 8–15.

Jacobs, H. (1970). *Mathematics: A human endeavor*. San Francisco: Freeman.

Johnson, E. (1981). Bar graphs for first graders. *Arithmetic Teacher, 29*(4), 30–31.

Kelly, M. (1986). Elementary school activity: Graphing the stock market. *Arithmetic Teacher, 33*(7), 17–20.

Nibbelink, W. (1982). Graphing for any grade. *Arithmetic Teacher, 30*(3). 28–31.

O'Neil, D., & Jensen, R. (1982). Looking at facts. *Arithmetic Teacher, 29*(8), 12–15.

Shaw, J. (1984). Dealing with data. *Arithmetic Teacher, 31*(9), 9–15.

Shulte, A. (1979). A case for statistics. *Arithmetic Teacher, 26*(6), 24.

Shulte, A. (Ed.). (1981). *Teaching statistics and probability: 1981 Yearbook*. Reston, VA: National Council of Teachers of Mathematics.

Shulte, A., & Choate, S. (1977). *What are my chances? Book A and Book B*. Palo Alto, CA: Creative Publications.

Silvey, L. (1978). *Polyhedra dice games for grades 5 to 10*. Palo Alto, CA: Creative Publications.

Silvey, L. & Smart, J. (Eds.). (1982). *Mathematics for the middle grades (5–9):1982 Yearbook*. Reston, VA: National Council of Teachers of Mathematics.

Slaughter, J. (1983). The graph examined. *Arithmetic Teacher, 30*(7), 41–45.

Souviney, R. (1976). Probability and statistics. *Learning Magazine, 5*(4), 51–52.

Souviney, R. (1977). Quantifying chance. *Arithmetic Teacher, 25*(3) 24–26.

Souviney, R. (1986). Problem solving tips for teachers: Conducting experiments. *Arithmetic Teacher*, (*33*)6, 56–57.

Vissa, J. (1987). Sampling treats from a school of fish. *Arithmetic Teacher, 34*(7), 36–37.

Woodward, E. (1983). A second-grade probability and graphing lesson. *Arithmetic Teacher, 30*(7), 23–24.

Zim, H. (1975). *Codes and secret writing*. New York: Scholastic.

Appendix:
Sample Elementary
Mathematics
Instructional Objectives

The following sample list of basic elementary mathematics instructional objectives is organized by topic and grade level. Though lists of performance objectives are by necessity incomplete, they do help with the sequencing of lessons. This continuum of objectives should be used as a resource for planning lessons rather than as a rigid guide and should be adapted to meet district requirements and the needs of individual students.

GEOMETRY

Kindergarten

1. The student will be able to identify circles, triangles, and squares.
2. The student will demonstrate an understanding of the common usage of the following geometric terms:

above	bottom	same
after	circle	shape
alike	different	square
before	down	top
below	left	triangle
beside	over	under
between	right	up

3. The student will be able to locate objects using positional words such as *bottom, above, below, up, between, down, on, under, inside, outside, beside, over, after, right,* and *left.*
4. The student will be able to participate in the group solution of one-step, graphic geometry story problems using role playing and materials.

Grade One

5. The student will be able to identify circles, triangles, and rectangles.
6. The student will be able to identify open and closed curves.
7. The student will be able to identify examples of a cube, sphere, and cylinder.
8. The student will be able to demonstrate an understanding of the common usage of the following geometric terms:

circle	cylinder	shape
closed curve	open curve	sphere
cube	rectangle	

9. The student will be able to distinguish between curved and straight paths.
10. The student will be able to solve one-step, graphic geometry story problems using materials and record the results by drawing a picture or constructing a graph.

Grade Two

11. The student will be able to identify the features of triangles, diamonds, rectangles, and squares, including vertex and side.
12. The student will be able to identify examples of rectangular solids, cubes, and pyramids.
13. The student will be able to solve one-step, graphic geometry story problems and record the results using pictures, graphs, and number sentences.

Grade Three

14. The student will be able to identify features of a circle including the diameter, center, and radius.
15. The student will be able to identify features of a cube including the vertex, edge, and face.
16. When given a coordinate pair, the student will be able to plot one or more points on a coordinate plane.
17. Working in a small group, the student will be able to apply the four-step problem plan employing one solution strategy (guess-and-test, conduct an experiment using materials, draw a sketch, make a graph, find a pattern) to solve simple geometry process problems.

Grade Four

18. The student will be able to identify simple geometric shapes (circles, triangles, quadrilaterals).

19. The student will be able to identify simple geometric solids (box, sphere, and cylinder) and their properties.

20. The student will demonstrate an understanding of the common usage of the following geometric terms:

circle	diameter	rectangle	square
cylinder	geometric figure	sphere	triangle

21. Working in a small group, the student will be able to apply the four-step problem plan employing one or more solution strategies (guess-and-test, work backward, conduct an experiment, draw a sketch, make a graph, find a pattern) to solve simple geometry process problems.

Grade Five

22. The student will be able to identify lines of symmetry in various geometric shapes (circles, triangles, quadrilaterals).

23. The student will be able to identify the common geometric features of polygons, including angle, side, diagonal, radius, and perimeter.

24. The student will be able to make a linear graph of ordered pairs on the coordinate plane from a table of (x,y) values.

25. The student will demonstrate an understanding of the common usage of the following geometry terms:

angle	diagonal	plane	region
circle	diameter	polygon	side
cone	geometric figure	quadrilateral	sphere
coordinate plane	line of symmetry	radius	symmetry
cylinder	ordered pair	rectangle	triangle

26. The student will be able to classify obtuse, acute, and right angles.

27. Working in a small group, the student will be able to apply the four-step problem plan employing one or more solution strategies (guess-and-test, work backward, solve a simpler case, conduct an experiment, use a table, find a pattern, construct a general rule) to solve geometry process problems.

Grade Six

28. The student will be able to identify geometric solids (prisms, pyramids, cylinders, cones, spheres), their parts (faces, edges, corners), and relationships between these features.

29. The student will be able to draw and identify the common geometric properties of points, lines, line segments, parallel lines, rays, and angles.
30. The student will be able to identify shapes, line segments, and angles that are congruent or have similar (proportional) features.
31. The student will be able to classify right, isosceles, and equilateral triangles.
32. The student will be able to compute the sum of the interior angles of a polygon and find the missing angle when all the remaining angles are given in a triangle.
33. The student will be able to carry out common geometric constructions using a straight edge and compass.
34. Working in a small group, the student will be able to apply the four-step problem plan employing one or more solution strategies (guess-and-test, work backward, solve a simpler case, conduct an experiment, use a table, find a pattern, construct a general rule, add elements to the problem situation) to solve geometry process problems.

MEASUREMENT

Kindergarten

1. The student will be able to match a nickel with five pennies.
2. The student will be able to compare similar objects to determine which are longer, larger, shorter, smaller, or taller.
3. The student will be able to order three similar objects according to their length.
4. The student will demonstrate an understanding of the common usage of the following measurement terms:

five cents	nickel	smaller
larger	penny	taller
longer	shorter	tallest
longest	shortest	

5. The student will be able to participate in the group solution of one-step, graphic measurement story problems using role playing and materials.

Grade One

6. The student will form sets of pennies, nickels, and dimes equivalent to twenty-five cents.
7. The student will be able to demonstrate understanding of comparative terms such as *longer, longest, shorter, shortest, taller, tallest, largest, smallest,* and the *same length* or *same size as* by identifying objects or pictures of objects that fit each description.

8. The student will have experience measuring length using nonstandard or traditional units.
9. The student will be able to tell time to the nearest hour using a digital and analog clock.
10. The student will demonstrate an understanding of the common usage of the following measurement terms:

dime	longer	shortest	taller
hour	longest	size	tallest
larger	quarter	smaller	unit
length	shorter	smallest	week

11. The student will be able to solve one-step, graphic measurement story problems using materials and record the results by drawing a picture or constructing a graph.

Grade Two

12. The student will form sets of pennies, nickels, dimes, and quarters equivalent to fifty cents.
13. The student will be able to measure length using nonstandard or traditional units and meters and/or feet and yards.
14. The student will have experience comparing the weights of objects using a balance scale.
15. The student will be able to tell time to the nearest half hour using a digital and analog clock.
16. The student will demonstrate an understanding of the common usage of the following measurement terms:

coins	half-hour	meter (m)	unit
dollar ($)	hour (hr)	pattern	week (wk)
foot (ft)	longer	shorter	weight
graph	longest	shortest	yard (yd)

17. The student will be able to solve one-step, graphic measurement story problems and record the results using pictures, graphs, and number sentences.

Grade Three

18. The student will be able to measure length to the nearest centimeter and inch.
19. The student will be able to tell time to the nearest minute using a digital and analog clock.
20. The student will be able to count a collection of coins and notes up to five dollars.

21. The student will demonstrate an understanding of the common usage of the following measurement terms:

centimeter (cm)
coins
degrees Celsius (°C)
degrees Fahrenheit (°F)
foot (ft)
gram (g)

inch (in)
meter (m)
same measure
temperature
weight
yard (yd)

22. Working in a small group, the student will be able to apply the four-step problem plan employing one solution strategy (guess-and-test, conduct an experiment using materials, draw a sketch, make a graph, find a pattern) to solve simple measurement process problems.

Grade Four

23. The student will be able to count a collection of coins and notes up to ten dollars.
24. The student will be able to count change for items less than one dollar.
25. The student will have experience measuring the weight (mass) of the objects in the environment using nonstandard and kilogram units.
26. The student will be able to measure area by counting square centimeter units and measure perimeter by counting the distance in units around rectangular figures.
27. The student will have experience finding the volume of rectangular prisms using various nonstandard and metric units (seeds, shells, blocks, cubic centimeters).
28. The student will have experience determining the capacity of common containers using liters.
29. The student will demonstrate an understanding of the common usage of the following measurement terms:

area
capacity
centimeter (cm)
gram (g)
kilogram (kg)
kilometer (km)

liter (L)
mass
meter (m)
mile (mi)
millimeter (mm)
perimeter

square centimeter (cm^2)
square foot (ft^2)
square inch (in^2)
square meter (m^2)
square unit
weight

30. Working in a small group, the student will be able to apply the four-step problem plan employing one or more solution strategies (guess-and-test, work backward, conduct an experiment, draw a sketch, make a graph, find a pattern) to solve simple measurement process problems.

Grade Five

31. The student will be able to measure length to the nearest centimeter and millimeter and/or inch and quarter-inch.

32. The student will be able to determine the area of a rectangle in square centimeter and/or square inch units and the perimeter in centimeters and/or inches.
33. The student will have experience finding the volume of common rectangular prisms using nonstandard and standard cubic units (cubic centimeters and/or cubic inches).
34. The student will have experience determining the capacity of common containers in liters and milliliters and/or quarts, pints, and cups.
35. The student will have experience determining the weight (mass) of objects from the environment in kilograms and grams and/or pounds and ounces.
36. The student will demonstrate an understanding of the common usage of the following measurement terms:

area	kilograms (kg)	quarts (qt)
area of a square	mass	second (s)
capacity	meter (m)	square centimeter (cm²)
centimeters (cm)	millimeter (mm)	square inch (in²)
cups (c)	minute (m)	square meter (m²)
degrees Celsius (°C)	ounces (oz)	temperature
degrees Fahrenheit (°F)	perimeter	volume
gram (g)	pints (p)	weight
hours (hr)	pounds (lb)	

37. Working in a small group, the student will be able to apply the four-step problem plan employing one or more solution strategies (guess-and-test, work backward, solve a simpler case, conduct an experiment, use a table, find a pattern, construct a general rule) to solve measurement process problems.

Grade Six

38. The student will be able to measure perimeter in inches, centimeters, meters, feet, and yards.
39. The student will be able to determine the area of a parallelogram or right triangle in square units.
40. The student will have experience finding the volume of rectangular prisms and will be able to calculate the volume in common and metric units.
41. The student will have experience estimating and measuring the capacity of containers and comparing the estimates with the actual measures.
42. The student will have experience estimating the weights of objects, weighing the objects in kilograms and grams and/or pounds and ounces, and comparing the estimates with the actual weights.
43. Working in a small group, the student will be able to apply the four-step problem plan employing one or more solution strategies (guess-and-test, work backward, solve a simpler case, conduct an experiment, use a table, find a pattern, construct a general rule, add elements to the problem situation) to solve measurement process problems.

NUMBER

Kindergarten

1. The student will be able to match zero to ten objects in one-to-one correspondence.
2. The student will be able to match objects by one-to-one correspondence to identify which of two sets with zero to ten members has more, fewer, or the same number of members.
3. The student will be able to count from zero to ten objects and sequence sets from zero to ten, matching with the correct numeral.
4. The student will be able to count sets of from zero to ten members and write the correct numeral symbol and words.
5. The student will demonstrate an understanding of the common usage of the following number and operation terms:

fewer	one, two, . . . , ten
more	same
most	set
number	

6. The student will demonstrate classification skills by grouping objects in terms of a common attribute.
7. The student will be able to describe objects by telling how they are alike and how they are different.
8. The student will be able to participate in the group solution of one-step, graphic number story problems using role playing and materials.

Grade One

9. The student will be able to count sets of objects to ten, match the correct numeral to the set, and write and sequence numerals from zero to ten.
10. The student will be able to identify which of two sets (zero to ninety-nine) has more, fewer, or the same number of members.
11. The student will demonstrate understanding of regrouping ones and tens using base-10 materials and coins.
12. The student will be able to represent a two-digit number using base-10 materials and coins.
13. The student will be able to count, read, write, and sequence numerals from zero to ninety-nine.
14. The student will be able to write a standard numeral for a given number of tens and ones and identify the number of tens and ones for any number from zero to ninety-nine.
15. The student will be able to demonstrate classification skills by grouping objects and pictures in terms of a common attribute.
16. The student will be able to solve one-step, graphic number story problems using materials and record the results by drawing a picture or constructing a graph.

Grade Two

17. The student will be able to name the ordinal numbers, first to tenth, by pointing to objects in a line.

18. The student will be able to write any two-digit number represented by base-10 materials and coins and vice versa.

19. The student will be able to count, read, write, sequence, and identify the place value of any numeral from zero through ninety-nine.

20. The student will be able to write any three-digit number represented by base-10 materials and coins and vice versa.

21. The student will demonstrate understanding of regrouping ones, tens, and hundreds using base-10 materials and coins.

22. The student will be able to count, write, and sequence numerals from zero to nine hundred and ninety-nine.

23. The student will be able to demonstrate classification skills by grouping objects and pictures according to two attributes ("and" (both)—intersection or conjunction; "or" (either or both)—union or disjunction).

24. The student will be able to solve one-step, graphic number story problems and record the results using pictures, graphs, and number sentences.

Grade Three

25. The student will be able to represent concretely, using base-10 materials and coins, count, read, write, sequence, and identify the place value of any numeral from 0 to 9999.

26. Working in a small group, the student will be able to apply the four-step problem plan employing one solution strategy (guess-and-test, conduct an experiment using materials, draw a sketch, make a graph, find a pattern) to solve simple number process problems.

Grade Four

27. The student will be able to count, read, write, sequence, and identify the place value of any numeral from 0 to 999,999.

28. Working in a small group, the student will be able to apply the four-step problem plan employing one or more solution strategies (guess-and-test, work backward, conduct an experiment, draw a sketch, make a graph, find a pattern) to solve simple number process problems.

Grade Five

29. The student will know the common properties of numbers and be able to read, write, sequence, and identify the place value of any numeral to the hundred millions place.

30. Working in a small group, the student will be able to apply the four-step problem plan employing one or more solution strategies (guess-and-test, work backward, solve a simpler case, conduct an experiment, use a table, find a pattern, construct a general rule) to solve number process problems.

Grade Six

31. The student will know common properties of numbers and be able to read, write, sequence, and identify the place value of any numeral through the billions place.
32. The student will be able to identify prime numbers (less than 100).
33. Working in a small group, the student will be able to apply the four-step problem plan employing one or more solution strategies (guess-and-test, work backward, solve a simpler case, conduct an experiment, use a table, find a pattern, construct a general rule, add elements to the problem situation) to solve number process problems.

ADDITION

Kindergarten

1. The student will be able to demonstrate the concept of addition by joining two sets whose sum is six or less.
2. The student will be able to participate in the group solution of one-step, graphic addition story problems involving addition using role playing and materials.

Grade One

3. The student will be able to demonstrate concretely and record addition with a set of base-10 materials, and then with coins, for sums of ten or less.
4. The student will be able to add concretely any two whole numbers whose sum is five or less and record the results in place value tables.
5. The student will be able to add concretely any two whole numbers with sums to ten, twelve, and eighteen, and record the results in place value tables (addition facts).
6. The student will be able to solve one-step, graphic addition story problems involving addition using materials and record the results by drawing a picture or constructing a graph.

Grade Two

7. The student will be able to add mentally any two one-digit whole numbers.
8. The student will be able to add concretely three or more addends whose sum is eighteen or less and record the results in place value tables.

9. The student will be able to add concretely any two whole numbers less than 100, regrouping as necessary, and record the results in place value tables.
10. The student will be able to solve one-step, graphic story problems involving addition and record the results using pictures, graphs, and number sentences.

Grade Three

11. The student will be able to add mentally up to three one-digit whole numbers whose sum is eighteen or less (addition facts mastery).
12. The student will be able to add up to three whole numbers less than 100 in place value tables, regrouping as necessary.
13. The student will be able to add concretely two whole numbers less than 1000, regrouping as necessary, and record the results in place value tables.
14. The student will be able to solve simple story problems involving addition of whole numbers and money.
15. Working in a small group, the student will be able to apply the four-step problem plan employing one solution strategy (guess-and-test, conduct an experiment using materials, draw a sketch, make a graph, find a pattern) to solve simple addition process problems.

Grade Four

16. The student will be able to add two whole numbers less than 10,000, regrouping as necessary, with and without place value tables.
17. The student will be able to solve story problems involving addition.
18. Working in a small group, the student will be able to apply the four-step problem plan employing one or more solution strategies (guess-and-test, work backward, conduct an experiment, draw a sketch, make a graph, find a pattern) to solve simple addition process problems.

Grade Five

19. The student will be able to add three whole numbers less than 100,000, regrouping as necessary, without place value tables.
20. The student will be able to solve story problems involving addition.
21. Working in a small group, the student will be able to apply the four-step problem plan employing one or more solution strategies (guess-and-test, work backward, solve a simpler case, conduct an experiment, use a table, find a pattern, construct a general rule) to solve addition process problems.

Grade Six

22. The student will be able to add four whole numbers, regrouping as necessary, using the standard addition algorithm.

23. The student will be able to solve multistep story problems involving addition.
24. Working in a small group, the student will be able to apply the four-step problem plan employing one or more solution strategies (guess-and-test, work backward, solve a simpler case, conduct an experiment, use a table, find a pattern, construct a general rule, add elements to the problem situation) to solve addition process problems.

SUBTRACTION

Kindergarten

1. The student will be able to demonstrate the concept of subtraction by removing objects from a set of six or less.
2. The student will be able to participate in the group solution of one-step, graphic story problems involving addition or subtraction using role playing and materials.

Grade One

3. The student will be able to demonstrate concretely and record subtraction with sets of base-10 materials, and coins, with minuends of ten or less.
4. The student will be able to subtract concretely two whole numbers with minuends and subtrahends of five or less and record the results in place value tables.
5. The student will be able to subtract concretely any two whole numbers with minuends to ten, twelve, and eighteen and differences of zero to nine, and record the results in place value tables (subtraction facts).
6. The student will be able to solve one-step, graphic story problems involving addition or subtraction using materials and to record the results by drawing a picture or constructing a graph.

Grade Two

7. The student will be able to subtract any whole numbers with minuends of eighteen or less using place value tables.
8. The student will be able to subtract concretely two whole numbers less than 100, regrouping as necessary, and to record the results in place value tables.
9. The student will be able to solve simple graphic story problems involving counting, addition, subtraction, length, time, or money.
10. The student will be able to solve one-step, graphic story problems involving addition or subtraction problems and to record the results using pictures, graphs, and number sentences.

Grade Three

11. The student will be able to subtract mentally any two whole numbers with minuends of eighteen or less (subtraction facts mastery).
12. The student will be able to subtract two whole numbers less than 100 in place value tables, regrouping as necessary.
13. The student will be able to subtract concretely two whole numbers less than 1000, regrouping as necessary and recording the results in place value tables.
14. The student will be able to solve simple story problems involving addition and subtraction of whole numbers.
15. Working in a small group, the student will be able to apply the four-step problem plan employing one solution strategy (guess-and-test, conduct an experiment using materials, draw a sketch, make a graph, find a pattern) to solve simple subtraction process problems.

Grade Four

16. The student will be able to subtract two whole numbers less than 10,000, regrouping as necessary, with and without place value tables.
17. The student will be able to solve story problems involving both addition and subtraction using whole numbers and money.
18. Working in a small group, the student will be able to apply the four-step problem plan employing one or more solution strategies (guess-and-test, work backward, conduct an experiment, draw a sketch, make a graph, find a pattern) to solve simple subtraction process problems.

Grade Five

19. The student will be able to subtract any two numbers less than 100,000, regrouping as necessary, with and without place value tables.
20. The student will be able to solve story problems involving both addition and subtraction.
21. Working in a small group, the student will be able to apply the four-step problem plan employing one or more solution strategies (guess-and-test, work backward, solve a simpler case, conduct an experiment, use a table, find a pattern, construct a general rule) to solve subtraction process problems.

Grade Six

22. The student will be able to subtract any two whole numbers using the standard subtraction algorithm.
23. The student will be able to solve story problems involving addition and subtraction.

24. Working in a small group, the student will be able to apply the four-step problem plan employing one or more solution strategies (guess-and-test, work backward, solve a simpler case, conduct an experiment, use a table, find a pattern, construct a general rule, add elements to the problem situation) to solve subtraction process problems.

MULTIPLICATION

Kindergarten

1. The student will be able to group objects in pairs.
2. The student will be able to count by twos to ten.
3. The student will be able to participate in the group solution of one-step, graphic story problems involving counting by twos using role playing and materials.

Grade One

4. The student will be able to group objects by groups of three and five.
5. The student will be able to count by threes to eighteen and fives to twenty-five.
6. The student will be able to solve one-step, graphic story problems involving addition, subtraction, or multiplication using materials and to record the results by drawing a picture or constructing a graph.

Grade Two

7. The student will be able to show concretely products to eighteen using arrays and repeated addition.
8. The student will be able to solve one-step, graphic story problems involving addition, subtraction, and multiplication and to record the results using pictures, graphs, and number sentences.

Grade Three

9. The student will be able to demonstrate concretely products to twenty-five using the repeated addition model of multiplication.
10. The student will be able to demonstrate concretely, with base-10 materials, products to twenty-five using the array model of multiplication.
11. The student will be able to multiply concretely two numbers with one factor less than six and the other less than ten and to record the results in place value tables.

12. The student will be able to multiply concretely two numbers less than ten and to record the results in place value tables (multiplication facts).
13. The student will be able to complete the (10 × 10) multiplication table with eighty percent accuracy.
14. Working in a small group, the student will be able to apply the four-step problem plan employing one solution strategy (guess-and-test, conduct an experiment using materials, draw a sketch, make a graph, find a pattern) to solve simple multiplication process problems.

Grade Four

15. The student will be able to multiply mentally any two numbers less than ten with ninety-five percent accuracy (multiplication facts mastery).
16. The student will be able to multiply concretely a number less than 100 by a number less than ten, regrouping as necessary, and to record the results in place value tables.
17. The student will be able to multiply a number less than 1000 by a number less than ten, regrouping as necessary, using place value tables.
18. The student will be able to multiply a number by a multiple of ten or 100 using place value tables.
19. The student will be able to multiply concretely any two numbers with products less than 1000, regrouping as necessary, and to record the results in place value tables.
20. The student will be able to solve story problems involving one or two of the basic operations (+ , − , ×) using whole numbers and money.
21. Working in a small group, the student will be able to apply the four-step problem plan employing one or more solution strategies (guess-and-test, work backward, conduct an experiment, draw a sketch, make a graph, find a pattern) to solve simple multiplication process problems.

Grade Five

22. The student will be able to multiply any two whole numbers that are less than 100, regrouping as necessary, with place value tables and without.
23. The student will be able to multiply any number less than 1000 by any number less than 100, regrouping as necessary, with and without place value tables.
24. The student will be able to solve story problems involving one or more of the basic operations (+ , − , ×).
25. Working in a small group, the student will be able to apply the four-step problem plan employing one or more solution strategies (guess-and-test, work backward, solve a simpler case, conduct an experiment, use a table, find a pattern, construct a general rule) to solve multiplication process problems.

Grade Six

26. The student will be able to multiply any number by a number less than 1000 using the standard multiplication algorithm.
27. The student will be able to solve story problems involving one or more of the basic operations.
28. Working in a small group, the student will be able to apply the four-step problem plan employing one or more solution strategies (guess-and-test, work backward, solve a simpler case, conduct an experiment, use a table, find a pattern, construct a general rule, add elements to the problem situation) to solve multiplication process problems.

DIVISION

Kindergarten

1. The student will be able to group objects in pairs.
2. The student will be able to participate in the group solution of one-step, graphic story problems involving grouping by two using role playing or materials.

Grade One

3. The student will be able to group objects by threes and fives.
4. The student will be able to solve one-step, graphic story problems ($+$, $-$, $\times$, $\div$) using materials and to record the results by drawing a picture or constructing a graph.

Grade Two

5. The student will be able to partition appropriate sets of up to twelve objects into equal groups and determine the number of groups.
6. The student will be able to solve one-step, graphic story problems ($+$, $-$, $\times$, $\div$) and record the results using pictures, graphs, and number sentences.

Grade Three

7. The student will be able to demonstrate the concept of division by separating appropriate sets of up to twenty-five objects into equivalent subsets.
8. The student will be able to demonstrate the array model of division (inverse of multiplication) using base-10 materials.
9. The student will be able to divide a number concretely, using basic facts and with the divisor less than six and the dividend less than twenty, and to record the results in place value tables.

10. The student will be able to divide a number concretely, using basic facts up to (81 ÷ 9), with no remainder and to record the results in place value tables.
11. The student will be able to solve simple story problems involving one operation (+, −, ×, ÷), using whole numbers and money.
12. Working in a small group, the student will be able to apply the four-step problem plan employing one solution strategy (guess-and-test, conduct an experiment using materials, draw a sketch, make a graph, find a pattern) to solve simple division process problems.

Grade Four

13. The student will be able to recall the one hundred division facts with ninety-five percent accuracy (division facts mastery).
14. The student will be able to divide concretely a number less than 100 by a number less than 10, first without, then with a remainder (64 ÷ 2 = 32; 87 ÷ 4 = 21 R3) and to record the results in place value tables.
15. The student will be able to divide a number less than 1000 by a number less than 10, requiring simple estimation, first without, then with a remainder (697 ÷ 3 = 232 R1) using place value tables.
16. The student will be able to divide concretely any number less than 1000 by any number less than 100, first without, then with remainders and to record the results in place value tables.
17. The student will be able to solve story problems involving one or two operations (+, −, ×, ÷), using whole numbers and money.
18. Working in a small group, the student will be able to apply the four-step problem plan employing one or more solution strategies (guess-and-test, work backward, conduct an experiment, draw a sketch, make a graph, find a pattern) to solve simple division process problems.

Grade Five

19. The student will be able to divide any number less than 1000 by any number less than 10, with and without a remainder, using place value tables and without.
20. The student will be able to divide any number less than 1000 by any number less than 100, with and without a remainder, using place value tables and without.
21. The student will be able to solve story problems involving one or more of the basic operations.
22. Working in a small group, the student will be able to apply the four-step problem plan employing one or more solution strategies (guess-and-test, work backward, solve a simpler case, conduct an experiment, use a table, find a pattern, construct a general rule) to solve division process problems.

Grade Six

23. The student will be able to divide any number greater than 1000 by a number less than 1000, with and without a remainder, using the standard division algorithm.
24. The student will be able to solve story problems involving one or more of the basic operations.
25. Working in a small group, the student will be able to apply the four-step problem plan employing one or more solution strategies (guess-and-test, work backward, solve a simpler case, conduct an experiment, use a table, find a pattern, construct a general rule, add elements to the problem situation) to solve division process problems.

PATTERNS AND FUNCTIONS

Kindergarten

1. The student will recognize a simple pattern based on shape or color and identify the member that comes next. For example:

    square, circle, square, circle, _____

Grade One

2. The student will be able to recognize and continue a simple pattern of shapes or colors. For example:

    square, diamond, circle, square, diamond, _____ , _____ , _____

Grade Two

3. The student will be able to count by twos and fives.
4. The student will be able to recognize and continue a simple number pattern. For example:

    5, 10, 15, _____ , _____ , _____

Grade Three

5. The student will be able to show concretely and record the patterns of even, odd, square, and triangle numbers.
6. Working in a small group, the student will be able to apply the four-step problem plan employing one solution strategy (guess-and-test, conduct an

experiment using materials, draw a sketch, make a graph, find a pattern) to solve simple pattern process problems.

Grade Four

7. The student will be able to record in tables the results of data collected in experiments (e.g., building structures with cubes) and to extend the pattern to the next value.
8. Working in a small group, the student will be able to apply the four-step problem plan employing one or more solution strategies (guess-and-test, work backward, conduct an experiment, draw a sketch, make a graph, find a pattern) to solve simple pattern process problems.

Grade Five

9. The student will be able to determine the rule (function) describing the linear relationship between two columns in a table.
10. Working in a small group, the student will be able to apply the four-step problem plan employing one or more solution strategies (guess-and-test, work backward, solve a simpler case, conduct an experiment, use a table, find a pattern, construct a general rule) to solve pattern and function process problems.

Grade Six

11. The student will be able to write a linear function relating the two columns of values in a table and plot ordered pairs to show a straight line graph.
12. Working in a small group, the student will be able to apply the four-step problem plan employing one or more solution strategies (guess-and-test, work backward, solve a simpler case, conduct an experiment, use a table, find a pattern, construct a general rule, add elements to the problem situation) to solve pattern and function process problems.

FRACTION AND DECIMAL OPERATIONS

Kindergarten

1. The student will be able to identify one-half of a region or object.
2. The student will be able to participate in the group solution of one-step, graphic story problems that involve appropriate rational numbers, using role playing and materials.

Grade One

3. The student will be able to identify halves and fourths of a region or object.
4. The student will be able to solve one-step, graphic story problems involving appropriate rational numbers using materials and to record the results by drawing a picture or constructing a graph.

Grade Two

5. The student will be able to construct and identify one-half, one-third, and one-fourth of a region or set.
6. The student will be able to solve one-step, graphic story problems involving appropriate rational numbers and to record the results using pictures, graphs, and number sentences.

Grade Three

7. The student will be able to identify halves, thirds, fourths, fifths, and sixths of a region or set and write the corresponding fraction numeral.
8. Working in a small group, the student will be able to apply the four-step problem plan employing one solution strategy (guess-and-test, conduct an experiment using materials, draw a sketch, make a graph, find a pattern) to solve simple process problems involving appropriate rational numbers.

Grade Four

9. The student will be able to identify halves, thirds, fourths, fifths, sixths, eighths, and tenths of a region or set and write the corresponding fraction numeral.
10. The student will be able to represent fractions concretely using an area model and to write an equivalent fraction for a given fraction (halves, thirds, fourths, fifths, sixths, and tenths).
11. The student will be able to add concretely fractions with like denominators and to record the results.
12. The student will be able to subtract concretely fractions with like denominators and to record the results.
13. The student will be able to represent concretely (using coins), say, and read decimals in tenths and hundredths written in place value tables.
14. The student will be able to solve one- or two-step story problems involving rational numbers.
15. Working in a small group, the student will be able to apply the four-step problem plan employing one or more solution strategies (guess-and-test, work backward, conduct an experiment, draw a sketch, make a graph, find a pattern) to solve simple process problems involving rational numbers.

Grade Five

16. The student will be able to write an equivalent fraction for a given fraction (halves, fourths, thirds, and sixths).
17. The student will be able to represent concretely and write fractions in simplest terms (halves, fourths, thirds, and tenths).
18. The student will be able to add concretely fractions with like denominators and to record the actions and results symbolically.
19. The student will be able to subtract concretely fractions with like denominators and to record the actions and results symbolically.
20. The student will be able to represent concretely, using base-10 materials and coins, say, read, and write decimals in tenths and hundredths using place value tables.
21. The student will be able to rename fractions in tenths and hundredths as decimal fractions and vice versa.
22. The student will be able to change a decimal expressed in tenths and hundredths to a percentage.
23. The student will be able to solve story problems involving one or more of the basic operations and percent.
24. Working in a small group, the student will be able to apply the four-step problem plan employing one or more solution strategies (guess-and-test, work backward, solve a simpler case, conduct an experiment, use a table, find a pattern, construct a general rule) to solve fraction and decimal process problems.

Grade Six

25. The student will be able to write an equivalent fraction for a given fraction and write the fraction in simplest terms.
26. The student will be able to rename mixed numbers as improper fractions and vice versa, renaming the answer in simplest terms.
27. The student will be able to add fractions, renaming the answer as a mixed numeral expressed in simplest terms.
28. The student will be able to subtract fractions, renaming the answer as a mixed numeral.
29. The student will be able to add mixed numbers with unlike denominators, renaming the answer in simplest terms.
30. The student will be able to subtract mixed numbers with unlike denominators, renaming the answer in simplest terms.
31. The student will be able to multiply fractions and mixed numerals.
32. The student will be able to divide fractions and mixed numerals.
33. The student will be able to say, read, and write decimals to hundredths and thousandths.
34. The student will be able to add decimals to hundredths, renaming as necessary.

35. The student will be able to subtract decimals to hundredths, renaming as necessary.
36. The student will be able to multiply a whole number and a number expressed in decimal form or two decimal numbers, both hundredths and tenths.
37. The student will be able to divide a number in decimal form by a whole number or by a number in decimal form, both hundredths and tenths.
38. The student will be able to convert a fraction to a decimal and vice versa.
39. The student will be able to solve problems involving ratio and proportion.
40. The student will be able to solve problems related to percentage, base, and rate.
41. Working in a small group, the student will be able to apply the four-step problem plan employing one or more solution strategies (guess-and-test, work backward, solve a simpler case, conduct an experiment, use a table, find a pattern, construct a general rule, add elements to the problem situation) to solve fraction and decimal process problems.

STATISTICS AND PROBABILITY

Kindergarten

1. The student will participate in the construction of an empirical graph.

Grade One

2. The students will contribute pertinent information about themselves to a class graphing project, participate in group decisions as to how to organize and record data, and discuss conclusions drawn from the graph including the terms *more than, less than*, and *the same*.

Grade Two

3. The students will contribute pertinent information about themselves to a class graphing project, participate in group decisions as to how to organize and record data, and discuss conclusions drawn from the graph including the terms *more than, less than, the same, base line* and *bar graph*.

Grade Three

4. The student will participate in a small group graphing activity that will include gathering statistical information, organizing the information in a bar graph, and interpreting the graph.

5. Working in a small group, the student will be able to apply the four-step problem plan employing one solution strategy (guess-and-test, conduct an experiment using materials, draw a sketch, make a graph, find a pattern) to solve simple statistics process problems.

Grade Four

6. The student will be able to plot points from number pairs on a coordinate plane to create a design.

7. The student will be able to construct and interpret a simple picture graph and a bar graph.

8. Working in a small group, the student will be able to apply the four-step problem plan employing one or more solution strategies (guess-and-test, work backward, conduct an experiment, draw a sketch, make a graph, find a pattern) to solve simple statistics and probability process problems.

Grade Five

9. The student will be able to plot ordered pairs on the coordinate plane and determine if the graph is linear.

10. The student will be able to construct and interpret a histogram, picture graph, and line graph using data collected from surveys or experiments.

11. The student will be able to use data to compute the mean (average) of a list of numbers.

12. The student will participate in a probability activity and be able to determine the experimental probability of an experiment.

13. Working in a small group, the student will be able to apply the four-step problem plan employing one or more solution strategies (guess-and-test, work backward, solve a simpler case, conduct an experiment, use a table, find a pattern, construct a general rule) to solve statistics and probability process problems.

Grade Six

14. The student will be able to graph ordered pairs of data collected from an experiment and determine if the relationship expressed is linear.

15. The student will be able to construct and interpret a circle graph, picture graph, histogram (bar graph), and line graph and to describe the advantages of using each.

16. The student will be able to find the mean, median, mode, and range for data collected in surveys and experiments.

17. The student will be able to determine the theoretical and experimental probability of an event.

18. Working in a small group, the student will be able to apply the four-step problem plan employing one or more solution strategies (guess-and-test, work backward, solve a simpler case, conduct an experiment, use a table, find a pattern, construct a general rule, add elements to the problem situation) to solve statistics and probability process problems.

Appendix:
Instructional Material
Suppliers and
Microcomputer
Resources

MATHEMATICS INSTRUCTIONAL MATERIALS SUPPLIERS

Activities Resources
P.O. Box 4875
Hayward, CA 94540

Addison-Wesley Publishing Company
2725 Sand Hill Road
Menlo Park, CA 94025

Burt Harrison & Company
P.O. Box 732
Weston, MA 02193

Creative Publications
P.O. Box 328
Palo Alto, CA 94303

Cuisenaire Company of America, Inc.
12 Church Street
New Rochelle, NY 10805

Dale Seymour Publications
PO Box 10888
Palo Alto, CA 94303

Developmental Learning Materials
7440 Natchez Avenue
Niles, IL 60648

Dick Blick
P.O. Box 1267
Galesburg, IL 61401

Didax, Inc.
Educational Resources
6 Doulton Place
Peabody, MA 01960

Educat
P.O. Box 2891
Clinton, IA 52735

Educational Teaching Aids
159 West Kinzie Street
Chicago, IL 60610

Good Year Books
Scott, Foresman and Company
1900 East Lake Ave.
Glenview, IL 60025

Ideal School Supply Company
11000 South Lavergne Street
Oak Lawn, IL 60453

591

Math House
Division of Mosaic Media, Inc.
Dept. C183
P.O. Box 711
Glen Ellyn, IL 60137

Nasco
901 Janesville Avenue
Fort Atkinson, WI 53538

Nasco West
1524 Princeton Avenue
Modesto, CA 95352

National Council of Teachers of Mathematics
1906 Association Drive
Reston, VA 22091

School Science and Mathematics
Arizona State University
203 Payne Hall
Tempe, AZ 85287

Selective Educational Equipment, Inc. (SEE)
3 Bridge Street
Newton, MA 02195

MATHEMATICS AND COMPUTER EDUCATION JOURNALS AND PERIODICALS

The following periodicals publish articles of interest to elementary mathematics teachers who use computers in their classrooms. Issues frequently contain reviews of software, discussions of current hardware developments, user-group columns, and classroom applications.

A+
Ziff-Davis Publishing Company
One Park Avenue
New York, NY 10016

Arithmetic Teacher
National Council of Teachers of
Mathematics
1906 Association Drive
Reston, VA 22901

Call-A.P.P.L.E.
A.P.P.L.E. Co-op.
290 S.W. 43rd Street
Renton, WA 98055

Classroom Computer Learning
19 Davis Drive
Belmont, CA 94002

Classroom Computer News
PO Box 266
Cambridge, MA 02138

Courseware Report Card
150 West Carob Street
Compton, CA 90220

The Computing Teacher
Dept. of Computer and Information
Science
University of Oregon
Eugene, OR 97403

CUE Newsletter
Computer-Using Educators
PO Box 18547
San Jose, CA 95158

Educational Computer
PO Box 535
Cupertino, CA 95015

Electronic Learning
902 Sylvan Ave.
Englewood Cliffs, NJ 07632

Microcomputers in Education
QUEUE
5 Chapel Hill Drive
Fairfield, CT 06432

Micro-Scope
Jem Research
Discovery Park, University of Victoria
PO Box 1700
Victoria, BC V8W 2Y2, Canada

Software Reports
10996 Torreyana Road
San Diego, CA 92121

SOFTWARE SUPPLIERS, DIRECTORIES, AND CATALOGS

The following suppliers, directories, and catalogs provide information on a wide range of education software for popular microcomputers.

Atari Program Exchange
Atari, Inc.
PO Box 427
155 Moffett Park Drive, B-1
Sunnyvale, CA 94086

Borg-Warner Educational Systems
600 West University Drive
Arlington Heights, IL 60004

CIE Software News
Computer Information Exchange
PO Box 159
San Luis Rey, CA 92068

Commodore Software Encyclopedia
Commodore Corporate Offices,
Education Dept.
487 Devon Park Drive
Wayne, PA 19087

Conduit Software
The University of Iowa
Oakdale Campus
Iowa City, IA 52244

Control Data Publishing Company
PO Box 261127
San Diego, CA 92126

CTB/McGraw-Hill
2500 Garden Road
Monterey, CA 93940

CUE Softswap
333 Main Street
Redwood City, CA 94063

DesignWare
185 Berry Street, Bldg. 3, Ste. 158
San Francisco, CA 94107

Developmental Learning Materials
One DLM Park
Allen, TX 75002

Edu-Ware Services, Inc.
P.O. Box 22222
Agoura, CA 91301

Encyclopaedia Britannica Educational Corporation
425 North Michigan Avenue
Chicago, IL 60611

EPIE Report
EPIE Institute
PO Box 620
Stony Brook, NY 11790

Hartley Courseware, Inc.
P.O. Box 431
Simindale, MI 48821

InterLearn Inc.
PO Box 342
Cardiff, CA 92007

The Learning Company
545 Middlefield Road
Suite 170
Menlo Park, CA 94025

Learning System, Ltd.
P.O. Box 9046
Fort Collins, CO 80525

Lightspeed Software
2124 Kittredge Street
Berkeley, CA 94704

Technology in the Curriculum: Mathematics Resource Guide
California State Department of Education
721 Capitol Mall
Sacramento, CA 95802

The Micro Center
Department JM
P.O. Box 6
Pleasantville, NY 10570

Milliken Publishing Company
1100 Research Blvd.
St. Louis, MO 63132

Minnesota Educational Computing Consortium
3490 Lexington Avenue, North
St. Paul, MN 55112

Opportunities for Learning, Inc.
8950 Lurline Avenue, Dept. JY
Chatsworth, CA 91311

Scholastic, Inc.
905 Sylvan Avenue
Englewood Cliffs, NJ 07632

Scott, Foresman and Company
Electronic Publishing
Glenview, IL 60025

Scott Resources, Inc.
P.O. Box 2121
Fort Collins, CO 80522

Sensible Software Inc.
210 South Woodward, Suite 229
Birmingham, MI 48011

Spinnaker Software
215 First Street
Cambridge, MA 02142

Sterling Swift Publishing Company
1600 Fortview Road
Austin, TX 78704

Sunburst Communications
39 Washington Avenue
Pleasantville, NY 10570

Weekly Reader
245 Longhill Road
Middletown, CT 06457

Index